PRINCIPLES OF APPLIED CLINICAL CHEMISTRY

CHEMICAL BACKGROUND AND MEDICAL APPLICATIONS

Volume 1
Maintenance of Fluid and Electrolyte Balance

PRINCIPLES OF APPLIED CLINICAL CHEMISTRY

CHEMICAL BACKGROUND AND MEDICAL APPLICATIONS

Volume 1
Maintenance of Fluid and Electrolyte Balance

SAMUEL NATELSON, Ph.D.

Director, Department of Biochemistry
Michael Reese Hospital and Medical Center
Chicago, Illinois

and

ETHAN A. NATELSON, M.D.

Baylor College of Medicine
Methodist Hospital
Houston, Texas

PLENUM PRESS · NEW YORK AND LONDON

Library of Congress Cataloging in Publication Data

Natelson, Samuel.
  Principles of applied clinical chemistry.

  Includes bibliographical references and index.
  CONTENTS: v. 1. Maintenance of fluid and electrolyte balance.
  1. Chemistry, Clinical. 2. Metabolism. 3. Metabolism, Disorders of. I.
Natelson, Ethan Allen, 1942-   joint author. II. Title. [DNLM: 1. Chem-
istry, Clinical. QY90 N273p]
RB27.N38   616.07'56   75-4798
ISBN 0-306-35231-1

© 1975 Plenum Press, New York
A Division of Plenum Publishing Corporation
227 West 17th Street, New York, N.Y. 10011

United Kingdom edition published by Plenum Press, London
A Division of Plenum Publishing Company, Ltd.
Davis House (4th Floor), 8 Scrubs Lane, Harlesden, London, NW10 6SE, England

Printed in the United States of America

To the Memory of
MORRIS NATELSON

Preface

"Clinical Chemistry encompasses the study of the fundamental principles of chemistry as applied to an understanding of the functioning of the *human* organism in health and disease."[1]

From its very definition, clinical chemistry is an applied science. Its scope includes the following:

1. Studies designed to elucidate the chemical mechanisms whereby the human normally functions.
2. The application of this information to an understanding of the disease process in the human.
3. The development of methodology and instrumentation in order to facilitate data gathering so as to apply the above principles to the diagnosis and treatment of disease in the human.

This book is an attempt to organize the information gathered relative to points 1 and 2 into a logical sequence so as to *define the areas of learning encompassed by the science of clinical chemistry*. It is constructed around the subject which is the target of this science, namely the human. The material is presented from the point of view of the clinical chemist, but since it is impossible to discuss a mechanism adequately without visualizing its parts, some schematic anatomical drawings are included to simplify the discussion of responses to chemical challenges.

The book is partly a curriculum which has been worked out by the authors for the training of clinical chemists and clinical pathologists. It should also be useful for the training of medical technologists.

[1] Chaney, A. L., McDonald, H. J., Natelson, S., Osterberg, A. E., and Routh, J. I., *Chem. Eng. News 32*: 4097 (1954).

This book is also designed to enable the practicing physician to utilize more fully the resources of the laboratory of clinical chemistry. For this reason summaries and tables are placed at the end of the various discussions correlating the chemical findings expected with certain disease processes. This should also be useful to the technically trained clinical chemist, as it will familiarize him with medical terminology.

The organization of the book follows the natural lines of the functioning human. It is assumed that the reader has an elementary background in organic and biological chemistry. For this reason, such standard headings as proteins, carbohydrates, enzymes, and vitamins do not appear in this book. These subjects are discussed in connection with the particular function at hand.

To cover so large a field in a small volume is obviously an insurmountable task. For this reason the literature citations have been selected to give preference to monographs and to the more general articles on the various subjects. In this way, this book stands supported by selected texts and journal publications.

Procedures for performing the various types of analyses required in a laboratory of clinical chemistry are not included in this book. The reader is referred to a list of textbooks on analytical clinical chemistry in the appendix.

Samuel Natelson, Ph.D.
Ethan Allen Natelson, M.D.

Chicago
Houston

June, 1975

Development of this text was supported in part by the training grant program in clinical chemistry of the National Institutes of Health, which support is gratefully acknowledged.

Contents

CHAPTER 7

Maintenance of Constant Ion Concentration in Body Fluids
Calcium, Magnesium, Phosphate, and Sulfate   141

CHAPTER 10
Fluid and Electrolyte Applications 247

SECTION II

THE KIDNEY AND SWEAT GLANDS IN FLUID AND ELECTROLYTE BALANCE

CHAPTER 11
The Kidney 285

SECTION III

APPENDIX AND INDEX

Introduction

Origin of the Term "Clinical Chemistry"

From the beginning of man's consciousness, he has centered his interests around his own well being. Every early civilization and culture gave a special position to the healer or medicine man. In turn, the healer has always explored and theorized about the workings of the human mechanism in health and disease. With the Renaissance and the spread of chemical information gathered in the Near East, medieval alchemy was adopted in the western world.

One of the main concerns of the alchemists was with finding the "elixir of life." Their studies on human and animal material laid the foundation for the development of the chemical sciences. Paracelsus, the son of a physician and a physician himself, proclaimed in 1526 that alchemy should not concern itself with the transmutation of elements but was the handmaiden of medicine. The alchemists were replaced by *iatrochemists*, the word *iatro* meaning pertaining to medicine. Thus we may say that Paracelsus is the Father of Clinical Chemistry.

Van Helmont, who was active during the middle of the first half of the 17th century, was the last prominent chemist to call himself an iatrochemist, because by this time those studying living organisms were calling themselves *organic chemists*. Van Helmont prepared carbon dioxide and studied the ferments (digestive enzymes) of the body. He introduced the analytical balance into the study of chemistry. This is in line with the fact that numerous analytical tools and techniques have originated in the clinical chemistry laboratory and continue to do so. For example, Pregl, who received the Nobel Prize in Microchemistry,

1

was a physician and was motivated by clinical problems. Modern instrumentation developed first by clinical chemists has subsequently been used in other disciplines.

Toward the end of the 18th century, Priestley and Lavoisier studied combustion and respiration, and Lavoisier, based on the work of Priestley, correctly interpreted the facts of oxidation, which replaced the phlogiston theory. It is of interest to note that Priestley went on to lecture at the University of Pennsylvania between 1795 and 1804 and to this day the University of Pennsylvania is an outstanding center for studies in pulmonary physiology and chemistry.

Berzelius and Gmelin reflected the thinking of the organic chemists of their time (1811) in proclaiming that a vital force was necessary to form an organic compound. This theory was abandoned when Wöhler and Liebig synthesized urea in 1828. It is of interest to note that Wöhler had a degree in medicine. From this time, organic chemistry came to mean the study of compounds of carbon. Those interested primarily in the chemistry of the human began to call themselves *physiological chemists*, with the publication of *Hoppe-Seyler's Zeitschrift für Physiologische Chemie*. Since most of these chemists were trained physicians, it was natural for these studies to stress physiology more than chemistry.

Feeling the need for a journal which was more closely related to their studies, with an emphasis on the chemistry of the living organism, the Biochemical Society was organized in England and the *Biochemical Journal* was first published in 1906.

Six years later the *Journal of Biological Chemistry* and *Biochemische Zeitschrift* were founded. These contained articles mainly on clinical chemistry by such authors as Van Slyke, Folin, Schoenheimer, Bloor, Somogyi, and Victor Meyer. The term biochemistry was almost synonymous with what we now call clinical chemistry. Courses in the medical schools in biochemistry were courses in clinical chemistry. The Johns Hopkins University was the first of the medical schools to employ a chemist to teach biochemistry early in the 20th century.

After the Second World War, stimulated by grants from the National Institutes of Health which were originally designed to encourage investigation into the chemistry of the human, numerous research projects were undertaken at universities in biochemistry. In the absence of a hospital environment, these investigators turned to detailed studies of the mechanisms of biochemical reactions. Soon these publications crowded out the publications more closely related to the human. Clinical chemical studies necessarily had to be published in medical journals.

To remedy this situation, the American Association of Clinical Chemistry was organized in 1948 and publication of a journal, *Clinical Chemistry*, followed in 1955. The word "clinical" literally means *bedside*. The term was probably chosen because of a book published on the subject in 1932 by Peters and Van Slyke entitled *Quantitative Clinical Chemistry*. Soon clinical chemistry journals appeared in many countries. *Clinica Chimica Acta* appeared in 1956. Similar journals in Scandinavia, Germany, and other countries followed.

The increasing utilization of the routine clinical chemistry laboratory by the physician resulted in the expansion of these laboratories and the introduction of extensive systems of automation. Certain clinical chemists then devoted their full time to the management of these laboratories and to the development of improved methodology and procedures and many of these chemists moved to commercial laboratories and companies devoted to supplying the analytical needs of the laboratories. The name clinical chemist began to mean, to some, an *analytical clinical chemist*. In England, a Society of Clinical Biochemists was formed to distinguish those who concerned themselves with activities designed to explore the chemistry of the human in health and disease and not methodology. A *Journal of Clinical Biochemistry* was published in Canada. Whether this term clinical biochemist will be generally accepted in other countries in this sense remains to be seen.

The clinical chemist has thus been called an alchemist, iatro-chemist, organic chemist, physiological chemist, biochemist, and now clinical chemist and clinical biochemist. Today many clinical chemists still call themselves biochemists and hospital laboratories are still called biochemical laboratories. *As pointed out in the preface, if one is motivated in seeking to improve our understanding of the chemical processes in the human in health and disease, then today he is a clinical chemist.*

Body Fluids and Electrolytes

Maintenance of the Steady State in the Human[1]

The human body is a highly complex servomechanism. Like an airplane on automatic pilot, it reacts to stimuli so as to maintain its balance. Like the automatic oil burner, it reacts to stimuli from within by taking more fuel (food). It adjusts its metabolic rate automatically to temperature changes in its environment, and adjusts its rate of blood flow automatically with changes in load. It maintains a constant temperature, pH, and salt concentration automatically in its components, which in the human are called organs, such as the kidney and the lungs.

Figure 1.1 indicates the complexity of what takes place when a simple stimulus is answered by a response. The stimulus arrives at the central nervous system, where it triggers the hypothalamus to release low molecular weight polypeptides. The hypothalamus simultaneously adjusts the blood flow, the respiration rate, the blood pressure, and the volume of blood passing through the kidneys and other organs. The pH and electrolyte concentrations are kept in adjustment within narrow limits.

The pituitary is then stimulated to secrete hormones which control the function of the various glands. In addition, protein synthesis is stimulated in order to carry out the innumerable functions necessary to eventually result in the necessary muscular activity to respond to the stimulus.

So sophisticated a machine requires a correspondingly sophisticated computer, which in the human is called the *central nervous system*. Just as certain servomechanisms made by man can learn by experience in a limited way, this servomechanism or robot can learn most compli-

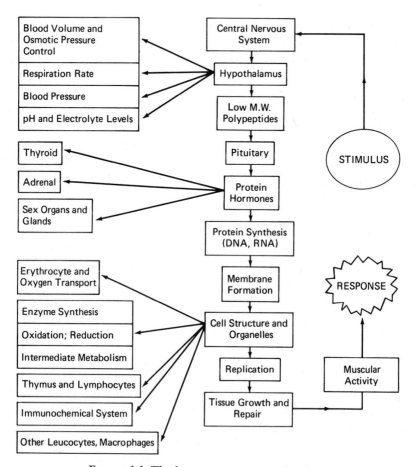

FIGURE 1.1 The human as a servomechanism.

cated tasks. While man has learned to build computers from minute electrical current modulators called transistors, these are massive in size compared to the relays in the brain. A computer constructed with present technology would occupy many square miles if it were to begin to simulate the function of the brain.

1.1   *THE STEADY STATE*

In order to discuss the various mechanisms for maintenance of an orderly system of metabolism in the human, it is important that the meaning of certain terms be clearly understood.

A system is said to be in *equilibrium* when it has reached a unique condition where change ceases.[2] Equilibrium is a state of rest. For example, if one half fills a bottle with water at room temperature, seals it, and places it in a constant-temperature bath at 37°C, then some of the water will evaporate. Now the pressure of water molecules leaving the water is called its *vapor tension*. The pressure of the water molecules condensing to form water is called the *vapor pressure*. Thus we have some water molecules leaving the liquid and others returning. At *equilibrium* these are equal. In other words, the vapor tension equals the vapor pressure. Note that this is a *dynamic equilibrium*, that is, activity is going on but a state of balance has been reached.

If one repeats the experiment by filling the bottle with hot water, sealing it, and then putting it in the 37°C water bath, it would cool and reach a point of equilibrium. If one now measures the partial pressure of the water vapor in this bottle and compares it with that in the first bottle, he will find them to be identical. We now see a characteristic of systems in equilibrium. *There is one and only one point of equilibrium* at constant temperature.

Another example which more closely resembles the problems at hand is the hydrolysis of methyl acetate. One places a 0.25 M solution of methyl acetate in one sealed container and equal volumes of 0.5 M methanol and 0.5 M acetic acid in another container. Both containers are placed in a constant-temperature bath. A few months later one opens the containers and titrates the acetic acid. Titration will be the same, volume for volume. In one case, the methyl acetate was hydrolyzed and in the second case, some methanol combined with acetic acid to reach the same ratio of hydrolyzed to unhydrolyzed product. If one then performs the same experiment adding an enzyme, esterase, again the same ratio is obtained, but in a shorter period of time. This experiment shows that in a chemical reaction that reaches equilibrium, there is one and only one point of equilibrium. The catalyst brings the reaction mixture to the equilibrium point more rapidly, *but does not change that point of equilibrium.*

In a constant-temperature bath we have a state of balance, but not equilibrium. We maintain a state of balance with a thermoregulator which turns the heat on and off as the bath cools or reaches a preset temperature. Adjusting the thermoregulator to another point would give us a different temperature in the bath. Thus there are an infinite number of points of balance and therefore we do not have a system in equilibrium. An automobile traveling at constant speed is in a state of balance since we can change speeds by stepping on the accelerator. Such systems are said to be in a *steady state*.[1,3] The human body is such a system. Body temperature is maintained at 37°C by temperature controls located in the central nervous system which regulate metabolic rate and oxygen consumption. Fuel (food) from the outside must be added to this system to maintain the steady state.

With the human system, the word *homeostasis*[4] is used to describe the steady state. This, from the Greek, literally means "staying in the same place." The use of this word is limited to organisms in physiological balance maintained by coordinated functioning of the brain and other organs of the body. It is generally not used outside of biological circles. *The human organism is in homeostasis; it is not in equilibrium.*

From the examples discussed above, one can see that all systems tend to move in the direction of equilibrium. This is really a statement of the second law of thermodynamics. All processes in the human tend toward equilibrium and *energy must be supplied to prevent this from being achieved.* It is interesting to note that this second law of thermodynamics stems particularly from the work of Helmholtz, a physician who, in his motivation, was a clinical chemist.

As one approaches equilibrium, the difference between the opposing forces is progressively reduced. For example, if a glass capillary tube is held vertically and touched to water, the liquid will enter the capillary and rise rapidly at first. As it approaches equilibrium it moves more and more slowly until its movement is barely perceptible.[5] It is for ease of control that systems in the human are kept a reasonable distance away from the equilibrium point.

In the past, some chemists have made the tacit assumption that the system being measured in the human is at equilibrium. Thus equations derived from equilibrium systems are erroneously applied directly to biochemical problems in the living organism. A few simple examples will illustrate the fallacy in this approach. The serum calcium level in the rabbit is of the order of 14 mg %, whereas it is 10 mg % in the human. The body temperature of women is somewhat higher than that of men. Sodium and chloride levels are lower in newborn infants than in adults. Obviously, these levels are maintained by servomechanisms and are not points of equilibrium.

1.2   SELECTED READING—EQUILIBRIUM, STEADY STATE, HOMEOSTASIS

Adkins, C. J., *Equilibrium Thermodynamics*. McGraw-Hill, New York (1968).

Blackburn, T. R., *Equilibrium: A Chemistry of Solutions*. Holt, Rinehart & Winston, New York (1969).

Hill, T. L., *Thermodynamics for Chemists and Biologists*. Addison-Wesley, Reading, Mass. (1968).

McIlwain, H. and Bachelard, H. S., Eds., *Biochemistry of the Central Nervous System*, 4th ed., Churchill-Livingstone, New York (1971).

1.3   THE NEED FOR CONSTANT pH IN BODY FLUIDS

The chemical reactions which take place in the human are pH dependent. If a plot is made of pH versus reaction velocity of any enzyme system, one finds that a pH exists at which this reaction proceeds most rapidly. This is called the optimum pH. Enzymes are proteins, and their properties are strongly affected also by a salt concentration and the nature of the salts present. Figure 1.2 illustrates the change in enzyme activity and optimum pH depending upon the nature of the buffer used.[6]

In the liver, for example, over 100 enzymes have been identified, of which approximately 30 have been isolated. These show different pH optima. Since in the liver they all work at the same pH, it is apparent that they are not all operating at maximum efficiency. This is necessary, since products formed need to be processed in an orderly manner, without pileup of intermediates at certain points. A change in pH could cause one system to speed up if it brought the system closer to its optimum pH, and another to slow down if it moved away from its optimum pH. With the numerous enzymes involved, chaos would result. It is therefore apparent why body pH must be maintained constant over a narrow range. In order to understand the manner in which this process proceeds, one must understand the significance of the expressions base, acid, and buffer, and these are discussed in Chapter 2.

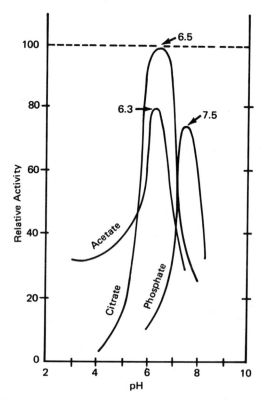

FIGURE 1.2   Effect of pH and the nature of the buffer on enzyme (urease) activity, illustrating the need for constant pH and salt concentration in the human.

1.4   *REFERENCES*

1. Walter, C., *Steady State Applications in Enzyme Kinetics*, Ronald Press, New York (1965).
2. Lewis, G. N., Randall, M., Pitzer, K. S., and Brewer, L., *Thermodynamics*, McGraw-Hill, New York (1961), p. 149.
3. Society for Experimental Biology, *Homeostasis and Feedback Mechanism*, Academic Press, New York (1964).
4. Langley, L. L., *Homeostasis*, Reinhold, New York (1965).
5. Mouquin, H., and Natelson, S., Equilibrium forces, acting on free drops in irregular capillaries, *Mikrochemie* **12**:293–302 (1932–3).
6. Howell, S. F. and Sumner, J. B., The specific effects of buffers on urease activity, *J. Biol. Chem.* **104**:619–26 (1934).

Acids and Bases

2.1 INTRODUCTION

There are three methods generally used for representing acids and bases. Arrhenius defined an acid as a substance that liberates hydrogen ions when dissolved in water.[2] A base, in accordance with his concept, is a substance that liberates hydroxyl ions. As examples, one can use HCl and NaOH:

$$\text{acid:} \qquad HCl \xrightarrow{\text{in water}} H^+ + Cl^-$$

$$\text{base:} \qquad NaOH \xrightarrow{\text{in water}} Na^+ + OH^-$$

It was soon found from the rate at which ions move in water under the influence of an electrical current (transference number) that the hydrogen ion exists in a hydrated form.[3] This is due to the fact that oxygen shares two of the six electrons in its outer shell with the two hydrogens in water. The other four are available for forming a bond with the hydrogen ion. Thus we can write

$$HCl + H_2O \rightleftharpoons H_3O^+ + Cl^-$$

H_3O^- is called the hydronium ion, but further studies have shown that more than one water is attached to the hydrogen ion and that this then is not a true representation. In this book, we will write H^+ for the hydrogen ion, keeping in mind that it is hydrated.

The Arrhenius system of acids and bases soon proved inadequate. For example, in anhydrous solvents, sodium acetate and piperidine will both catalyze the condensation of aldehydes with malonic acid. What do they have in common?

Brönsted then proposed his system of representing acids and bases to explain the above and other phenomena.[4,5] In his system, an acid is always associated with its conjugate base. The acid is the substance that can dissociate to liberate a hydrogen ion or proton. The base is the substance that can readily combine with the hydrogen ion or proton. A few examples will suffice:

(acid) $HAc \rightleftharpoons H^+ + Ac^-$ (conjugate base)

(acid) $NH_4^+ \rightleftharpoons H^+ + NH_3$ (conjugate base)

(acid) $H_2O \rightleftharpoons H^+ + OH^-$ (conjugate base)

(acid) $H_2CO_3 \rightleftharpoons H^+ + HCO_3^-$ (conjugate base)

Note that bicarbonate ion (HCO_3^-), NH_3, OH^-, and Ac^- are all bases and if a particular reaction requires a base for catalysis, these could substitute for each other. On the other hand, one must keep in mind that all bases are not of equal strength. The OH^- ion is a much stronger base than the bicarbonate ion.

The Brönsted system, although originally designed to explain reactions in nonaqueous media, has been most widely applied in clinical chemistry for aqueous solutions.

The most general way of representing acids and bases is the Lewis System, which focuses on the electron exchange instead of the proton. A base is defined as a substance *which can donate a pair of electrons* to complete the stable configuration of another atom or radical.[5] The acid is the acceptor of the pair of electrons, or is two electrons short of having a complete valence shell:

$$
\begin{array}{c} H \\ | \\ H-N^x_x \\ | \\ H \end{array} + H^+ \rightleftharpoons \left[\begin{array}{c} H \\ | \\ H-N^x_x H \\ | \\ H \end{array}\right]^+
$$

$$
H-\overset{xx}{\underset{xx}{O}}{}^x{}^- + H^+ \rightleftharpoons H-\overset{xx}{\underset{xx}{O}}{}^x H
$$

$$
\begin{array}{cc} H & F \\ | & | \\ H-N^x_x & B-F \\ | & | \\ H & F \end{array} \rightleftharpoons \begin{array}{cc} H & F \\ | & | \\ H-N^x_x B-F \\ | & | \\ H & F \end{array}
$$

Note that in these examples the hydrogen ion is the acceptor of two electrons and therefore is the acid, while NH_3 and OH^- are bases, or donors of the two electrons. This system is useful in explaining reactions between organic compounds in anhydrous solvents where hydrogen is not even involved in the reaction. These are referred to as Lewis acids or bases, depending upon whether they accept or donate the pair of electrons in forming the addition compound (salt).

The Brönsted concept will be used in our discussions.

2.2 SELECTED READING—ACIDS AND BASES

Kolthoff, I. M. and Bruckenstein, S., *Acid–Bases in Analytical Chemistry*, Wiley, New York (1959).

Davenport, H. C., *The ABC of Acid–Base Chemistry*, 5th rev. ed., University of Chicago Press (1969).

Masoro, E. J. and Sigel, P. D., *Acid–Base Regulation: Its Physiology and Pathophysiology*, Saunders, Philadelphia, Pennsylvania (1971).

Frisell, W. R., *Acid–Base Chemistry in Medicine*, Macmillan, New York (1968).

2.3 BUFFERS

A buffer is a solution in which changes in pH are small when acid or alkali is added to the solution.[6] The reason certain solutions have this property derives from the very meaning of the symbol pH.[7]

The ionization of acetic acid can be represented as follows:

$$HAc \rightleftharpoons H^+ + Ac^-$$

At equilibrium, the velocity of the forward reaction must equal the velocity of the backward reaction. The rate of the forward reaction is proportional to the concentration of acid. The velocity of the backward reaction is proportional to the product of the concentration of hydrogen and acetate ion. According to equilibrium laws[8,11] (the law of mass action) if the hydrogen ion concentration is tripled, the backward reaction would go three times as fast. If at the same time the acetate ion concentration is tripled, then the reaction rate backward would again increase threefold, or nine times altogether. This can be shown in symbols, using brackets to indicate concentrations:

$$\text{velocity (forward)} \propto [HAc]$$
$$\text{velocity (backward)} \propto [H^+][Ac^-]$$

From elementary algebra, if a is proportional to b, then a is equal to a constant times b. Then

$$V_f = K_1[\text{HAc}], \qquad V_b = K_2[\text{H}^+][\text{Ac}^-]$$

As pointed out above, these reaction velocities are equal at equilibrium. Then

$$K_1[\text{HAc}] = K_2[\text{H}^+][\text{Ac}^-]$$

and by transposing, we obtain

$$\frac{[\text{H}^+][\text{Ac}^-]}{[\text{HAc}]} = \frac{K_1}{K_2} = K_{eq} \qquad (2.1)$$

Carrying out the same derivation for water yields

$$\frac{[\text{H}^+][\text{OH}^-]}{[\text{H}_2\text{O}]} = K_{eq}$$

Water has a molecular weight of 18. In 1 liter of water there are $1000/18 = 55.6$ mol of water (mol will be used as the abbreviation for mole in this book). The amount of water ionized is $\sim 10^{-7}$ mol/liter. Thus the value of the denominator remains essentially unchanged after ionization, or is a constant. Multiplying the equilibrium constant by 55.6, we obtain the ionization constant for water:

$$[\text{H}^+][\text{OH}^-] = 10^{-14} \qquad (2.2)$$

When the solution is neutral, we have a value of 10^{-7} for the hydrogen and hydroxyl ion concentrations. This number being inconvenient, Sörensen devised the symbol pH to represent the number in the exponent.[7] At neutral pH, then, we say the pH is 7 and the pOH is also 7. At pH 6 the pOH^- must be 8, since the sum must always be 14. Taking the logarithm of the hydrogen ion concentration at neutral pH, we have

$$\log[\text{H}^+] = -7 \log 10$$

but the logarithm of 10 is 1, and changing signs, we see that

$$-\log[\text{H}^+] = 7 = \text{pH} \quad \text{(neutral)}$$

Thus we say that *the pH is the negative logarithm of the hydrogen ion concentration*. For example, what would be the pH of a 0.01 N HCl solution assuming 100% ionization. Here the hydrogen ion concentration would be 0.01 M or 10^{-2} M. The pH is then 2.

Referring back to equation (2.1), we noted that

$$\frac{[H^+][Ac]}{[HAc]} = K_{eq}$$

Solving for the hydrogen ion concentration, we obtain

$$[H^+] = \frac{K_{eq}[HAc]}{[Ac^-]}$$

Taking the negative logarithm of both sides, remembering that $\log ab = \log a + \log b$ and $\log (a/b) = -\log (b/a)$, we obtain

$$pH = -\log [H^+] = -\log K_{eq} - \log \frac{[HAc]}{[Ac^-]}$$

Substituting pK for $-\log K_{eq}$ and inverting the fraction to change signs, we obtain

$$pH = pK + \log \frac{[Ac^-]}{[HAc]} \qquad (2.3)$$

If the concentration of the acetate ion is equal to the concentration of acetic acid, then the value of the fraction is one. The logarithm of one being equal to zero, at this point the pH is equal to pK. This permits us to obtain the pK for an acid by a simple procedure. All we need do is titrate half the acid and measure the pH to obtain the pK. For example, let us assume that we have an unknown solid acid. We need only dissolve, say, 1 g in 100 ml of water. We take 50 ml and find that 26 ml of 0.1 M alkali neutralizes the solution. We then take the other 50 ml and add 13 ml of 0.1 M alkali and measure the pH to obtain the pK. In this experiment we are assuming that the salt of the acid is completely ionized. This is not true generally and a more exact method will be shown below.

Let us assume that the pK of a particular acid is 5. We can now, in our experiment, make the numerator of equation (2.3) ten times the denominator. This means that 90% of the acid has been neutralized. The value of the fraction will be 10. The logarithm of 10 is 1, and the pH will move to $5 + 1 = 6$. If now we titrate only 10% of the acid, then the value of the fraction will be 1/10. The log of 1/10 is -1. Then the pH will be $5 - 1 = 4$. Thus a wide change in the amount of alkali added produces a small change in pH. *This is what is meant by a buffer.*

Equation (2.3) can be plotted to observe the buffering action

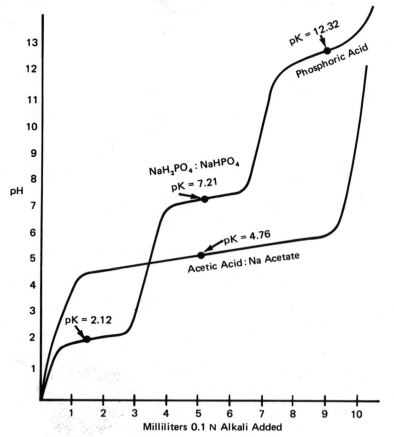

FIGURE 2.1   Titration curve of 10 ml of 0.1 N acetic acid and 10 ml of 0.1 N H₃PO₄ with 0.1 N NaOH.

graphically (Figure 2.1). Equation (2.3) is valid only for weak acids with strong bases, or strong acids reacted with weak bases. If any strong acid like sulfuric or hydrochloric acid is titrated with NaOH, a plateau does not occur as in Figure 2.1 except at pH values below 2.0.

From the titration curve in Figure 2.1, the pK is the midpoint of the flat portion of the curve or the point of inflection. The experimental pK values can be obtained from this point. These values are usually referred to as pK', if determined in a particular system such as human serum, as discussed above.

The buffering action of an acid–base system is best in the proximity of its pK value.[10] For acetic acid the pK value is 4.76. Phosphoric acid, being tribasic, has three pK values, one for H_3PO_4, one for NaH_2PO_4, and one for Na_2HPO_4, although the latter is a poor buffer. Phosphoric acid, having one pK value at 7.21, is often used as a 7.4 buffer to simulate blood pH. The pK values for a few common acids are listed in Table 2.1 for comparison. Extended tables of pK values are given in the handbooks of chemistry and physics.

The data in Table 2.1 permit us to select a buffer for a particular purpose. For example, if one wished to prepare a buffer at pH 5, then acetic acid (pK 4.75) would be a good choice. At pH 7, phosphoric acid would be satisfactory. At pH 9, boric acid would be chosen.

Bases are also useful as buffers and tables of their pK values are listed in the handbooks. In this case, one must keep in mind that the pK value given for bases is equal to $-\log pOH^-$ at the half titration point. Thus a pK value for NH_4OH is given as 4.75. This means that at half titration point the pOH is 4.75 and the pH is $14 - 4.75 = 9.25$. This is the pH range in which a buffer prepared from ammonia would be used. The pK value for triethylamine is at 3.24. Buffers prepared from this base would be useful in the vicinity of pH 10.8. Not all bases are used on the alkaline side. For example, pyridine, with a pK value of 8.77, is useful near pH 5.23. A base commonly used as a buffer is *Tris* (trishydroxymethyl aminomethane), because of its strong buffering

TABLE 2.1

Dissociation Constants of Acids in Water[9]

Acid		pK'	Acid		pK'
Acetic		4.75	Lactic		3.08
Barbituric		4.01	Malic	(1)	3.4
Benzoic		4.19		(2)	5.11
Boric	(1)	9.14	Methyl orange		3.80
	(2)	12.74	Oxalic	(1)	1.23
	(3)	13.80		(2)	4.19
Butyric		4.81	Phenol		9.89
Carbonic	(1)	6.37	Phenolphthalein		9.70
	(2)	10.25	Phenol red (P.S.P.)		7.40
Citric	(1)	3.08	Phosphoric	(1)	2.12
	(2)	4.74		(2)	7.21
	(3)	5.40		(3)	12.67
Fumaric	(1)	3.03	Picric	(1)	0.38
	(2)	4.44	Tartaric	(2)	2.98
Glycine		9.87		(2)	4.34

TABLE 2.2

Dissociation Constants of Bases in Water

Base	pK_{OH^-}	pK_{H^+} [a]	Base	pK_{OH^-}	pK_{H^+} [a]
Ammonia	4.75	9.25	Piperazine	4.19	9.81
				8.43	5.57
Amylamine	3.36	10.64	Piperidine	2.79	11.21
Aniline	9.42	4.58	Pyridine	8.77	5.23
Dimethyl aniline	8.94	5.04	Quinoline	9.20	4.80
Ethanolamine	4.56	9.44	Semicarbazide	10.57	3.43
Hydroxylamine	7.97	6.03	Triethanolamine	6.24	7.76
Hydrazine	5.77	8.23	Triethyl amine	3.24	10.76
α-Picoline	7.52	6.48	Trimethyl amine	4.26	9.74

[a] pK_{H^+} when titrated with a strong acid such as HCl. pH of midpoint of buffer range.

action at pH 7.4. Table 2.2 lists the pK_{OH^-} values for some common bases used as buffers.

If a crystal of sodium chloride is examined by x-ray diffraction it exhibits a cubic lattice with each sodium atom equidistant from six chlorine atoms and each chlorine atom surrounded by six sodium atoms.[12] Thus we cannot identify a chlorine atom with any particular sodium atom. In solution, each sodium ion is surrounded by a cluster of chloride ions and each chloride ion is surrounded by a cluster of sodium ions. Thus we cannot say that a sodium chloride molecule exists in solution and sodium chloride is completely ionized. When we measure the *colligative properties* such as vapor pressure, freezing point depression, and osmotic pressure of 0.1 N NaCl solution, we find that the solution acts as though the NaCl were about 78% ionized. We say then that the activity coefficient is 0.78.

Colligative properties are those based on the number of particles present. For example, for 0.1 M NaCl the solution would behave as though we had 0.078 M Na^+, 0.078 M Cl^-, and 0.022 M NaCl, the solution acting as if it were 0.178 M in its colligative properties as compared to an undissociated substance such as glucose. Thus the effective concentration or *activity* of a 0.1 M NaCl solution is 0.078 M with respect to Na^+ and 0.078 M with respect to Cl^- as far as its ionic behavior is concerned. As the solution is diluted, the *apparent* degree of ionization increases, so that at infinite dilution (by extrapolation) its activity coefficient would be 1. That is, the concentration and activity would be the same for each ion.

For salts like sodium acetate or sodium bicarbonate, the degree of ionization is of the order of only 50%. Thus equation (2.3) would seem to be invalid. We correct this by substituting an experimentally obtained pK value as from Figure 2.2 and call it pK'. A different pK' would be obtained at different concentrations of acid. For example, the pK' for carbonic acid in water (0.1 M) is 6.37, as can be seen in Table 2.1. At the concentrations occurring in human blood, the value of 6.1 is obtained experimentally.[14]

2.4 *INDICATORS*[13]

Equations similar to equation (2.3) can be written for acids like the ordinary acid–base indicators:

$$pH = pK + \log \frac{[Ind^-]}{[HInd]}$$

Suppose the indicator is yellow and its ionized form is blue. When the numerator equals the denominator of the fraction, we will have a combination of blue and yellow, or green. This indicator would go from yellow in acid, through green, and finally to blue in base, when most of it has been ionized. The pH span through which color changes take place is called the pH range of the indicator. Indicators, like other acids, have pK values as shown in Table 2.1. For example, when half of an amount of phenolphthalein is titrated, the pH is 9.7. When 10% has been converted (pH 8.7), we can see the pink color easily. The color change of methyl orange (pK 3.8), on the other hand, can be observed at pH 2.9. A table of the range of pH change for indicators is available in chemical handbooks.

2.5 *TITRATION*

In titrating an acid, *we want to reach the pH of the fully titrated salt.* For example, NH_4Cl solution has a pH of 4. In titrating HCl solutions with ammonia, we therefore want a color change at pH 4 (methyl orange). On the other hand, if we are titrating acetic acid, we want a pH change at about 9 because that is the pH of sodium acetate solutions (phenolphthalein).

By using a mixture of indicators, we can titrate a weak acid and a strong acid simultaneously. This is the situation when we titrate gastric juice. We have organic acids like lactic acid and an inorganic acid, hydrochloric acid. We use phenolphthalein for the organic acids whose salt solutions are at a pH of approximately 8, and Töpfer's reagent (dimethylaminoazobenzene) for the hydrochloric acid.

To see how this operates, the HCl concentration in gastric juice is approximately 0.015 N. Töpfer's reagent begins to change color at pH 3–4. The concentration of HCl (assuming 100% ionization) would be 10^{-3} at pH 3 and 10^{-4} at pH 4. Thus the HCl concentration when Töpfer's reagent first shows a change would be between 0.001 and 0.0001 N, depending upon the color discrimination of the operator. Assuming some point in between like pH 3.5, then the HCl remaining would be 0.0005 N. This is 5/150, or about 3.3% of the original acid present. For the purposes of gastric analyses this may be ignored, since approximately 97% of the hydrochloric acid has been titrated.

One then goes on to titrate the weak acids with phenolphthalein to a pH of approximately 8.5. If the weak acid is acetic acid with a pK value at 4.75, then very little has been titrated at pH 3.5, and both results are approximately correct. However, for lactic acid (pK 3.08), a serious error has been introduced because at the color change of Töpfer's reagent, half this acid would already have been neutralized. Lactic acid and other organic acids are formed by the action of microorganisms on food remaining in the stomach. With food retention, the lactic acid level rises significantly and half of this is interpreted as HCl. For this reason the test should ideally be performed on a subject in a fasting state with a previously emptied stomach. A chloride level in gastric juice should be obtained for more accurate interpretation of the results.

2.6   *SELECTED READING—BUFFERS*

Bates, Roger G., *Determination of pH, Theory and Practice*, 2nd ed., Wiley, New York (1973).

Christensen, H. N., *pH and Dissociation*, 2nd ed., Saunders, Philadelphia, Pennsylvania (1964).

Ricci, J. E., *Hydrogen Ion Concentration*, Princeton University Press, Princeton, New Jersey (1952).

2.7 *THE BICARBONATE BUFFER SYSTEM*

Referring again to equation (2.3), we can write a similar equation for the $H_2CO_3:HCO_3^-$ system, recalling that the pK value for this system in blood is 6.1,[15-17]

$$pH = 6.1 + \log \frac{[HCO_3^-]}{[H_2CO_3]} \qquad (2.4)$$

This is the well-known Henderson–Hasselbalch equation.[18,19] Now H_2CO_3 decomposes to form CO_2 and water:

$$H_2CO_3 \rightleftharpoons CO_2 + H_2O$$

Writing the equilibrium equation for this system, we obtain

$$\frac{[CO_2][H_2O]}{[H_2CO_3]} = K_{eq}$$

Since the amount of water reacting with CO_2 is negligible compared to the 55.6 mol of water in a liter, we can divide the constant by 55.6 and obtain a new constant

$$\frac{[CO_2]}{[H_2CO_3]} = K_1$$

or

$$[CO_2]_{dissolved} = K_1[H_2CO_3] \qquad (2.5)$$

The blood at the lungs is in equilibrium with the gases in the alveoli. The escaping tendency of the CO_2 in this blood (CO_2 tension) is normally of the order of 40 mmHg. The partial pressure of the CO_2 in the alveolar gas is also 40 mmHg, since a steady state exists. Now according to Henry's law, the amount of a gas dissolving in water is proportional to the partial pressure of that gas above the water and in equilibrium with it.[20] The symbol used for CO_2 partial pressure, and also (erroneously) for CO_2 tension, is p_{CO_2}. *Note that the symbol p_{CO_2} has no relationship to the symbol pH since it is not a logarithm.*

We can then write

$$p_{CO_2} \propto [CO_2]_{dissolved}$$

Therefore,

$$p_{CO_2} = K_2[CO_2]_{dissolved} \qquad (2.6)$$

Combining equation (2.5) with equation (2.6), we see that

$$p_{CO_2} = K_2[CO_2]_{\text{dissolved}} = K_1[H_2CO_3]$$

and

$$p_{CO_2}/K_3 = [H_2CO_3] \qquad (2.7)$$

$1/K_3$ has been experimentally determined to be from 0.0301 to 0.0324 by different investigators.[15-17,21] This value can be estimated from equations (2.4), (2.7), and the fact that the total CO_2 (which can be measured readily) is equal to the sum of the numerator and denominator of equation (2.4):

$$\text{total } CO_2 = [H_2CO_3] + [HCO_3^-] = (1/K_1)p_{CO_2} + [HCO_3^-] \quad (2.8)$$

Total CO_2, p_{CO_2}, and pH can all be measured directly. From equations (2.4) and (2.8) we then have two equations and two unknowns $[HCO_3^-]$ and $1/K_1$, from which $1/K_1$ can be calculated.

An estimate of $1/K_1$ can also be made as follows: In 1 ml of water, 0.545 ml of CO_2, corrected to 0°C, will dissolve at 38°C when the p_{CO_2} is 760 mmHg. The value of 0.545 has been corrected to indicate the volume at 0°. This is called the *solubility coefficient* (α) of CO_2 in water.

For serum the value of α has been determined to be 0.5108. By introducing serum into a Van Slyke gasometer, evacuating it a few times, and ejecting the supernatant gases, one has serum without CO_2 except as $NaHCO_3$. The pH of this serum would be approximately 8.5. If anhydrous CO_2 is now added in a measured amount, this system is shaken at 760 mmHg, and a measurement is taken, then the amount of CO_2 which dissolves per milliliter (α) can now be determined. This is 0.5108 ml of CO_2 corrected to 760 mmHg and 0°C. From Henry's law, the amount dissolved at a particular p_{CO_2} would be proportioned to the partial pressure and thus

$$0.5108 \times \frac{p_{CO_2}}{760} = \text{ml dissolved } CO_2 \text{ in 1 ml at } p = p_{CO_2}$$

and multiplying by 1000 to convert to one liter, we have

$$510.8 \times \frac{p_{CO_2}}{760} = \text{ml dissolved/liter}$$

Dividing by 22.2 to convert to mmol/liter, since 22.2 is the gram molecular volume of CO_2, we obtain

$$\frac{510.8}{760} \times \frac{1}{22.2} \times p_{CO_2} = 0.03028 p_{CO_2} = \frac{\text{mmol dissolved } CO_2}{\text{liter}}$$

This is the amount of CO_2 dissolved *other than the bicarbonate*, and to obtain the total CO_2 we must add this to the bicarbonate value:

$$\text{total } CO_2 = 0.03028 p_{CO_2} + [HCO_3{}^-] \tag{2.8a}$$

But this is the same as equation (2.8). Therefore $1/K_1$ is equal to 0.03028. Substituting in equation (2.4), we obtain the important equation

$$pH = 6.1 + \log \frac{[HCO_3{}^-]}{0.03 p_{CO_2}} \tag{2.9}$$

In the routine laboratory we can measure the blood pH, total CO_2, and p_{CO_2}. The total CO_2 is equal to the sum of the numerator and denominator of the fraction in equation (2.9), as shown in equation (2.8a).

From equations (2.8a) and (2.9) it is apparent that we have two equations and only one unknown, namely the bicarbonate ion concentration. Thus we can measure any two of the variables and calculate the third. Let us solve a problem to see how this operates. The laboratory finds that the pH of a patient's blood is 7.30, with a total CO_2 of 18 mmol/liter and a p_{CO_2} of 35 mmHg. Substituting the value of the pH in equation (2.9), we have

$$7.3 = 6.1 + \log \frac{[HCO_3{}^-]}{0.03 p_{CO_2}}; \quad \text{or} \quad \log \frac{[HCO_3{}^-]}{0.03 p_{CO_2}} = 1.2$$

From an antilog table, we find that the antilog of 1.2 is 15.85. Then

$$\frac{[HCO_3{}^-]}{0.03 p_{CO_2}} = 15.85$$

But from equation (2.10) and the experimentally determined value of the total CO_2 obtained (18 mmol/liter), we also know that

$$[HCO_3{}^-] + 0.03 p_{CO_2} = 18$$

We have two equations and two unknowns,

$$x/y = 15.85 \quad \text{and} \quad x + y = 18$$

Solving for y we obtain,

$$y = 0.03 p_{CO_2} = 1.068 \quad p_{CO_2} = 35.6$$

The measured value checks within experimental error and the results are acceptable. If they did not check, we would repeat the measurements to find the source of error.

TABLE 2.3

p_{CO_2} (in mmHg) as a Function of pH and Total CO_2

pH

CO_2 mmol/liter	6.90	6.95	7.00	7.05	7.10	7.15	7.20	7.25	7.30	7.35	7.40	7.45	7.50	7.55	7.60	7.65	7.70	7.75
4	18.2	16.5	14.9	13.5	12.1	10.9	9.80	8.80	7.91	7.10	6.36	5.70	5.10	4.57	4.09	3.66	3.27	2.92
5	22.8	20.6	18.6	16.8	15.2	13.6	12.3	11.0	9.89	8.88	7.96	7.13	6.38	5.71	5.11	4.57	4.10	3.65
6	27.4	24.8	22.4	20.2	18.2	16.4	14.7	13.2	11.9	10.6	9.5	8.6	7.66	6.85	6.13	5.48	4.90	4.38
7	31.9	28.9	26.1	23.5	21.2	19.1	17.2	15.4	13.8	12.4	11.1	10.0	8.9	8.0	7.2	6.4	5.7	5.11
8	36.5	33.0	29.8	26.9	24.2	21.8	19.6	17.6	15.8	14.2	12.7	11.4	10.2	9.1	8.2	7.3	6.5	5.83
9	41.0	37.1	33.5	30.3	27.3	24.5	22.1	19.8	17.8	16.0	14.3	12.8	11.5	10.3	9.2	8.2	7.4	6.57
10	45.6	41.3	37.3	33.6	30.3	27.3	24.5	22.0	19.8	17.7	15.9	14.3	12.8	11.4	10.2	9.1	8.2	7.30
11	50.2	45.4	41.0	37.0	33.3	30.0	27.0	24.2	21.8	19.5	17.5	15.7	14.0	12.6	11.2	10.1	9.0	8.03
12	54.7	49.5	44.7	40.4	36.4	32.7	29.4	26.4	23.7	21.3	19.1	17.1	15.3	13.7	12.3	11.0	9.8	8.76
13	59.3	53.6	48.5	43.7	39.4	35.5	31.9	28.6	25.7	23.1	20.7	18.5	16.6	14.8	13.3	11.9	10.6	9.49
14	63.8	57.8	52.2	47.1	42.4	38.2	34.3	30.8	27.7	24.8	22.3	20.0	17.9	16.0	14.3	12.8	11.4	10.2
15	68.4	61.9	55.9	50.4	45.5	40.9	36.8	33.0	29.7	26.6	23.9	21.4	19.1	17.1	15.3	13.7	12.3	11.0
16	73.0	66.0	59.6	53.8	48.5	43.6	39.2	35.2	31.7	28.4	25.5	22.8	20.4	18.3	16.4	14.6	13.1	11.7
17	77.5	70.1	63.4	57.2	51.5	46.4	41.7	37.5	33.6	30.2	27.0	24.2	21.7	19.4	17.4	15.5	13.9	12.4
18	82.1	74.3	67.1	60.5	54.5	49.1	44.2	39.7	35.6	31.9	28.6	25.7	23.0	20.6	18.4	16.4	14.7	13.1
19	86.6	78.4	70.8	63.9	57.6	51.8	46.6	41.9	37.6	33.7	30.2	27.1	24.2	21.7	19.4	17.4	15.5	13.9
20	91.2	82.5	74.5	67.3	60.6	54.6	49.1	44.1	39.6	35.5	31.8	28.5	25.5	22.8	20.4	18.3	16.3	14.6
21	95.8	86.6	78.3	70.6	63.6	57.3	51.5	46.3	41.5	37.3	33.4	29.9	26.8	24.0	21.5	19.2	17.2	15.8
22	100	90.8	82.0	74.0	66.7	60.0	54.0	48.5	43.5	39.0	35.0	31.4	28.1	25.1	22.5	20.1	18.0	16.1
23	104	94.9	85.7	77.3	69.7	62.7	56.4	50.7	45.5	40.8	36.6	32.8	29.3	26.3	23.5	21.0	18.8	16.8
24	109	99.0	89.5	80.7	72.7	65.5	58.9	52.9	47.5	42.6	38.2	34.2	30.6	27.4	24.5	21.9	19.6	17.5
25	114	103	93.2	84.1	75.8	68.2	61.3	55.1	49.5	44.4	39.8	35.6	31.9	28.6	25.6	22.8	20.4	18.3

26	19.0	21.2	23.8	26.6	29.7	33.2	37.1	41.4	46.1	51.4	57.3	63.8	70.9	78.8	87.4	96.9	107	118
27	19.7	22.1	24.7	27.6	30.8	34.5	38.5	43.0	47.9	53.4	59.5	66.2	73.6	81.8	90.8	100	111	123
28	20.4	22.9	25.6	28.6	32.0	35.7	39.9	44.5	49.7	55.4	61.7	68.7	76.4	84.8	94.2	104	115	127
29	21.2	23.7	26.5	29.6	33.1	37.0	41.3	46.1	51.5	57.4	63.9	71.1	79.1	87.9	97.5	108	119	132
30	21.9	24.5	27.4	30.7	34.3	38.3	42.8	47.7	53.2	59.3	66.1	73.6	81.8	90.9	100	111	123	136
31	22.6	25.3	28.3	31.7	35.4	39.6	44.2	49.3	55.0	61.3	68.3	76.0	84.6	93.9	104	115	127	141
32	23.4	26.1	29.2	32.7	36.5	40.8	45.6	50.9	56.8	63.3	70.5	78.5	87.3	97.0	107	119	132	145
33	24.1	27.0	30.2	33.7	37.7	42.1	47.0	52.5	58.6	65.3	72.7	80.9	90.0	100	111	123	136	150
34	24.8	27.8	31.1	34.7	38.8	43.4	48.5	54.1	60.3	67.3	74.9	83.4	92.7	103	114	126	140	155
35	25.6	28.6	32.0	35.8	40.0	44.7	49.9	55.7	62.1	69.2	77.1	85.8	95.5	106	117	130	144	159
36	26.3	29.4	32.9	36.8	41.1	45.9	51.3	57.3	63.9	71.2	79.3	88.3	98.2	109	121	134	148	164
37	27.0	30.2	33.8	37.8	42.3	47.2	52.7	58.9	65.7	73.2	81.5	90.8	100	112	124	137	152	168
38	27.7	31.0	34.7	38.8	43.4	48.5	54.2	60.5	67.4	75.2	83.7	93.2	103	115	127	141	156	173
39	28.5	31.9	35.6	39.9	44.5	49.8	55.6	62.0	69.2	77.1	85.9	95.7	106	118	131	145	160	177
40	29.2	32.7	36.5	40.9	45.7	51.0	57.0	63.6	71.0	79.1	88.1	98.1	109	121	134	149	165	182
41	29.9	33.5	37.5	41.9	46.8	52.3	58.4	65.2	72.8	81.1	90.3	100	111	124	137	152	169	187
42	30.7	34.3	38.4	42.9	48.0	53.6	59.9	66.8	74.5	83.1	92.5	103	114	127	141	156	173	191
43	31.4	35.1	39.3	43.9	49.1	54.9	61.3	68.4	76.3	85.1	94.7	105	117	130	144	160	177	196
44	32.1	35.9	40.2	45.0	50.3	56.2	62.7	70.0	78.1	87.0	97.0	108	120	133	148	164	182	201
45	32.8	36.8	41.1	46.0	51.4	57.4	64.1	71.6	79.9	89.0	99.0	110	123	136	151	168	186	205
46	33.6	37.6	42.0	47.0	52.6	58.7	65.6	73.2	81.7	91.0	101	113	126	139	155	172	190	210
47	34.3	38.4	43.0	48.0	53.7	60.0	67.0	74.8	83.4	92.9	104	115	128	142	158	175	194	214
48	35.0	39.2	43.9	49.1	54.8	61.3	68.4	76.4	85.2	94.9	106	118	131	145	161	179	198	219
49	35.7	40.0	44.8	50.1	56.0	62.5	69.8	78.0	87.0	96.9	108	120	134	149	165	183	202	224
50	36.5	40.8	45.7	51.1	57.1	63.8	71.2	79.6	88.7	98.9	110	123	136	152	168	186	206	228
51	37.2	41.7	46.6	52.1	58.3	65.1	72.6	81.1	90.5	101	112	125	139	155	172	190	210	233
52	37.9	42.5	47.5	53.1	59.4	66.4	74.1	82.7	92.3	103	115	128	142	158	175	194	215	237
53	38.7	43.2	48.4	54.2	60.5	67.6	75.5	84.3	94.1	105	117	130	145	161	178	198	219	242
54	39.4	44.1	49.3	55.2	61.7	68.9	77.0	85.9	95.9	107	119	133	147	164	181	201	223	246
55	40.1	44.9	50.3	56.2	62.8	70.2	78.4	87.5	97.6	109	121	135	150	167	185	205	227	251
56	40.9	45.7	51.2	57.2	64.0	71.5	79.8	89.1	99.4	111	123	137	153	170	188	209	231	255
57	41.6	46.6	52.1	58.3	65.1	72.7	81.2	90.7	101	113	126	140	156	173	192	212	235	260
58	42.3	47.4	53.0	59.3	66.3	74.0	82.7	92.3	103	115	128	142	158	176	195	216	239	265

The p_{CO_2} and pH could have been used in a similar manner to calculate the total CO_2, or the total CO_2 and p_{CO_2} could have been used to calculate the pH. These calculations have been made for a wide range of values and are shown in Table 2.3. Tables A.3 and A.4 in the appendix use this equation to relate the p_{CO_2} and pH to the bicarbonate value in plasma. p_{CO_2} and p_{O_2} are usually given in mmHg at 0°C. For mmHg many journals require the designation torr, or Torr, or Tors. In this book we will use the more descriptive mmHg.

2.8 SELECTED READING—THE BICARBONATE SYSTEM

Peters, J. P. and Van Slyke, D. D., *Quantitative Clinical Chemistry*, Williams & Wilkins, Baltimore, Maryland (1931), pp. 868–1018.

Siggaard-Andersen, O., *Acid–Base Status of the Blood*, 3rd ed., Williams & Wilkins, Baltimore, Maryland (1974).

Christensen, H. N. *Body Fluids and the Acid–Base Balance*, Saunders, Philadelphia, (1964).

2.9 REFERENCES

1. Howell, S. F. and Sumner, J. B., The specific effects of buffers on urease activity, *J. Biol. Chem.* **104**:619–26 (1934).
2. Arrhenius, S. A., The influence of neutral salts on reaction velocity of the hydrolysis of ethylacetate. Theory of isohydric solutions, *Z. Physik. Chem.* **1**:110–133 (1887); **2**:284–295 (1888).
3. Bagster, L. S. and Cooling, G., Electrolysis of hydrogen bromide in liquid sulfur dioxide, *J. Chem. Soc.* **117**:693–696 (1920).
4. Brönsted, J. N., The conception of acids and bases, *Rec. Trav. Chim.* **42**:718–728 (1923).
5. Brönsted, J. N., Acid and base catalysis, *Chem. Rev.* **5**:231–338 (1928).
6. Clark, W. M., *The Determination of Hydrogen Ions*, Williams and Wilkins, Baltimore, Maryland (1920), p. 19.
7. Sörensen, S. P. L., On the measurement and significance of hydrogen ion concentration in enzymatic processes, *Biochem. Z.* **21**:131–200 (1909).
8. Guldberg, C. M. and Waage, P., Etudes sur les affinités chimiques, Brogger and Christie, Christiana (1867).
9. Weast, R. C. and Selby, S. M., eds., *Handbook of Chemistry and Physics*, 46th Ed., Chemical Rubber Publ. Company, Cleveland, Ohio (1965–66), p. D-78.
10. Van Slyke, D. D., On the measurement of buffer values and on the relation-

ship of buffer values to the dissociation constant of the buffer and the concentration and reaction of the buffer solution, *J. Biol. Chem.* **52**:525–570 (1922).

11. Van't Hoff, J. H., The role of osmotic pressure in the analogy between solutions and gases, *Z. Physik. Chem.* **1**, 481–508 (1887).

12. Cullity, B. D., *Elements of X-ray Diffraction*, Addison-Wesley, Reading, Mass. (1956), p. 47.

13. Kolthoff, I. M. *et al.*, *Volumetric Analysis*, Vol. I (1942), Vol. II (1947), Vol. III (1957), Wiley, New York.

14. Hastings, A. B. and Sendroy, J., Jr., The value of pK' in the Henderson–Hasselbalch equation for blood serum, *J. Biol. Chem.* **79**:183–192 (1928).

15. Rispens, P., Dellebarre, C. W., Eleveld, D., Helder, W., and Zijlstra, W. G., The apparent first dissociation constant of carbonic acid in plasma between 16 and 42.5°C, *Clin. Chim. Acta*, **22**:627–637 (1968).

16. Severinghaus, J. W., Stupfel, M., and Bradley, A. F., Variations of serum carbonic acid pK' with pH and temperature, *J. Appl. Physiol.* **9**:197–200 (1956).

17. Siggaard–Andersen, O., The first dissociation exponent of carbonic acid as a function of pH, *Scand. J. Clin. Lab. Invest.* **14**:587–597 (1962).

18. Henderson, L. J., The theory of neutrality regulation in the animal organism, *Am. J. Physiol.* **21**:427 (1908).

19. Hasselbalch, K. A., The calculation of blood pH from free and bound CO_2 as a function of pH, *Biochem. Z.* **78**:112 (1916), **78**:251 (1917).

20. Sackur-Breslau, O. and Stern, O., The osmotic pressure of concentrated solutions of carbon dioxide, *Z. Elektrochem.* **18**:641–644 (1912).

21. Van Slyke, D. D., and Sendroy, J., Jr., Studies of gas and electrolyte equilibrium in blood; line charts for graphic calculations by the Henderson–Hasselbalch equation, and for calculating plasma carbon dioxide content from whole blood content, *J. Biol. Chem.* **79**:781–798 (1928).

22. Van Slyke, D. D., Sendroy, J., Jr., Hastings, A. B., and Neill, J. M., Studies of gas and electrolyte equilibria in blood. X. The solubility of carbon dioxide at 38°C in water, salt solution, serum and blood cells, *J. Biol. Chem.* **78**:765–799 (1928).

Partial Pressures $\left(p_{O_2} \text{ and } p_{CO_2}\right)$

3.1 INTRODUCTION

The total pressure of a mixture of gases is equal to the sum of the pressures of all the gases present as though each were there alone. For example, if a gas in a sealed container is at 1 atm pressure and consists of 20% oxygen and 80% nitrogen, then if the oxygen is removed at constant volume, we have 0.8 atm of pressure. If we remove the nitrogen, keeping the volume constant, then we have 0.2 atm. We say, then, that the *partial pressures* of oxygen p_{O_2} and nitrogen p_{N_2} are 0.2 and 0.8 atm, respectively.

The composition of dry air is shown in Table 3.1 in mmHg,

TABLE 3.1

Composition of Dry Air at 0°C and 760 mmHg[1]

Component	Content, %		Partial pressure, mmHg	
Nitrogen	78.08		593.4	
Oxygen	20.95		159.2	
CO_2	0.033		0.25	
Argon	0.934		7.10	
Neon	18	$\times 10^{-4}$	13.7	$\times 10^{-3}$
Helium	5.3	$\times 10^{-4}$	4.0	$\times 10^{-3}$
Krypton	1.2	$\times 10^{-4}$	0.9	$\times 10^{-3}$
Xenon	0.087	$\times 10^{-4}$	0.07	$\times 10^{-3}$
Hydrogen	0.5	$\times 10^{-4}$	0.4	$\times 10^{-3}$
Methane	2	$\times 10^{-4}$	1.5	$\times 10^{-3}$
N_2O	0.5	$\times 10^{-4}$	0.4	$\times 10^{-3}$

TABLE 3.2
Vapor Pressure (vp) of Water at Various Temperatures[a]

Temp., °C	vp, mmHg	Temp., °C	vp, mmHg	Temp., °C	vp, mmHg	Temp., °C	vp, mmHg
16	13.6	23	21.1	30	31.8	37	47.1
17	14.5	24	22.4	31	33.7	38	49.7
18	15.5	25	23.8	32	35.7	39	52.4
19	16.5	26	25.2	33	37.7	40	55.3
20	17.5	27	26.7	34	39.9	41	58.3
21	18.7	28	28.3	35	42.2	42	61.5
22	19.8	29	30.0	36	44.6	43	64.8

[a] Abridged from table in Ref. 2.

assuming a total pressure of 760 mmHg. In this table the partial pressure of the gas is obtained by taking the percentage of the total pressure P (760 mmHg). In air, water vapor is always present in variable amounts. Neglecting the rare gases other than argon, we can represent the total pressure p of air as[1]

$$P = p_{N_2} + p_{O_2} + p_{CO_2} + p_A + p_{H_2O}$$

The vapor pressure of water is given by a hygrometer in *relative humidity*, that is, in percentage saturation of air. Thus a value of 60% at 25°C means 60% of a p_{H_2O} of 23.8 mmHg (from Table 3.2), or 14.3 mmHg. Thus, if one reads the total air pressure from a manometer as 759 mmHg at that temperature and humidity, then the dry air pressure is 744.7 mmHg. Taking the corresponding percentages of 744.7 from Table 3.1, we find

$$P = 759 = 581.5 \ (p_{N_2}) + 156.0 \ (p_{O_2}) + 14.3 \ (p_{H_2O}) \\ + 0.24 \ (p_{CO_2}) + 7.0 \ (p_A)$$

If air is inhaled by a normal individual, held for 1 min, and partially expelled, and the residual gas is blown into a rubber balloon, we find that the gas is saturated with water at 37°C: $p_{H_2O} = 47$ mmHg, $p_{CO_2} = 40$ mmHg and p_{O_2} is approximately 100 mmHg. These values, then, are the partial pressures of the gases in the alveoli. But this gas, when in the lungs, was in equilibrium with the arterial blood in the

---

[1] A capital P symbolizes the total pressure. A lowercase p symbolizes a partial pressure. Thus the symbol p_{O_2} and not P_{O_2} to symbolize the partial pressure of oxygen is the convention accepted by chemists.

lungs; therefore the corresponding tensions of the gas are the same as the partial pressures of the gas. As explained above, the gas pressure is the partial pressure of the gas phase of a particular component, while the gas tension is the escaping tendency of the gas into the gas phase from the solution. When p_{CO_2} is being measured in the Severinghaus type of electrode, tensions are being measured and these symbols have come to mean the tensions of the gases in blood. If the CO_2 and O_2 tensions are measured for *venous blood* with these electrodes, then one obtains, for the normal, values of the order of $p_{CO_2} = 46$ mmHg and $p_{O_2} = 40$ mmHg. For arterial blood, the values obtained are $p_{CO_2} = 40$ mmHg and $p_{O_2} = 100$ mmHg. Table 3.3 lists p_{O_2} and p_{CO_2} values found in arterial blood, venous blood, and mixed alveolar gas.

If no chemical reaction takes place when a gas dissolves in water, then the amount dissolved is proportional to the partial pressure of the gas in contact with the water (Henry's law). Thus, in water, at p_{O_2} of 100 mmHg, twice as much is dissolved as at p_{O_2} of 50 mmHg. This is not true for either O_2 or CO_2 in blood. From the equation

$$pH = 6.1 + \log \frac{[HCO_3^-]}{0.03 p_{CO_2}}$$

one can readily see that the amount of total CO_2 which would dissolve in a buffered solution under various values for p_{CO_2} is not a linear function of the partial pressure of the gas.

In the case of oxygen, Henry's law is also not followed because of the reaction between oxygen and hemoglobin to form oxyhemoglobin. Since oxyhemoglobin is more acidic than hemoglobin, the amount of oxygen dissolved will also vary somewhat with pH.

Figure 3.1 shows a plot of percentage HbO_2 in blood (oxygen

TABLE 3.3

Partial Pressures of Blood and Alveolar Gases at 37°C[3]

	Alveolar gas	Arterial blood	Venous blood
p_{O_2}	104	100	40
p_{CO_2}	40	40	46
p_{H_2O}	47	47	47
$p_{N_2} + p_A$	569	573	573
P (total)	760	760	706

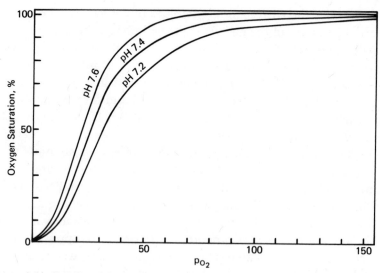

FIGURE 3.1   Relationship between percentage oxygen saturation and p_{O_2} at different pH values.[4] (The Bohr effect.)

saturation) versus change of p_{O_2}. If Henry's law were followed, this would be a straight line. Obviously, chemical reaction is taking place between the hemoglobin and the oxygen. The hemoglobin molecule contains four heme residues and combines with four oxygen molecules. Since the nature of the molecule changes as each heme reacts, we have four different equilibrium constants. For solutions of hemoglobin, the following formula has been derived and seems to apply at low Hb concentrations (less than 4%) and up to 75% saturation:

$$\frac{\% \text{HbO}_2}{100} = \frac{K_1 p + 2K_1 K_2 p^2 + 3K_1 K_2 K_3 p^3 + 4K_1 K_2 K_3 K_4 p^4}{4(1 + K_1 p + K_1 K_2 p + K_1 K_2 K_3 p^3 + K_1 K_2 K_3 K_4 p^4)}$$

where K_1, K_2, K_3, and K_4 are constants related to the equilibrium constants of the four intermediate reactions involved and p represents p_{O_2}.[5] From this equation we can average and factor out the constants to yield a new constant, average out the exponents of p, and simplify the equation to read

$$\frac{\% \text{HbO}_2}{100} = \frac{K p^n}{1 + K p^n} \qquad (3.1)$$

This simplified equation seems to fit the curve (Figure 3.1) fairly well. It was originally developed by Hill and has been used for many years.[6] It is in effect an empirical equation. The exponent of p is not 2.5 but varies from 1.5 to 3, the lower values being obtained at low salt concentrations. Thus at pH 7.1 and in 0.1 M phosphate buffer, for whole blood, $n = 2.8$.

From Figure 3.1, one can readily see that at p_{O_2} of 100 mmHg, 97% of the hemoglobin has reacted with oxygen. Even at a p_{O_2} of 50 mmHg, 85% of the hemoglobin has reacted. Thus the human can survive at various altitudes, and if the p_{O_2} is as low as 60 mmHg, the %HbO_2 will be 90%.

From the curve shown in Figure 3.1 one can see that the dissociation curve varies with pH. The p_{O_2} of the blood also increases with temperature, as would be expected. This is considered in preparing the nomogram of Figure 3.2. Figure 3.2 is based on the relationship connecting p_{O_2}, pH, and temperature developed by Astrup *et al.*[8] from the data of Van Slyke and others:

$$\log p_{O_2} = 0.0244\, \Delta T - 0.05\, \Delta pH$$

Since pH and p_{O_2} are routine measurements in the clinical laboratory, this relationship yields a convenient way of obtaining percentage oxygen saturation, which is technically more difficult to measure. ΔT represents the number of degrees away from 37 and ΔpH the units different from 7.4.

To find the percentage of hemoglobin saturated with oxygen (O_2 sat %) given the temperature, the measured p_{O_2} in mmHg, and the pH, connect the point on the temperature scale (scale A) to the p_{O_2} value on scale C. This intersects scale D giving a p_{O_2} value corrected for the temperature. Now draw a straight line from the measured pH value on scale B and the temperature-corrected p_{O_2} value, now read on scale C. Scale D will yield a p_{O_2} value now corrected for both temperature and pH, corresponding to a percent oxygen saturation value on scale E. For a sample with p_{O_2} measured at 60 mmHg and pH at 7.45, at 30°C, p_{O_2} corrected for temperature is 89.4 and corrected for pH is 94.1. This corresponds to an oxygen saturation value of 97.2%. *Oxygen saturation* means the amount of oxygen combined with hemoglobin as compared to the maximum capacity of the blood.

Using a second example more closely related to measurements taken in the routine laboratory, at 37°C, we find $p_{O_2} = 30$ mmHg and a blood pH of 7.25. The p_{O_2} value needs no temperature correction since we are at 37°C and the temperature-corrected p_{O_2} is 30 mmHg.

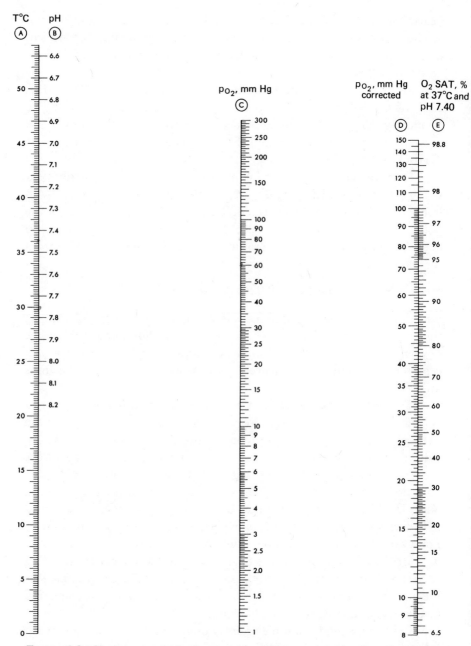

FIGURE 3.2 Nomogram to obtain percentage oxygen saturation for whole blood given the pH, temperature, and p_{O_2}.[7,8] (Radiometer Co.)

For a pH of 7.25, corrected p_{O_2} on the D scale is 24 mmHg. Connecting these two values (pH = 7.25 and p_{O_2} = 24 mmHg) on scales B and C, we read 52% oxygen saturation on scale E.

3.2   MEASUREMENT OF p_{CO_2} AND p_{O_2} [7]

Plastics, including rubber, are porous to gases but not to liquids. Evidence for this is the relatively rapid rate with which the helium in a balloon purchased at the zoo soon exchanges with the air and ceases to float. Advantage is taken of these phenomena to sense the CO_2 and oxygen tension of blood. For CO_2 the blood is placed in contact with a thin Teflon film. On the other side of the membrane is a weak bicarbonate solution (0.01 M). A glass electrode senses the pH change and records it as p_{CO_2}.

For oxygen, a weak current passes through a KCl solution 0.01 M with respect to KOH. Hydrogen collects at the cathode and reduces the current by acting as an insulator (polarizes the electrode). The solution is contained in polypropylene and the blood contacts the outer side of the polypropylene. As oxygen diffuses through the membrane, it reacts with the hydrogen and permits an increased current to flow (depolarizes the electrode). This degree of depolarization is a function of the oxygen tension of the blood.

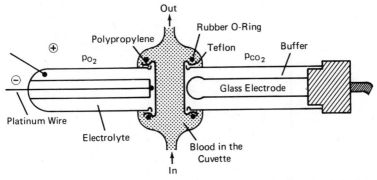

FIGURE 3.3   Representation of the p_{CO_2}, p_{O_2} electrode. The blood is injected between the two membranes. Diffusion of CO_2 changes the pH of the 0.01 M bicarbonate buffer, which is sensed by the pH electrode. Diffusing oxygen from the blood depolarizes the platinum electrode, causing an increase in flow of current, which is sensed by a sensitive ammeter.

Usually, p_{CO_2} and p_{O_2} are measured simultaneously as shown in Figure 3.3.

3.3 EFFECT OF OXYGENATION OF HEMOGLOBIN ON BLOOD p_{CO_2}, TOTAL CO_2, AND pH

Hemoglobin, like proteins in general, is amphoteric in nature. Its isoelectric point is 6.8 when fully reduced and 6.65 when fully oxygenated. At physiological pH (7.4–7.7) it is therefore an anion and acts as an acid.

At pH 7.4, oxyhemoglobin has five acid groups per mole of oxygen uptake or per mole of Fe or heme. These are therefore capable of buffering the blood. On titration, 2.5 Eq (equivalents) of alkali are needed to be added to fully oxygenated hemoglobin to bring a change of one pH unit per mole of Fe. Since there are four heme residues per molecule of hemoglobin, this indicates that a molecule of oxyhemoglobin has the titration equivalent of ten carboxyl groups. *For this reason, oxygenation of hemoglobin has the effect of adding acid to the blood, since it serves to increase the degree of ionization of the carboxyl groups.*

The pK of hemoglobin is 6.6 when oxidized and 7.85 when reduced. Conversion of hemoglobin to the oxidized form results in a loss of CO_2 from the blood and a drop in p_{CO_2}, if the same pH is to be

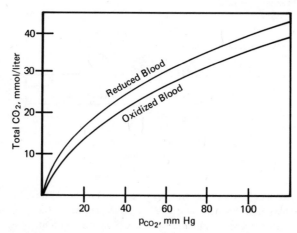

FIGURE 3.4   Effect of oxygenation of hemoglobin on CO_2 retention at constant pH. (The Haldane effect.)

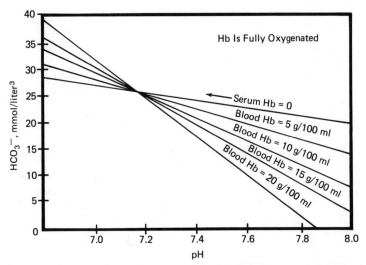

FIGURE 3.5　Effect of oxyhemoglobin on pH and bicarbonate content of horse blood. [Adapted from D. D. Van Slyke *et al.*, *J. Biol. Chem.* **56**:765 (1923)].

maintained. The loss of bicarbonate ion is 0.36 mEq/liter for each gram of hemoglobin oxygenated, as determined by Van Slyke. Thus, for 14 g of hemoglobin per 100 ml, one would experience an approximately $14 \times 0.36 = 5.0$ mEq/liter decrease in the CO_2 content of the blood when fully oxygenated, as compared to fully reduced blood. Since venous blood is partially oxygenated, the moving from venous to arterial blood results in less than a 2 mEq/liter decrease in bicarbonate level. From Figure 3.4 one can see this effect of oxygenating the blood on the p_{CO_2} and total CO_2 levels. This is known as the "Haldane effect." As Figure 3.4 shows, reduced blood at any given p_{CO_2} level binds more CO_2 than does oxygenated blood. At 40 mmHg this difference amounts to approximately 3 mmol/liter, as can be seen from the figure.

　　Figure 3.5 demonstrates that increasing the concentration of oxyhemoglobin significantly decreases the bicarbonate level, if constant pH is to be maintained. Referring to Figure 3.1, one sees the complementary effect, namely that the affinity of hemoglobin for oxygen decreases with decreasing pH.

　　From the above consideration it is apparent that *the CO_2 content of whole blood, even in the presence of a particular CO_2 tension, is variable and depends upon the pH, the hemoglobin concentration, and the extent of oxygenation.* The nomogram in Figure 3.6 demonstrates this relationship. It calcu-

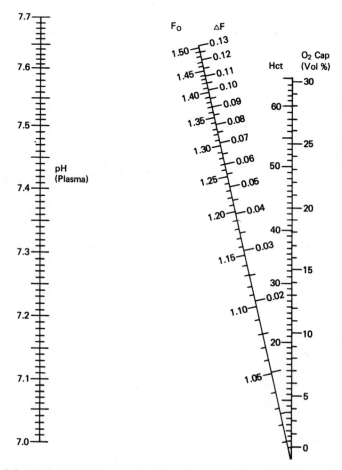

FIGURE 3.6   Relationship between extent of oxygenation of whole blood and the factor to convert blood CO_2 content to the CO_2 content of the plasma therefrom. [Adapted from D. D. Van Slyke and Sendroy, *J. Biol. Chem.* **79**:788 (1925).] (vol %, O_2 Cap) × 0.736 = Hb (g/100 ml); F × whole blood CO_2 = plasma CO_2; $F_0 = F$ for oxygenated blood; To obtain F for reduced blood, subtract ΔF from F_0; To obtain F for partially reduced blood, subtract from F_0, ΔF × % O_2 unsaturation/100.

lates a factor by which the whole blood CO_2 content is multiplied in order to obtain the plasma or serum CO_2 content. In the figure, O_2 capacity (O_2 Cap) signifies the oxygen content of the *fully oxygenated* blood in volume percent (milliliters of oxygen liberated per 100 ml of blood). Where the erythrocytes are not hypochromic or hyperchromic, this figure correlates with the percentage of cell volume or hematocrit value of the blood. A line drawn from the hematocrit to the pH line intersects the factor line F_0. Multiplying this number by the whole blood CO_2 level yields the plasma CO_2 level when the blood is fully oxygenated. If the blood is fully reduced, then one subtracts the ΔF value from the F_0 value to obtain the factor. For example, for a hematocrit value of 40% and pH of 7.4, the factor for oxygenated blood F_0 is 1.19. For fully reduced blood it is $1.19 - 0.0375 = 1.1525$. If the total CO_2 for the fully oxygenated blood is 23.5, then the plasma CO_2 content is $23.5 \times 1.19 = 28$ mmol/liter.

For an actual blood sample as received from the same patient, the blood pH and p_{O_2} are measured and the percentage oxygen saturation is first obtained from Figure 3.2. Let us assume that the patient's percentage oxygen saturation is 60%. Then we have from the equation for mixed blood,

$$\text{factor to be subtracted} = \Delta F \times \frac{\% \ O_2 \ \text{saturation}}{100}$$

$$= 0.0375 \times \frac{60}{100} = 0.0225$$

and thus

$$F_0 = 1.19 - 0.0225 = 1.1675$$

so that

$$\text{plasma } CO_2 \text{ content} = 23.5 \times 1.1675 = 27.4 \text{ mEq/liter}$$

The factor may be used in the inverse sense. That is, given the plasma CO_2 content, we can find the whole blood CO_2 content. One then divides the plasma CO_2 content by the factor to obtain the whole blood CO_2 content.

In calculating base excess or deficit in the patient, it is necessary, from the above, to take into account the hemoglobin level and extent of oxygenation. This will be discussed in Chapter 9 on acidosis and alkalosis.

3.4   OXYGEN CONTENT, CAPACITY, AND PERCENTAGE SATURATION

Oxygen content refers to the concentration in volume of oxygen per 100 ml of blood or vol %. The *oxygen capacity* is the concentration of oxygen in whole blood when the blood is saturated with oxygen, usually by equilibrating with air at p_{O_2} of approximately 150 mmHg, and is also expressed in vol %. Dividing the oxygen content by the oxygen capacity and multiplying by 100 gives the oxygen saturation, which is the amount of oxygen that has combined with hemoglobin as compared to the oxygen capacity, and is expressed in percent.

Oxygen content and capacity can be measured directly with a gasometer and the percentage oxygen saturation can be calculated from them.[9,10] Since hemoglobin and oxyhemoglobin have different absorption spectra, spectrophotometric procedures have been applied in the visible[11] and infrared[12] to estimate the percentage of oxygen saturation. Coupled with an estimation of the hemoglobin content, the oxygen content and capacity are also calculated. Since there is a direct relationship between the hematocrit value and hemoglobin content in the normal individual, some measure the hematocrit value instead of the hemoglobin content for the same purpose. Instrumentation is available for measuring oxygen content directly by amperometric titration. Figures 3.1, 3.2, and 3.4–3.6 interrelate the percentage oxygen saturation with p_{O_2}, pH, p_{CO_2}, bicarbonate, total CO_2, and hematocrit. These data are integrated into one nomogram in Figure 9.5 for practical application.

3.5   *HEMOGLOBIN AFFINITY FOR OXYGEN AND 2.3-DIPHOSPHOGLYCERATE*

It has been pointed out that a change in blood pH will change the affinity of oxygen for hemoglobin and the shape of the oxygen saturation curve (Figure 3.1). It has also been shown that at constant pH, the affinity of oxygen for hemoglobin will change with the concentration of 2.3-diphosphoglyceric acid (3,5-DPG or simply DPG) within the erythrocyte.[13–14a] DPG originates from the metabolism of glucose by the erythrocyte.

Normally, blood contains $15.7 \pm 1.4 \, \mu$mol of DPG per gram of hemoglobin.[15] This calculates to approximately equimolar quantities of DPG and hemoglobin. Inside the red cell the concentration of DPG is approximately 5 μmol/ml.

Approximately 30% of the DPG in the erythrocyte is bound to the cell wall, and the other 70% is presumably bound to hemoglobin.

When DPG binds to hemoglobin the curve is shifted to the right as though one had acidified the blood. This shift is usually described in terms of the oxygen tension that will yield a percentage oxygen saturation value of 50%. From Figure 3.2, this can be seen to be 26.5 mmHg. The symbol used for this point on the curve is usually P_{50}.[16]

An increase in the DPG concentration in the erythrocyte will result in a shift of the P_{50} point to the right. This can be seen in Figure 3.7. Since 1 g of hemoglobin reacts with 1.39 ml of oxygen, with a hemoglobin value of 14.4 g one would have an oxygen capacity of $14.4 \times 1.39 = 20$ vol %. The P_{50} value would then be that pressure where the oxygen content is 10 vol %, or at 27 mmHg. In the patient with cardiac disease, the DPG concentration may be increased[17,18] so that $P_{50} = 37$ mmHg. The shape of the curve is a similar sigmoid curve but shifted to the right. Let us see what results from this shift.

The p_{O_2} of arterial blood is approximately 97 mmHg and is approximately 97% saturated. Venous blood, on the other hand, is at

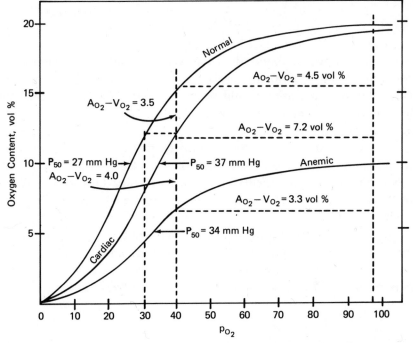

FIGURE 3.7 Oxygen dissociation curves in the normal as compared to a patient with decreased cardiac output due to disease and a patient with sickle cell anemia. The figures demonstrate the increased release of oxygen when the curve shifts to the right. It also shows that this effect decreases markedly at low oxygen tension, such as in lung disease.

p_{O_2} of 40 mmHg. The dashed vertical line at 40 mmHg and the one at 97 mmHg each intercepts both curves. Horizontal lines from the 40 mmHg intercepts permit us to calculate how much oxygen was liberated at the tissues per 100 ml of blood. For the normal curve this value is 4.5 ml of oxygen/100 ml of blood. For the particular patient with cardiac disease, with the curve shifted to the right, this calculates to 7.5 ml of oxygen/100 ml of blood. One can readily see the advantage of this in that metabolism may be maintained in the tissues with a substantial reduction in blood flow. This suggests that *control of erythrocyte DPG levels permits compensation for reduced blood flow in the patient with cardiac disease.*

Another example shown in Figure 3.7 is that observed in sickle cell anemia [19,20] where the hemoglobin concentration is reduced to 7.5 g. This is approximately one-half that shown in the normal and an arterial–venous difference of $4.5/2 = 2.25$ vol % would be expected. Actually the potential yield to the tissues is 3.3 ml of oxygen/100 ml of blood, an approximately 30% increase over that which would be expected from the hemoglobin concentration.

The minimal p_{O_2} required for survival is estimated to be somewhere between 20 and 30 mmHg. This is called the "critical p_{O_2}." Below this value the oxygen tension would be so low that it could not reach the mitochondria in adequate amounts. The minimum pressure required at the mitochondria is approximately 1 mmHg of oxygen but a pressure of at least 20–30 mmHg in the blood is required, depending upon the tissue, to attain this level. The shift of the P_{50} value to the right permits this value to be maintained with less hemoglobin. This is particularly true with certain abnormal hemoglobins such as hemoglobin Seattle.[21,22]

The alternative to this shift of the hemoglobin curve would be to increase cardiac output. In patients with anemia, cardiac output changes are small,[23] *the shift of the hemoglobin dissociation curve being a more practical method of compensation for low blood hemoglobin and thus low oxygen capacity.*

3.5.1 *Effect of Change in p_{O_2}*

To this point we have assumed that the arterial and venous p_{O_2} would remain constant. In pulmonary disease this is not the case, both values being decreased. Referring to Figure 3.7, the oxygenation of the patient's blood would be described by the Hb–oxygen dissociation curve at the steep portion of the curve. Let us assume an arterial p_{O_2} of 40 mmHg (the vertical dashed line) and a venous p_{O_2} of 30 mmHg. For the normal curve one finds an A–V oxygen content difference of approximately 3.5 vol %. For the curve shifted to the right and labeled "cardiac," this difference is approximately 4 vol % in this area as compared to 7.2 vol % at saturation. *Thus the advantage from the shift when the blood is well oxygenated is lost as the blood becomes less well oxygenated.*[24] If the p_{O_2} drops low enough, it is also possible for a shift of the P_{50} value to reach a level where a decrease in the oxygen availability from the hemoglobin takes place. The advantages of the DPG mechanism are thus compromised with lung disease.[25,26]

3.5.2   *Effect of Exercise*

With exercise the requirement for increased oxygen exchange can be met by two mechanisms, one by increased cardiac output and the second by a shift of the P_{50} value to the right. With pyruvate kinase deficiency, metabolism of DPG is inhibited and the DPG piles up, causing a shift to the right. With hexokinase deficiency less DPG is formed causing a shift to the left. Comparing two such patients doing exercise, the one with the shift to the left (hexokinase deficiency) was found to double cardiac output during exercise over that with the pyruvate kinase deficiency.[27] This demonstrates rather dramatically the two alternatives for compensating for increased oxygen demand.

3.5.3   *Shift of P_{50} by Drugs*

Certain drugs, such as propranolol, relieve the symptoms of angina pectoris. It has been shown that propranolol causes release of DPG from the wall of the erythrocyte and makes more DPG (30% more) available for combination with hemoglobin, causing a shift to the right.[15] This suggests that part of the mechanism of propranolol action may be due to its action in making more oxygen available to cardiac tissue. Epinephrine blocks the effect of propranolol on the erythrocyte membrane. This implies an additional mechanism for control of DPG concentration within the erythrocyte, involving the secretions of the adrenal medulla.

3.5.4 *Control of DPG Concentration*

The effect of DPG is controlled by two mechanisms. First, DPG is synthesized within the erythrocyte and does not pass the cell wall. Accumulation of DPG will cause acidification within the erythrocyte and a shift of the P_{50} value due solely to a pH effect as seen in Figure 3.2.

More important, DPG binds with hemoglobin at the beta chain of hemoglobin involving the terminal amino acid valine and the beta H-21 histidine group. These sites are located at the entrance to the central cavity of the hemoglobin molecule when hemoglobin is deoxygenated. If oxygenation takes place, then a conformational change in the molecule ensues, compressing the central cavity so that less DPG can bind. *Once DPG has been bound, the molecule resists this conformational change and an increased oxygen tension is required to oxygenate the blood.*[22] This is the significance of the change in the curve with a shift to the right as shown in Figure 3.7.

In normal glycolysis in the erythrocyte, 1,3-diphosphoglycerate is converted directly to 3-phosphoglycerate. Alternatively, 1,3-diphosphoglycerate converts to 2,3-diphosphoglycerate (DPG) and then to 3-phosphoglycerate. Thus a pathway to DPG and also a bypass are provided so that the DPG formed can be controlled without interfering with normal metabolism. Small increases in intracellular pH will stimulate DPG formation. Thus in alkalosis the DPG formation will be increased and the dissociation curve will shift to the right. In acidosis, DPG formation is decreased and the curve shifts to the left. *Thus variation in DPG concentration serves to neutralize the effect of pH on the shift of the dissociation curve and helps to maintain a normal dissociation curve in the face of acidosis or alkalosis.*[28–30]

3.5.5 *Stored Blood*

DPG concentration decreases in stored blood, resulting in a shift of the P_{50} value to the left and an increase in oxygen affinity for stored blood as compared to normal blood.[31,32] When administered to a patient, the DPG concentration will return to normal after 24 hr.[33] Addition of various substances, such as inosine, has been proposed to prevent this decrease, but except for those patients where an acute defect in oxygen delivery to the tissues exists, this decrease in DPG level is usually not considered of serious significance in clinical practice.

3.5.6   *Other Factors Affecting the Shift of the Oxygen Dissociation Curve*

Factors other than 2,3-DPG concentration and blood pH also can affect a shift of the oxygen dissociation curve to the right or left. For example, ATP has an effect resembling that of 2,3-DPG.[24,25] While DPG will cause a shift of the curve to the right for hemoglobins *A*, *S*, and *C*, this is not true for all hemoglobins. The effect of DPG on the major component of hemoglobin *F* (F_{11}) is approximately one-third that of hemoglobin *A*.[23,36] *However, it has no effect on the minor component* F_1. Thus the conformation of the particular hemoglobin plays a major role in the shift of the percentage dissociation curve.[14a]

3.5.7   *Methods for Obtaining P_{50} from the Oxygen Saturation Curve*

The hemoglobin concentration can be readily obtained. Multiplying this value by 1.39 yields the oxygen content when saturated. A partial pressure in mmHg to produce one-half this value (Figure 3.7) is the P_{50} value. The P_{50} value is inversely proportional to the hemoglobin binding force for oxygen. Obviously, the more pressure it requires to cause the oxygen to bind, the lower the binding force and the easier for oxygen release.

The P_{50} value can be obtained by measuring the oxygen content at various points and interpolating. This is done since just above and below the P_{50} point (Figure 3.7) we have essentially a straight line. Instrumentation has been proposed for continuously monitoring the oxygen content spectrophotometrically while increasing p_{O_2} values, a continuous plot being so obtained. The P_{50} value can readily be determined from these data.[37,38] A method which has been utilized by several derives from the Hill equation (3.1), which can be arranged as follows:

$$\frac{\%\ Hb\ sat'n}{100} = \text{fractional oxygen saturation} = y = \frac{Kp^n}{1 + Kp^n}$$

where K is the binding constant, n is the result of interaction of the four heme residues on each other's equilibrium constant and is of the order of 2.8, and p is the partial pressure of oxygen. Rearranging this equation, we have

$$y/(K - Ky) = p^n \qquad (3.2)$$

Substituting the value of P_{50} ($y = 0.5$) in equation (3.2), we obtain

$$0.5/(K - 0.5K) = 1/K = P_{50}^n \qquad (3.3)$$

Note from equation (3.3) that P_{50}^n is inversely proportional to the equilibrium or binding constant and is therefore an inverse measure of oxygen hemoglobin bond strength.

Dividing equation (3.3) into equation (3.2), we have

$$y/(1 - y) = (p/P_{50})^n$$

and

$$\log \frac{y}{1 - y} = n \log \frac{p}{P_{50}} \qquad (3.4)$$

A plot of the logarithm of p vs. the logarithm of $y/(1 - y)$ will yield a straight line. When $y = 0.5$, the value of p is P_{50} and $\log[0.5/(1 - 0.5)]$

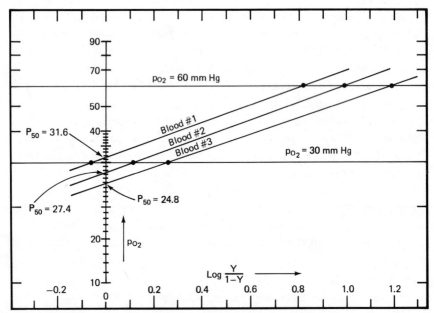

FIGURE 3.8   Plot on semilog paper of p_{O_2} vs. log $[y/(1 - y)]$, in order to obtain the P_{50} value. Blood is equilibrated with gas with p_{O_2} at 60 and 30 mmHg to obtain the two points for each curve.

TABLE 3.4

Calculations for Plotting log p_{O_2} *vs.* log $[y/(1 - y),]$ *from Fractional Oxygen Saturation Measurement y at Two p_{O_2} Values*[a]

Blood #		$p_{O_2} = 30$ mm			$p_{O_2} = 60$ mm		
	y	$\dfrac{y}{1-y}$	$\log \dfrac{y}{1-y}$	y	$\dfrac{y}{1-y}$	$\log \dfrac{y}{1-y}$
1	0.47	0.8865	-0.0522	0.87	6.692	0.8255
2	0.57	1.326	0.1225	0.91	10.11	1.057
3	0.67	1.826	0.2615	0.94	15.67	1.196

[a] Fractional oxygen saturation is calculated from the measured oxygen contents divided by the oxygen capacity (hemoglobin value × 1.39). $y \times 100 = \%$ oxygen saturation.

$= \log 1 = 0$. An intercept at the ordinate for the line will then yield P_{50}. This can be seen in Figure 3.8. In Figure 3.8 the line is obtained by equilibrating the blood with gas to two known p_{O_2} values. The oxygen content is determined at these two points, spectrophotometrically, by gasometer, or by amperometric titration and the value is used to obtain the logarithm of $y/(1 - y)$. Table 3.4 shows the calculations which resulted in the lines drawn for Figure 3.8 in order to obtain the P_{50} values. The blood sample #1 shows a shift to the right and blood sample #3 shows a shift of the curve to the left. Blood sample #2 is essentially normal.

For the above purposes one can determine the oxygen content by adding 50 μl of whole blood to 5 ml of 0.2% sodium ferricyanide solution and measuring the p_{O_2} of the solution containing the liberated oxygen.[39] With a desk top minicomputer one can readily program for obtaining the P_{50} value from the data above and equation (3.4).[39a]

3.6 SELECTED READING—PARTIAL PRESSURES p_{O_2} and p_{CO_2}

Astrup, P. and Rørth, M., Eds., *Oxygen Affinity of Hemoglobin and Red Cell Acid Base Status*, Academic Press, New York (1972).

Présent, R. D., *Kinetic Theory of Gases*, McGraw-Hill, New York (1958).

Roughton, F. J. W. and Kendrew, J. C., *Haemoglobin*, Interscience, New York (1949), p. 311.

Dittmer, D. S. and Grebe, R. M., *Handbook of Respiration*, Saunders, Philadelphia, Pennsylvania (1958).

3.7   *BLOOD AND ALVEOLAR* p_{CO_2} *AND* p_{O_2}

The blood p_{CO_2} and p_{O_2} values are functions of several factors. These include the rate and depth of breathing, the rate of blood flow to the lungs and the rate of gas exchange taking place in the alveoli of the lungs.[40] All of these factors are subject to change in the human.

Voluntarily, one can increase or decrease both the depth and rate of breathing. Blood flow to and from the alveoli increases with exercise. With cardiac disease, this blood flow may be impaired. In lung disease, such as emphysema, destructive changes in the alveoli can result in impaired gas exchange between the blood and the gases in the lungs.[41] In order to discuss this problem, certain terms need to be defined.

1. The volume of air inspired or expired during each cycle of normal breathing is called the *tidal volume*. This is of the order of 450–600 ml, when at rest.

2. If one inspires maximally, more air is inspired than at rest. The maximal amount of gas that can be inspired from the end of the normal inspiratory position is called the *inspiratory reserve volume*.

3. After one has expired normally, one can forcefully eject more air from the lungs. The volume of gas that can be expired from the end of the normal expiratory level is called the *expiratory reserve volume*.

4. Even after one forcefully expires as much as he can, there is still air left in the air spaces in the lungs. The volume of gas remaining in the lungs at the end of maximal expiration is called the *residual volume*.

The maximum amount of air capable of being moved by the lungs by forced expiration after maximal inspiration is called the *vital capacity*. This is the sum of the tidal volume, the inspiratory reserve volume, and the expiratory reserve volume. It excludes the dead space or residual volume.

The *total lung capacity* is the sum of the vital capacity plus the residual capacity. The sum of the expiratory reserve volume and the residual volume is called the *functional residual capacity*. The *inspiratory capacity* is the sum of the tidal volume and the inspiratory reserve volume.

The vital capacity, inspiratory capacity, and expiratory reserve volume are easily measured by the clinician and serve for preliminary diagnostic purposes. The mean volumes for the various factors discussed

TABLE 3.5

Mean Lung Volumes for Normal Adults, 20–30 y of Age[a]

Volume measured, ml	Male	Female
Vital capacity	4800	3200
Total lung capacity	6000	4200
Inspiratory capacity	3600	2400
Expiratory residual capacity	1200	800
Functional residual capacity	2400	1800
Residual volume	1200	1000

[a] Volumes are considered normal when $\pm$ 20% of the figures listed.[3]

above in the adult male and female are listed in Table 3.5. For children the volumes change with growth. The formulas in Table 3.6 are used to calculate volumes for vital capacity in boys and girls from ages 4 to 19.

Alveolar ventilation (V_A) is defined as the volume of air entering the alveoli per minute. In the normal, this is approximately four liters. Total ventilation is the volume of air entering the nose and mouth per minute. This signifies the sum of all the air inspired during the minute. It is apparent that alveolar ventilation is less than total ventilation since some air stays in the nose, oral cavity, trachea, and bronchi, which never reaches the alveoli before being expired. The difference between total and alveolar ventilation depends upon the tidal volume, the dead space volume, and the frequency of breathing. The p_{CO_2} in the alveoli is an inverse function of the rate of alveolar ventilation. This is apparent

TABLE 3.6

Vital Capacity in Normal Children[a]

Sex	Age	Vital capacity
Male	4–9	$(193 \times \text{age}) + 88$
	11–12	$(194 \times \text{age}) + 83$
	13–19	$(338 \times \text{age}) - 1720$
Female	4–11	$(191 \times \text{age}) - 62$
	12–16	$(200 \times \text{age}) - 121$
	17–18	$(154 \times \text{age}) + 608$

[a] The normal range is considered $\pm 30\%$ of the values calculated.[4]

from the fact that practically all of the CO_2 is derived from the lungs, very little arising from the air:

$$p_{CO_2}(\text{alveolar}) \propto \frac{CO_2 \text{ production per minute}}{\text{alveolar ventilation } (V_A)}$$

Another factor which must be considered is the rate of blood flow to the alveoli, called the *perfusion rate* $\dot{Q}$, where Q represents the quantity of blood flowing, and per unit time we have Q/T. The perfusion rate at any instant is dQ/dT, or in the dot symbolism $\dot{Q}$. In the same way the rate of alveolar ventilation at any instant is symbolized by $\dot{V}_A$. The ratio of rate of air flow in the alveoli divided by the perfusion rate is called the *ventilation–perfusion ratio* or *ventilation–blood flow ratio*. Since the flow of air into and out of the alveoli is approximately four liters/min, and the blood flow to the lungs is approximately five liters/min, this ratio, in the normal, is approximately $4/5 = 0.8$,[44]

$$\text{ventilation–blood flow ratio} = \dot{V}_A/\dot{Q} = 0.8$$

If the ventilation rate increases above that required to maintain a p_{CO_2} value of 40 mmHg in the alveoli, we call this *hyperventilation*. This can occur voluntarily, or with damage to the central nervous system such as in cerebrovascular accident or in salicylate poisoning.[45,46] The normal increased rate of ventilation, in proportion to increased metabolic rate, such as in exercise, is called *hyperpnea*.[47] Hyperpnea occurs to maintain a p_{CO_2} level of 40 mmHg in the face of increased CO_2 production.

If ventilation decreases below that required to maintain a p_{CO_2} value of 40 mmHg in the alveoli, the condition is called *hypoventilation*. In this case the p_{CO_2} in the alveoli rises and blood pH falls.[48]

3.8   RELATIONSHIP BETWEEN ALVEOLAR AND BLOOD p_{CO_2} AND p_{O_2}

The p_{CO_2} and p_{O_2} of the gases in the alveoli will determine the p_{CO_2} and p_{O_2} of the blood, since, in the normal, the membranes of the alveoli are thin and porous to the flow of gases. This can be seen from Table 3.3.

The percentage of oxygen in dry air is 20.95% (Table 3.1). The p_{O_2} of air saturated with water vapor is 20.95% of the atmospheric pressure in mmHg minus the p_{H_2O}. At body temperature (37°C), p_{H_2O} is 47 mmHg (Table 3.2). For the p_{O_2} value of moist air we have 0.2095(760 − 47) = 149 mmHg. This is the p_{O_2} of air as it would exist after it passes the moist trachea. This air, when introduced into the alveoli, mixes with residual air, which has some of its oxygen removed and replaced with CO_2. The p_{O_2} in the alveoli will be a function, there-fore, of the ventilation of the alveoli, which depends on the depth and the rate of breathing.[49] By means of controls located in the central nervous system and in the carotid artery and aorta, which are sensitive to slight changes in pH and p_{O_2} levels, breathing rate is carefully adjusted so that the blood p_{CO_2} remains at approximately 40 mmHg and the p_{O_2} at approximately 100 mmHg.[50–53] The respiratory rate and depth of breathing or the pulmonary ventilation are then adjusted to the rate of CO_2 production in the body (Figure 4.2). With exercise, there is an increased rate of CO_2 production. Lung ventilation also increases, resulting in the p_{O_2} and p_{CO_2} remaining essentially constant, both in the blood and the alveoli. With constant vigorous exercise, the complete adjustment takes some time and one says, when balance is reestablished, that "he got his second wind."

One mole of oxygen yields 1 mol of CO_2. Then one would expect that if the p_{CO_2} in the alveoli were 40 mmHg, then the p_{O_2} would be 149 − 40 = 109 mmHg. The gram molecular volume of oxygen is 22.4 liters, while for CO_2 it is 22.2 liters. This is so because there is greater interaction between CO_2 molecules than O_2 molecules. This requires a greater number of CO_2 molecules to exert the same pressure as a certain number of oxygen molecules. In addition, some of the CO_2 formed is excreted in the sweat, urine, and through other pathways. Lesser amounts of the O_2 absorbed are also excreted through the urine, sweat, etc. The ratio of the number of milliliters of CO_2 excreted through the lungs to the oxygen absorbed is therefore not equal to one. This ratio, called the *respiratory exchange ratio R*, is 0.8 in the normal individual. To correct for this, the following simple formula can be derived:[54]

$$p_{O_2}(\text{alveolar}) = F(P - 47) - p_{CO_2}(\text{alveolar})\left(F + \frac{1 - F}{R}\right)$$

where F is the fraction of oxygen in the dry gas being breathed; from Table 3.1 we can see that in ordinary air it is 0.2095. P is the atmospheric pressure and $P - 47$ is the partial pressure of dry air; for air at 760 mmHg saturated with water, it is 760 − 47 = 713 mmHg. R is the respiratory exchange ratio, or 0.8 normally. Substituting these values in the equation, we obtain

$$p_{O_2}(\text{alveolar}) = 0.2095(760 - 47)$$
$$- 40\left(0.2095 + \frac{1 - 0.2095}{0.8}\right) = 101.5$$

If the ratio R were 1, then we would have $149 - 40 = 109$, as we calculated above.

Let us now apply this equation to a practical problem. A patient in an oxygen tent is breathing a gas containing 40% oxygen, the remainder being nitrogen. The laboratory reports a $p_{CO_2} = 50$ mmHg. This then is also the mean p_{CO_2} in the alveoli. *This may not always be exactly correct, but we make this assumption.* Atmospheric pressure reads 750 mmHg. What is the alveolar p_{O_2} and what should the blood p_{O_2} be? We have

$$p_{O_2}(\text{alveolar}) = 0.4(750 - 47) - 50\left(0.4 + \frac{1 - 0.4}{0.8}\right) = 223.7$$

The blood p_{O_2} should then be approximately 224 mmHg. We measure the blood p_{O_2} and find 180 mmHg. Assuming that the laboratory reports are reliable, and assuming that the gas in the tent is 40% oxygen and not a lesser value due to leakage, this discrepancy is significant and suggests that R is not 0.8. By rearranging the equation, we can solve for R:

$$R = \frac{(1 - F)p_{CO_2}}{F(P - 47 - p_{CO_2}) - p_{O_2}} = \frac{\text{ml } CO_2 \text{ excreted}}{\text{ml } O_2 \text{ absorbed}}$$

In the practical example cited above, the blood p_{CO_2} is 50 mmHg and p_{O_1} is 180 mmHg. The patient is breathing 40% oxygen at 750 mmHg atmospheric pressure. Assuming that the blood values are the mean alveolar values for p_{O_2} and p_{CO_2}, we have

$$R = \frac{(1 - 0.4)50}{0.4(750 - 47 - 50) - 180} = 0.37$$

In order to interpret this low respiratory exchange ratio, let us first examine the alveoli in a normal lung (Figure 3.9). No two alveoli are the same even in the normal lung, and the values listed above for the rate of blood flow, ventilation rate, and respiratory exchange ratio, and the volumes listed in Tables 3.5 and 3.6 are *mean values*.[55]

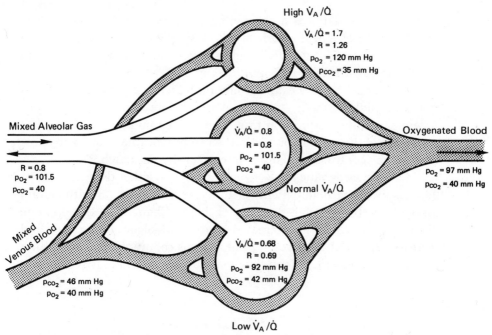

FIGURE 3.9 Effect of variations in the ventilation–blood flow ratio $\dot{V}_A/\dot{Q}_A$ on p_{O_2}, p_{CO_2}, and respiratory exchange ratio R in different alveoli in an individual.

Figure 3.9 compares three alveoli in a lung, one with a normal ventilation–blood flow ratio, another where the ratio is high, and a third where the ratio is low. Note that for a particular alveolus, the respiratory exchange ratio (CO_2 removed/oxygen dissolved) varies with the ventilation–blood flow ratio. This means, simply, that if an alveolus is highly ventilated with respect to the blood flow, the p_{O_2} in the alveolus will approach that of tracheal air (149 mmHg).

The rate of CO_2 removal is then greater than the rate at which it is being produced, which is also the rate at which oxygen is being consumed. The value of the numerator is therefore increased with respect to the denominator when solving for R. This is shown in Figure 3.9 as an increased value for R. In contrast, if the respiration rate is decreased, the rate of CO_2 removal will be decreased and the value of R will decrease.

When the value of R differs significantly from 0.8 it is often referred to as a "mismatch" between respiration rate to remove CO_2 and oxygen consumption. This condition is usually transient. If the condition causing the mismatch is a chronic one, readjustment will take place and the value of R will return to 0.8. A simple example is that of the distance runner who gradually adjusts ventilation and blood flow to match increased rate of CO_2 production.

A practical clinical example of a mismatch is a patient in acidosis who is rapidly brought to normal pH before the central nervous system can adjust respiration rate. This can be minimized by gradual correction of the acidosis or alkalosis when practicable in order to allow time for the control mechanisms to correct the mismatch.[55a]

An increased value for $V_A/\dot{Q}$ may be due to either decreased blood flow (as in the cardiac patient) or to hyperventilation (as with certain cases of cerebrovascular disease), or both. Changes in the opposite direction will result in decreased ventilation–blood flow ratios and increased p_{CO_2} and R values.[56] The various conditions that result in a decreased or increased blood flow to the lungs are summarized below.

3.9 EFFECT OF DECREASED BLOOD FLOW TO THE LUNGS ON p_{O_2} AND p_{CO_2}

Decreased blood flow to the lungs can result from any of the following causes[57]:

1. *Embolic* occlusion of pulmonary capillaries due to blood clot, fat, parasites, amniotic fluid, tumor, or gas.
2. *Obliteration* of pulmonary capillaries due to arteritis accompanying disseminated collagen diseases.
3. *Congenital* pulmonary arterial hypoplasia or pulmonary arterial branch stenosis.
4. *Mechanical* compression of pulmonary vessels by pneumothorax, pleural effusion, or space-occupying lesions.
5. *Degeneration* of the pulmonary vascular bed accompanying such chronic granulomatous processes as tuberculosis and histoplasmosis.
6. *Congestion* of pulmonary capillaries leading to chronic elevations of pulmonary arterial pressures and subsequent thickening and obliteration of the pulmonary arteriolar walls, occurring with such cardiac diseases as mitral stenosis and large atrial septal defects.

7. *Closure* of some pulmonary capillaries and an increase in intrapulmonary arterial to venous shunting, occurring when circulating blood volume is markedly reduced and during circulatory shock. This is similar to the creation of "Trueta shunts" discussed in the chapter on the kidney.

8. *Anastamotic arterial to venous channels* may persist within pulmonary hemangiomas. In this case mixed venous blood bypasses the alveoli and is not oxygenated.

In each of the above cases, since the rate of blood flow to the alveoli $\dot{Q}$ is decreased, the ventilation-to-blood flow ratio ($\dot{V}_A/\dot{Q}$) will be high for the alveoli involved if the ventilation rate is unchanged or increased.

While the level of p_{CO_2} will be low in the alveoli, since the alveoli are well ventilated, it will be higher in the blood, since in these cases blood is perfusing the alveoli slowly. The elevation of p_{CO_2} in the blood causes a lowering of blood pH, since p_{CO_2} is in the denominator of the pH equation [equation (2.9)]. This will result in activating the chemo-receptors in the brain to stimulate respiration and increase ventilation. This may or may not reach a rate to compensate for poor blood circulation. In any case, the value of the respiratory exchange ratio will be low.

In summary, the effect of decreased blood flow to the alveoli will generally result in elevated blood p_{CO_2} values.

3.10 EFFECT OF DECREASED VENTILATION ON p_{CO_2} AND p_{O_2}

Decreased ventilation of the lungs can occur for any of the following reasons[58]:

1. *Reduction* in the number of functioning alveoli such as occurs in advanced emphysema.

2. *Obstruction* of the smaller airways, as occurs in bronchial asthma and with chronic bronchitis; also with larger airway obstruction due to intrabronchial neoplasms or by simple mechanical obstruction due to foreign objects lodged in the trachea or mainstem bronchi.

3. *Collapse* of the terminal bronchioles leading to the formation of

"check valves" at the entrance to the alveolus such as occurs in obstructive emphysema.

4. *Interference* with the expansion of the alveoli caused by fluid or exudate in the alveoli or interstitial spaces to the lung. This type of restrictive defect occurs with pulmonary congestion, atelectasis, and tumor and is characteristic of pulmonary fibrosis.

In the areas of the lung affected by the above defects, the mean value of $\dot{V}_A/\dot{Q}$ is decreased due to a reduction in the value of the numerator, as measured with expired air. This results in a decrease in the mean respiratory exchange ratio with an increase in p_{CO_2} in the expired air of the alveoli and in the blood. The increased p_{CO_2} in the blood results in an acidosis and a signal to the chemoreceptors in the brain to stimulate respiration. The increased ventilation which results often does not compensate for the anatomical defect and these patients are in chronic respiratory acidosis, with high arterial p_{CO_2} levels and low p_{O_2} levels.

Defects in ventilation and in perfusion are often regional, and can both occur in different segments of the same lung. Their net effect on respiration depends on the extent of involvement. In general, both defects lead to an initial reduction in the arterial p_{O_2}. In the case of airway obstruction, this is due to alveolar hypoventilation. With pulmonary arteriolar blockage, redirection of nonoxygenated blood toward other alveolar capillaries is not entirely efficient and intrapulmonary arterial to venous shunting occurs.[61] The decrease in systemic arterial p_{O_2} stimulates the drive to respiration in an attempt to restore normal p_{O_2} tension. Thus the patient with moderate emphysema and regional defects in both perfusion and ventilation will have both somewhat reduced p_{O_2} and p_{CO_2} levels in the systemic arterial blood.

When lung disease is severe, with substantial bronchiolar obstruction and alveolar destruction, the p_{CO_2} level will rise as high as 50–60 mmHg. This condition is called *hypercapnia*, which merely means elevated CO_2 levels in the blood.[59] This individual will have an arterial p_{O_2} level, on room air, of 45–55 mmHg. If one now administers oxygen and raises the p_{O_2} level significantly above 100 mmHg and maintains it at that level, *his respiratory rate will decrease, resulting in further increases of CO_2 levels in the blood.*

The reason for this phenomenon is twofold. Located in the carotid and aortic arch are chemoreceptors[60] which stimulate respiration in response to lowered oxygen tension. Up to an arterial p_{O_2} of 100 mmHg these centers still serve to stimulate respiration. With increasing oxygen tension above this value due to oxygen administration, the respiratory rate will be lowered due to decreased stimulation from these centers. The p_{CO_2} tensions will now proceed to rise higher than their originally elevated values, reaching values as high as 70–90 mmHg.

Simply stated, oxygen administration designed to lower elevated p_{CO_2} levels will actually cause a rise in these levels since oxygen tensions above 100 mmHg result in respiration being suppressed.

The high oxygen tensions also keep a high percentage of the hemoglobin in the venous blood in the form of oxyhemoglobin, thus reducing its CO_2-carrying capacity. This results in an increased amount of the CO_2 accumulating at the tissue as dissolved CO_2, interfering with normal metabolism.

The highly elevated arterial p_{CO_2} levels also cause constriction of the pulmonary capillaries and dilation of the cerebral capillaries. This may result in increased intracraneal pressure. Approximately 10% of patients with chronic lung disease exhibit papilledema from time to time and mental confusion. Chronic pulmonary constriction can also lead to pulmonary hypertension and eventual damage to the pulmonary vascular bed and heart failure.[64]

3.11 SHUNT EFFECTS

Of importance to the clinician is the amount of "wasted" pulmonary blood flow or "shunting" that goes on in a diseased lung. If an alveolus is full of liquid, then blood around it will not be oxygenated. If flow through a portion of the lung is blocked, then some blood would flow through a blood vessel bypassing this area. Note also that blood passing through the capillaries of an alveolus which is not functioning is in effect a shunt. Resistance to a set of alveoli that causes extra flow to other alveoli is also a shunt. Since the blood is equilibrated with the alveolar oxygen in a normal alveolus, extra blood flow will not increase oxygen uptake significantly. This can be shown as in Figure 3.10. In the figure, the blood with a hemoglobin concentration of 14.8 g/100 ml is being oxygenated. If it were fully oxygenated, one would obtain $14.8 \times 1.39 = 20.57$ vol % oxygen capacity, since 1.39 ml of oxygen

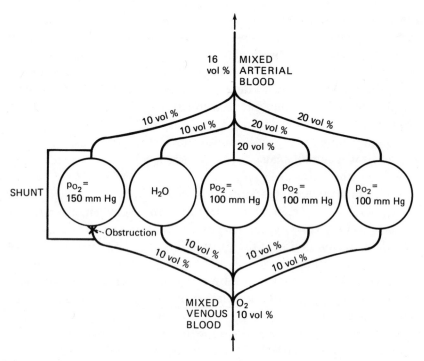

Figure 3.10   Schematic representation of the effect of shunting. In the illustration one alveolus is blocked by an obstruction and another is nonfunctioning owing to a pulmonary edema.

is taken up for each gram of hemoglobin. However, the blood is saturated to an extent of 97% at 100 mmHg and one obtains $20.57 \times 0.97 = 19.95$ or approximately 20 vol % for the normal alveoli. For the nonfunctioning alveoli the oxygen content is the same before and after passing the alveolus. Adding, we obtain $10 + 10 + 20 + 20 + 20 = 80$, and dividing by 5 we obtain a mixed arterial blood of 16 vol %.

In Figure 3.10 the ratio of venous admixture $\dot{Q}_{va}$ to total blood flow $\dot{Q}_t$ would then be

$$\frac{\dot{Q}_{va}}{\dot{Q}_t} = \frac{C_{iO_2} - C_{aO_2}}{C_{iO_2} - C_{\bar{v}O_2}}$$

where C_{aO_2} is the measured arterial oxygen content (16 vol %), $C_{\bar{v}O_2}$ is the measured mixed venous oxygen content (10 vol %), and C_{iO_2} is the "ideal" arterial oxygen content if all the alveoli were operating perfectly, which is 20% (see above):

$$\frac{\dot{Q}_{va}}{\dot{Q}_t} = \frac{20 - 16}{20 - 10} = \frac{4}{10} = 0.4 = \text{fraction "shunted"}$$

This signifies that 40% of the blood that flows to the lung is not being oxygenated. This can be confirmed from Figure 3.10 in that two of five (40%) alveoli are inoperative. This could occur due to shunts or inoperative alveoli but it could also be a result of partial obstruction to alveoli, the extra flow going to other areas of the lung. Thus the figure is an average value describing the overall efficiency in oxygenating the blood brought to the lung.

One must be cautioned that if blood is drawn from the artery and vein of the arm (e.g., radial artery and median cubital vein) for O_2 content, the extent of shunting will be overestimated in the normal since the blood from the vein is not the same as the mixed venous blood entering the lung, the blood from the arm having a somewhat higher O_2 content. In some cases it will be underestimated if blood flow to the arm is impaired.

Since it is difficult to obtain a sample of true mixed venous blood as presented to the lungs, various procedures have been devised to evaluate the shuntlike effects. A common procedure is to assume an A–V difference of 4.5 mmHg. The patient is given 100% oxygen and arterial content is measured. From this measurement and the hemoglobin and pH value that gives the "ideal" oxygen content, the shunt effect is estimated. The test is repeated with the patient breathing air of a lower oxygen tension and the shuntlike effect again calculated. If these values agree, then one is dealing with a true shunt effect. If they do not agree, then some other physiologic cause is involved. This procedure has been described in detail.[62]

For measuring oxygen content the Van Slyke apparatus can be used. In a more recent approach the oxygen content can be rapidly obtained with microliter quantities of blood by measuring the current generated as the oxygen reacts with a cadmium electrode[63] (Lexington Instrument Co., Waltham, Massachusetts).

3.12 RECAPITULATION

By means of controls located in the aorta and carotid artery and the central nervous system, changes in pH and p_{O_2} are sensed and adjustment made in the rate of ventilation of the lungs in order to maintain a constant p_{CO_2} level of approximately 40 mmHg and p_{O_2} level of approximately 100 mmHg in the arterial blood. In disease, this mechanism may be impaired. Determination of p_{CO_2}, p_{O_2}, and pH in the laboratory, with the knowledge of the composition of the gases being breathed by the patient, permits the calculation of the *respiratory exchange ratio R*. This measures the ratio of the CO_2 removed to the O_2 consumed. Significant variation from the normal ratio ($R = 0.8$) is referred to as a "mismatch" between CO_2 removal and CO_2 production. In correcting an acidosis or alkalosis the rate of pH adjustment should be designed to minimize the possibility of a mismatch.

Several factors, such as blood pH, 2,3-diphosphoglycerate (DPG) concentration in the erythrocyte, adrenaline, ATP, pyridoxal phosphate and certain drugs, cause a shift in the oxygen dissociation curve. The extent of the shift can be indicated by obtaining the oxygen tension when the blood is half saturated (P_{50} value). A shift to the right, as with increased DPG concentration or in acidosis, permits a greater exchange of oxygen between oxyhemoglobin and the tissues at a particular blood flow. A shift to the left has the opposite effect.

This mechanism is of significant value in providing adequate supply of oxygen to the tissue when cardiac output is diminished. Adjustment of 2,3-DPG concentration also helps to maintain a normal oxygen dissociation curve in acidosis and alkalosis.

The extent of wastage of blood circulation to the lungs (shunting effect) can be estimated from the oxygen content of venous and arterial blood provided that one can obtain a sample of the mixed venous blood presented to the lungs. Since this presents some difficulty, the shunting effect is usually estimated from the oxygen content of arterial blood with the patient breathing 100% oxygen and respiratory gases of lesser oxygen content, assuming an arterial venous difference of 4.5 vol %. If the several values agree, then one is dealing with a true shunting effect.

3.13   SELECTED READING—BLOOD AND ALVEOLAR p_{CO_2} AND p_{O_2}

Comroe, J. H., Jr., Forster, R. E. II, Dubois, A. B., Briscoe, W. A., and Carlsen, E., *The Lung, Clinical Physiology and Pulmonary Function Tests*, Year Book, 2nd ed., Chicago, Illinois (1967).

Pace, W. R., Jr., *Pulmonary Physiology in Clinical Practice*, 2nd ed., F. A. Davis Company (1970).

Woolner, R. F., Ed., *A Symposium on pH and Blood Gas Measurement*, Little, Brown, and Company, Boston (1959).

Torrance, R. W., Ed., *Arterial Chemoreceptors*, Blackwell, Oxford (1968).

Comroe, J. H., Jr. *Physiology of Respiration*, Year Book, Chicago, Illinois (1965).

3.15 REFERENCES

1. Standard atmosphere, in *Handbook of Chemistry and Physics*, 46th ed., Chemical Rubber Publ. Co., Cleveland, Ohio (1965–6), p. F-116.

2. *Handbook of Chemistry and Physics*, Chemical Rubber Publ. Co., Cleveland, Ohio (1965–6), pp. D-94, D-95.

3. Arndt, H. and Dolle, W., Comparative investigations with a quickly recording CO_2 electrode and the Astrup equipment in the determination of CO_2 pressure in human blood, *Klin. Wochenschr.* **44**, 511–15 (1966).

4. Sveringhaus, J. W., Blood gas calculator, *J. Appl. Physiol.* **21**:1108–1116 (1966).

5. Roughton, F. J. W., Recent work on carbon dioxide transport of the blood, *Physiol. Rev.* **15**:241–296 (1935).

6. Hill, A. V., The possible effects of the aggregation of the molecules of haemoglobin on its dissociation curves, *J. Physiol.* **40**:IV (1910).

7. Severinghaus, J. W., Design of a capillary pH electrode incorporating an open liquid junction and reference electrode in a single unit, *Scand. J. Clin. Lab. Invest.* **17**:614–616 (1965).

8. Astrup, P., Engel, K., Severinghaus, J. W., and Munson, E., The influence of temperature and pH on the dissociation curve of oxyhemoglobin of human blood, *Scand. J. Clin. Lab. Invest.* **17**:515–523 (1965).

9. Natelson, S. and Menning, C. M., Improved methods of analysis for oxygen, carbon monoxide and iron on fingertip blood, *Clin. Chem.* **1**:167–179 (1955).

10. Van Slyke, D. D. and Neil, J., The determination of gases in blood and other solutions by vacuum extraction and manometric measurement, *J. Biol. Chem.* **61**:523–573 (1924).

11. Maas, A. H. J., Hamelink, M. L., and De Leeuw, R. J. M., An evaluation of the spectrophotometric determination of HbO_2, HbCO and Hb in blood with the co-oximeter I. L. 182, *Clin. Chim. Acta* **29**:303–309 (1970).

12. Cole, J. S., Martin, W. E., Cheung, P. W., and Johnson, C. C., Clinical studies with a solid state fiberoptic oximeter, *Am. J. Cardiology* **29**:383–388 (1972).

13. Benesch, R. and Benesch, R. E., The effect of organic phosphates from the human erythrocyte on the allosteric properties of hemoglobin, *Biochem. Biophys. Res. Commun.* **26**:162–167 (1967).

14. Chanutin, A. and Curnish, R. R., Effect of organic and inorganic phosphates on the oxygen equilibrium of human erythrocytes, *Arch. Biochem. Biophys.* **121**:96–102 (1967).

14a. Mansouri, A. and Winterhalter, K. H., Nonequivalence of chains in hemoglobin oxidation and oxygen binding. Effect of organic phosphates, *Biochemistry* **13**:3311–3314 (1974).

15. Oski, F. A., Miller, L. D., Delivoria-Papadopoulos, M., Manchester, J. H., and Shelburne, J. C., Oxygen affinity in red cells: Changes induced in vivo by propranolol, *Science* **175**:1372–1373 (1972).

16. Klocke, R. A., Oxygen transport and 2,3-diphosphoglycerate (DPG), *Chest* **62** (suppl., part 2):79S–85S (1972).

17. Rosenthal, A., Mentzer, W. C., Eisenstein, E. B., Nathan, D. G., Nelson, N. M., and Nadas, A. S., The role of red blood cell organic phosphate in adaptation to congenital heart disease, *Pediatrics* **47**:537–547 (1971).

18. Woodson, R. D., Torrance, J. D., Shappell, S. D., and Lenfant, C., The effect of cardiac disease on hemoglobin–oxygen binding, *J. Clin. Invest.* **49**:1349–1356 (1970).

19. Charache, S., Grisolia, S., Fiedler, A. J., and Hellegers, A. E., Effect of 2,3-diphosphoglycerate on oxygen affinity of blood in sickle cell anemia, *J. Clin. Invest.* **49**:806–812 (1970).

20. Bromberg, P. A. and Jensen, W. N., Blood oxygen dissociation curves in sickle cell disease, *J. Lab. Clin. Med.* **70**:480–488 (1967).

21. Stamatoyannopoulos, G., Parer, J. T., and Finch, C. A., Physiologic implications of a hemoglobin with decreased oxygen affinity (hemoglobin Seattle), *N. Eng. J. Med.* **281**:915–919 (1969).

22. Bunn, H. F. and Briehl, R. W., The interaction of 2,3-diphosphoglycerate with various human hemoglobins, *J. Clin. Invest.* **49**:1088–1095 (1970).

23. Brannon, E. S., Merrill, A. J., Warren, J. V., and Stead, E. A., Jr., The cardiac output in patients with chronic anemia as measured by the techniques of right atrial catheterization, *J. Clin. Invest.* **24**:332–336 (1945).

24. Lenfant, C. A., Ways, P., Aucutt, C., and Cruz, J., Effect of chronic hypoxic hypoxia on the O_2-Hb dissociation curve and respiratory gas transport in man, *Respir. Physiol.* **7**:7–29 (1969).

25. Edwards, M. J. and Cannon, B., Normal levels of 2,3-diphosphoglycerate in red cells despite severe hypoxemia of chronic lung disease, *Chest* **61**:258–268 (1972).

26. Oski, F. A., Gottlieb, A. J., Miller, W. W., and Delivoria-Papadopoulos, M., The effect of deoxygenation of adult and fetal hemoglobin on the synthesis of red cell 2,3-diphosphoglycerate and its *in vivo* consequences, *J. Clin. Invest.* **49**:400–407 (1970).

27. Oski, F. A., Marshall, B. E., Cohen, P. J., Sugerman, H. J., and Miller, L. D., Exercise with anemia. The role of the left-shifted or right-shifted oxygen hemoglobin equilibrium curve, *Ann. Intern. Med.* **74**:44–46 (1971).

28. Bellingham, A. J., Detter, J. C., and Lenfant, C., Regulatory mechanisms of hemoglobin oxygen affinity in acidosis and alkalosis. *J. Clin. Invest.* **50**:700–706 (1971).

29. Bellingham, A. J., Detter, J. C., and Lenfant, C., The role of hemoglobin affinity for oxygen and red cell 2,3-diphosphoglycerate in the management of diabetic ketoacidosis, *Trans. Assoc. Amer. Physicians* **83**:113–120 (1970).

30. Asakura, T., Sato, Y., Minakami, S., and Yoshikawa, H., pH dependency

of 2,3-diphosphoglycerate content in red blood cells, *Clin. Chim. Acta* **14**:840–841 (1966).

31. Bunn, H. F., May, M. H., Kocholaty, W. F., and Shields, C. E., Hemoglobin function in stored blood, *J. Clin. Invest.* **48**:311–321 (1969).

32. Duhm, J., Deuticke, B., and Gerlack, E., Complete restoration of oxygen transport function and 2,3-diphosphoglycerate concentration in stored blood, *Transfusion* **11**:147–151 (1971).

33. Valeri, C. R. and Hirsch, N. M., Restoration *in vivo* of erythrocyte adenosine triphosphate, 2,3-diphosphoglycerate, potassium ion, and sodium ion concentrations following the transfusion of acid-citrate-dextrose-stored human red blood cells *J. Lab. Clin. Med.* **73**:722–733 (1969).

34. Garby, L, Gerber, G., and De Verdier, C. H., Binding of 2,3-diphosphoglycerate and adenosine triphosphate to human haemoglobin A, *Eur. J. Biochem.* **10**:110–115 (1969).

35. Benesch, R. E., Benesch, R., and Yu, C. I., The effect of pyridoxal phosphate on the oxygenation of hemoglobin, *Fed. Proc.* **28**:604 (1969).

36. Tyuma, I. and Shimizu, K., Different response to organic phosphates of human fetal and adult hemoglobins, *Arch. Biochem.* **129**:404–405 (1969).

37. Duvelleroy, M. A., Buckles, R. G., Rosenkaimer, S., Tung, C., and Laver, M. B., An oxyhemoglobin dissociation analyzer, *J. Appl. Physiol.* **28**:227–233 (1970).

38. Herman, C. M., Rodkey, F. L., Valeri, C. R., and Fortier, N. L., Changes in the oxyhemoglobin dissociation curve and peripheral blood after acute red cell mass depletion and subsequent red cell mass restoration in baboons, *Ann. Surg.* **174**:734–743 (1971).

39. Laver, M. B., Blood O_2 content measured with the pO_2 electrode: A modification, *J. Appl. Physiol.* **22**:1017–1019 (1967).

39a. Aberman, A., Cavanilles, J. M., Trotter, J., Erbeck, D., Weil, M. H., and Shubin, H., An equation for the oxygen hemoglobin dissociation curve, *J. Appl. Physiol.* **35**:570–571 (1973).

40. Armstrong, B. W., Hurt, H. H., Blide, R., and Workman, J., Relation of pulmonary ventilation to CO_2 partial pressure and hydrogen ion concentration in mixed venous and arterial blood, *Clin. Res.* **8**:252 (1960).

41. Barach, A. L. and Bickerman, H. A., *Pulmonary Emphysema*, Williams & Wilkins, Baltimore, Maryland (1956).

42. Kaltreider, N. L., Fray, W. W., and Hyde, H. V., Effect of age on total pulmonary capacity and its subdivisions, *Am. Rev. Tuberc.* **37**:662–689 (1938).

43. Stewart, C. A., Vital capacity of lungs of children in health and disease, *Am. J. Dis. Child.* **24**:451–496 (1922).

44. Briscoe, W. A., Forster, R. E., and Comroe, J. H., Jr., Alveolar ventilation at very low tidal volumes, *J. App. Physiol.* **7**:27–30 (1954).

45. Tenney, S. M. and Miller, R. M., Respiratory and circulatory actions of salicylate, *Am. J. Med.* **19**:498–508 (1955).

46. Lewis, B. I., Hyperventilation syndromes; a clinical and physiological observation, *Postgrad. Med. J.* **21**:259 (1959).

47. Comroe, J. H., Jr., The hyperpnea of muscular exercise, *Physiol. Rev.* **24**:319–339 (1944).

48. Eichenholz, A., Mulhausen, R. O., Anderson, W. E., and MacDonald, F. M., Primary hypocapnia: a cause of metabolic acidosis. *J. Appl. Physiol.* **17**:283–288 (1962).

49. Riley, R. L., Lilienthal, J. L., Jr., Proemmel, D. D., and Franke, R. E., On the determination of the physiologically effective pressures of O_2 and CO_2 in alveolar air, *Am. J. Physiol.* **147**:191–198 (1946).

50. Mitchell, R. W., Loeschke, H. H., Massion, W. H., and Severinghaus, J. W., Respiratory responses mediated through superficial chemosensitive areas on the medulla, *J. Appl. Physiol.* **18**:523–533 (1963).

51. Katz, R. L., Ngai, S. H., Nahas, G. G., and Wang, S. C., Relationship between acid base balance and the central respiratory mechanisms, *Am. J. Physiol.* **204**:867–872 (1963).

52. Schmidt, C. F. and Comroe, J. H., Jr., Functions of the carotid and aortic bodies, *Physiol. Rev.* **20**:115–157 (1940).

53. Liljestrand, A., Neural control of respiration, *Physiol. Rev.* **38**:691–708 (1958).

54. Riley, R. L., Cournand, A., and Donald, K. W., Analysis of factors affecting partial pressures of O_2 and CO_2 in gas and blood of lungs, *J. Appl. Physiol.* **4**:77–101 (1951).

55. Rahn, H., A concept of mean alveolar air and the ventilation-blood flow ratio relationships during pulmonary gas exchange, *Am. J. Physiol.* **158**:21–30 (1940).

55a. Goldring, R. M., Cannon, P. J., Heinemann, H. O., and Fishman, A. P., Respiratory adjustments in chronic metabolic acidosis in man, *J. Clin. Invest.* **47**:188 (1968).

56. West, J. B., Dollery, C. T., and Hugh-Jones, P., The use of radioactive CO_2 to measure regional blood flow in the lungs of patients with pulmonary disease, *J. Clin. Invest.* **40**:1–12 (1961).

57. de Reuk, A. V. S. and O'Connor, M., Eds., *Problems of the Pulmonary Circulation*, Little, Brown, Boston, Massachusetts (1961).

58. Gordon, B. L. and Kory, R. C., Ed, *Clinical Cardiopulmonary Physiology*, Grune & Stratton, New York (1960).

59. Schwartz, W. B., Hays, R. M., Polak, A., and Haynie, G. D., Effects of chronic hypercapnia on electrolyte and acid–base equilibrium. II. Recovery with special reference to the influence of chloride intake, *J. Clin. Invest.* **40**:1238–1249 (1961).

60. Dripps, R. D., Jr. and Comroe, J. H., Jr., The clinical significance of carotid and aortic bodies, *Am. J. Med. Sci.* **208**:681–694 (1944).

61. Wilson, J. E., III, Harrell, W. R., Mullins, C. B., Winga, E. R., Johnson, R. L., Jr., and Pierce, A. K., Hypoxia in pulmonary embolism, *Clin. Res.* **19**, 81 (1969).

62. Druger, G. L., Simmons, D. H., and Levy, S. E., The determination of shunt-like effects and its use in clinical practice, *Am. Rev. Resp. Dis.* **108**:1261–1265 (1973).

63. Valeri, C. R., Zaroulis, C. G., Marchionni, L., and Patti, K. J., A simple method for measuring oxygen content in blood, *J. Lab. Clin. Med.* **79**:1035–1040 (1972).

Maintenance of Constant pH in the Human

The problem the body faces is created by the fact that it obtains its energy for existence by oxidation of foods. Two reactions are involved in the overall scheme:

$$2H_2 + O_2 \rightarrow 2H_2O \tag{4.1}$$

$$C + O_2 \rightarrow CO_2 + H_2O \rightarrow H_2CO_3 \tag{4.2}$$

The first reaction produces water and does not affect the pH. The second reaction, however, produces carbon dioxide, which reacts with water to produce carbonic acid. The objective of the pH regulatory system is to maintain an alkaline pH (7.4) in the face of the production of massive quantities of CO_2. *The human system is then designed to contend with a tendency toward acidosis.* To do this, *a major problem is to conserve base* to be used in a cyclic manner for transporting the CO_2 from the cells where it is produced to the lungs where it can be expelled. The CO_2 transport system must operate in such a manner that constant pH is maintained throughout the fluids bathing the cells where metabolism is taking place. How this is done will now be discussed.

4.1 THE LUNGS

The most important system for maintenance of constant pH in the human is the combination of the respiratory centers in the brain, the diaphragm, and the lungs.[1] The importance of this system can be readily seen from a few simple facts.

The human breathes, at rest, approximately 11–14 times per minute. In each exchange, approximately 450–600 ml of air is taken in and the same volume of a gas containing approximately 5.3% of CO_2 by volume is expelled.[2] One can then calculate how much CO_2 is expelled per minute as follows, using 500 ml and 5% as round numbers:

500 ml × 0.05 = 25 ml

25 ml × 14 = 350 ml of pure CO_2 expelled per minute

Even at rest, 350 ml of pure CO_2 gas is expelled per minute. With exercise this can be many times higher. It is no wonder then that if an individual is prevented from breathing for 5 minutes, the pH of his blood will go from 7.4 to approximately 6.3 and he will soon die.

The respiratory centers of the brain are sensitive to pH changes and will adjust breathing rate to expel more CO_2 gas, to attempt to correct acidosis by breathing deeply (Kussmaul respiration).[3] To correct an alkalosis, as much CO_2 as compatable with minimum oxygen requirements is retained by shallow breathing (alkalotic respiration).

4.2   *THE HEMOGLOBIN–OXYHEMOGLOBIN SYSTEM*

There are two locations where a marked change in pH of the blood could take place. One is at the cells, where CO_2 is being produced, and the other is at the lungs, where CO_2 is being expelled. In order to correct for these sudden changes, a corresponding change takes place at the lungs and the tissues in the hemoglobin–oxyhemoglobin system.

The pK of oxyhemoglobin is 6.60, and the pK of hemoglobin is 7.85; thus oxyhemoglobin is a much stronger acid than hemoglobin. At the lungs, where CO_2 is lost, a weak acid, hemoglobin, is being changed to a stronger acid, oxyhemoglobin, in a mole-to-mole (stoichiometric) relationship. At the cells, for every mole of CO_2 liberated, a mole of O_2 leaves the oxyhemoglobin to form hemoglobin, whose pK is on the alkaline side, balancing the acidity produced by the CO_2. We thus have a strong buffering action based on a reversible change of an acid to a base,[4]

$$HbO_2(pK = 6.60) \rightleftharpoons Hb(pK = 7.85) + O_2$$
$$\text{(lungs)} \qquad\qquad \text{(cells)}$$

Hemoglobin has free amino groups which are capable of combining with carbonic acid in the following manner:

$$\underset{\substack{|\\R-N-H}}{\overset{H}{|}} + \underset{\substack{\|\\HO-C-OH}}{\overset{O}{\|}} \underset{\xrightarrow{-H_2O}}{\rightleftharpoons} \underset{\substack{|\ \|\\RN-COH}}{\overset{H\ O}{|\ \|}}$$

These are called carbamino compounds and 18–20% of carbon dioxide in the blood is in this form.[4a,4b] Actually, proteins in general contain free amino groups, and some carbon dioxide in the plasma is combined as carbamino compounds with the plasma proteins. It is of interest to note that *carbamino compound formation is of great significance in fixing CO_2 in sea water for growth of algae and plants.*

4.3 *THE BICARBONATE SYSTEM*

The bicarbonate system, along with other buffering systems discussed below, acts as the medium maintaining the pH of the extracellular fluid bathing the cells and thus their well being. The pK of this system is 6.1 and therefore at pH 7.4 it is somewhat mobile. As discussed above, this permits delicate adjustment so that the pH can be maintained precisely by controls in the central nervous system within a narrow range of a few hundredths of a pH unit.[5]

How this is done is best seen from a mechanical model of the servomechanism working in the human (Figure 4.1). Figure 4.1 illustrates a model system set up to simulate the bicarbonate system in the human. Into the container is placed 0.05 N sodium bicarbonate solution. This solution is at a pH of 8.6 when first dissolved. To this solution is added phenol red (P.S.P.), which is pink at a pH of 8.6. The air inlet is attached to a source of compressed air. The CO_2 inlet is attached to a tank of CO_2. By means of the two valves the rate of CO_2 and air flow can be regulated. The pH electrodes record the pH on a pH meter. An intermediate valve creates a constriction between the two containers.

The whole instrument is placed in a 37°C water bath. The capacity of the container is 50 ml, and 25 ml of the buffer is added. The stirrer stirs vigorously to assure even mixing. *Carbonic anhydrase (25 mg) is added so that equilibrium is brought about rapidly in the system.* The reason for this will be discussed below.

If air is blown rapidly through the bicarbonate buffer, CO_2 is lost and the $NaHCO_3$ becomes Na_2CO_3, the pH becoming about 10. If this

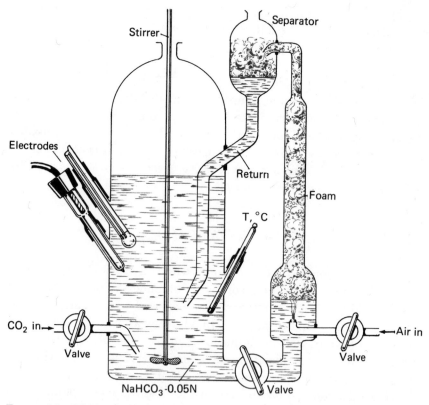

FIGURE 4.1 Mechanical model of the servomechanism utilizing the bicarbonate system for CO_2 transfer from the cells to the lungs while maintaining constant pH.

process is continued with CO_2-free air, a mixture of Na_2CO_3 and NaOH is formed and the pH will become approximately 11–12. If CO_2 is blown rapidly through the bicarbonate system, the pH will drop to approximately 6.1 and the phenol red will become colorless.

In the system of Figure 4.1, both air and CO_2 are blown through the system. The air flow is rapid and the solution foams. The foam is carried over to the separator, which is coated with silicone defoaming agent. It then returns to the main container. The foam and separator simulate the lungs. The CO_2 pouring into the solution represents the CO_2 from the tissues. By adjusting the rate of air flow and rate of flow of CO_2, the pH in the container can be maintained at 7.4 ± 0.01 pH units indefinitely. By increasing the flow of air, the pH can be maintained at any pH value up to approximately 8.0. By keeping the air flow constant and increasing the CO_2 flow, the pH can be maintained at any pH from 7.4 to approximately 6.4. If samples are taken from time to time and the pH, total CO_2, and p_{CO_2} are measured, they follow the relations

$$pH = 6.1 + \log \frac{[HCO_3^-]}{0.03 p_{CO_2}}$$

$$\text{total } CO_2 = [HCO_3^-] + 0.03 p_{CO_2}$$

Of importance is the fact that in this system the CO_2 is driving toward an acidosis which is prevented by a conservation of OH^- ions which serve to balance the constant sodium ion concentration electrically. This can be visualized in the following manner:

$$NaHCO_3 \rightleftharpoons Na^+ + HCO_3^-$$

$$H_2O \rightleftharpoons \underset{(1)}{\underline{OH^-}} + H^+$$

$$\Updownarrow$$

$$H_2CO_3$$

$$\Updownarrow$$

$$H_2O + \underset{(3)}{\underline{CO_2}}$$

$$\text{(2)}$$

The three products of the reaction are OH^-, CO_2, and H_2O. But the H_2O formed is equal to the H_2O we started out with, so that the net products of the reaction are OH^- and CO_2. The sodium content of the solution is the same at the end as the beginning. The hydroxyl ion and CO_2 originate from HCO_3^-. The net reaction taking place is then

$$HCO_3^- \rightarrow OH^- + CO_2$$

in the simulated lung of Figure 4.1. The solution returns to the container with a higher concentration of OH^- ions after being defoamed.

In the container to which CO_2 is being added, simulating the tissues, the reverse reaction takes place,

$$OH^- + CO_2 \rightarrow HCO_3^-$$

It is apparent then that the OH^- ion is being circulated and can be considered the carrier of CO_2. Now in the equation

$$HCO_3^- \rightleftharpoons OH^- + CO_2$$

we can see that *the bicarbonate, which is a base* according to the Brönsted theory, is being replaced by a *stronger base, namely the OH^- ion.*

This concept can be illustrated as follows: if blood from a patient in metabolic acidosis, which assays for pH = 7.2; p_{CO_2} = 24 mmHg; $[HCO_3^-]$ = 9 mEq/liter, is equilibrated with a gas where the pCO_2 = 40 mmHg, then one obtains pH = 7.05; p_{CO_2} = 40 mmHg; and $[HCO_3^-]$ = 11 mEq/liter (see Table A.3, Appendix). Thus, an increase of bicarbonate ion without a concomitant increase in sodium ion replaces OH^- ion and acidifies the blood. When the expression *bicarbonate base* is used, it is usually understood that sodium ion has increased along with bicarbonate ion.

We can also write an equation for air blowing through $NaHCO_3$ in the following manner, providing we stop the reaction before all of the CO_2 is blown off:

$$2NaHCO_3 \xrightarrow[\text{air}]{\text{blown}} Na_2CO_3 + H_2O + CO_2$$

and say that HCO_3^- base is being replaced by the stronger CO_3^{2-} base.

In either case, one must keep in mind that the major problem in maintaining constant pH in the human is conservation of base.[6] This can be shown schematically as it occurs in the human in Figure 4.2. The cells generate CO_2 and water, and the lungs expel the CO_2. The oxyhemoglobin generated neutralizes the liberated hydroxyl ions. The brain maintains the system in balance by regulating the metabolic rate (rate of CO_2 production) and the respiratory rate and tidal volume. The action of the excretory organs like the kidney is long range.

The OH^- ion shown in Figure 4.2 must not be taken literally for the simple reason that it is immediately neutralized by conversion of Hb to HbO_2 at the lungs as it forms. The CO_2 is removed by ventilation. Thus the reaction $HCO_3^- \rightarrow CO_2 + OH^-$ *tends to completion at the lungs because the products of the reaction are being removed*, the OH^- ion by oxygenated Hb and the CO_2 by aeration.

At the cell, the reaction taking place is decarboxylation of organic acids, such as pyruvate going to acetate. This reaction can be simulated in a test tube by dissolving phenylpyruvic acid in alkali. The phenylpyruvic acid is chosen for demonstration purposes since its alkaline solutions are a pink color. On addition of hydrogen peroxide, the pink color is bleached, producing phenyl acetic acid and sodium bicarbonate.

$$C_6H_5CH_2 \cdot \overset{O}{\overset{\|}{C}} - \overset{O}{\overset{\|}{C}} - ONa \xrightarrow[\text{NaOH}]{H_2O_2} C_6H_5 \cdot CH_2 \cdot \overset{O}{\overset{\|}{C}}ONa + NaHCO_3$$

In the human, the bicarbonate ion generated needs to pass the cell

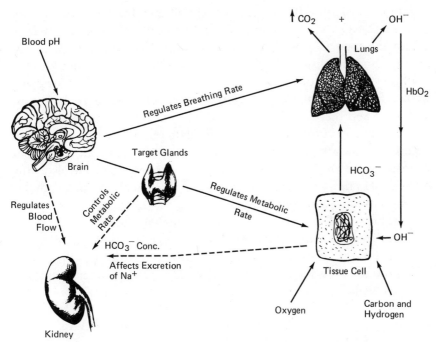

FIGURE 4.2   Conservation of base to maintain blood pH.

membrane to get to the blood stream. At the intracellular pH, which is somewhat less than pH 7.4 of the blood, a significant amount of bicarbonate is converted to dissolved CO_2, which passes the membrane more rapidly than bicarbonate ion itself.

At the pH within the cell (~ 7.2), the distribution of bicarbonate and dissolved CO_2 will be given by the Henderson–Hasselbalch equation, assuming that the H_2CO_3 breaks up to give dissolved CO_2:

$$7.2 = 6.1 + \log \frac{[HCO_3^-]}{[H_2CO_3]}$$

$$1.1 = \log \frac{[HCO_3^-]}{[H_2CO_3]}$$

$$\frac{[HCO_3^-]}{[H_2CO_3]} = \text{antilog } 1.1$$

$$\frac{[HCO_3^-]}{[H_2CO_3]} = 12.6$$

The molar ratio of $[HCO_3^-]/[H_2CO_3]$ will then be 12.6/1. Since each mole of H_2CO_3 breaks up to yield $H_2O + CO_2$, then approximately 8% of the total CO_2 will be in the form of dissolved CO_2. Both the HCO_3^- and the dissolved CO_2 can diffuse out of the cell and into the erythrocyte, where their effect in lowering pH is promptly neutralized by the $HbO_2 \rightarrow Hb$ reaction. Note that HCO_3^- ion is replacing an OH^- ion and therefore is acidic relative to this ion (Figure 4.2).

The action of the brain can be simulated in the mechanical model Figure 4.1 by attaching the pH electrodes to a pH-stat instrument coupled to the oxygen and CO_2 valves. By setting the controls to the desired pH, the CO_2 valve and oxygen valve will automatically be adjusted to maintain constant pH.

In Figure 4.2 the role of other organs such as the kidney and sweat glands in maintaining constant pH are long range and are discussed later. The target gland regulating metabolic rate, and thus the rate of CO_2 production, is indicated as the thyroid, although other glands are also involved.

4.4 *PROTEIN AS A BUFFER*

Proteins exhibit buffering properties for several reasons.[7] The relationship between protein concentration and pH follows the same law as for organic acids:

$$pH = pK + \log \frac{[Pr^-]}{[HPr]}$$

There is approximately 70 g of protein per liter of serum, or 70,000 mg. One can assume an average molecular weight of 60,000 for purposes of calculation. This is approximately the molecular weight of albumin and much lower than the globulins. Even with this figure one can see that the concentration is of the order of 1 mmol/liter. Titration of dialyzed serum (protein = 7 g/100 ml) with alkali to a pH of 8 requires 16 mEq/liter of acid. Compared to the bicarbonate concentration (~ 28 mEq/liter), one can see that this value is significant. This high value is due to the fact that the protein molecules possess numerous free carboxyl groups, due to the presence of glutamic and aspartic acid residues in the molecule.

Proteins are amphoteric, having in addition to the carboxyl groups numerous basic groups due to the presence of arginine, lysine, and histidine in the molecule. The presence of polypeptide linkages and other groups capable of hydrogen bonding permits them to adsorb acids, like lactic acid, acetoacetic acid, and acids of the citric acid cycle. With cations like potassium, calcium, and magnesium, proteins will also form both salts and chelates.[8] *Thus their buffering action on both sides of pH 7.4 goes beyond their behavior as an anion.*[9] As pointed out above, proteins also combine with bicarbonate to form carbamino compounds. This reaction becomes of greater significance in alkalosis.

If blood is drawn from a patient with a mild diabetic acidosis, one might find a pH of 7.3. If one dilutes the serum from this blood with normal saline to five times its volume, the pH drops rapidly to the acid side of the pH scale. This is due to the dissociation of organic acids, like acetoacetic acid, adsorbed to the protein by hydrogen bonding. This is not observed with normal serum, which moves to a pH of ~ 7.5 on dilution. *In acidosis, the buffering action of protein is greater than would be indicated by considering it as the sodium salt of a weak acid.*

Van Slyke used a factor of 2.43 to convert total protein to mEq/liter of anion.[10] Thus for a serum protein of 7, the anion equivalent would be $7 \times 2.43 = 16.92$ mEq/liter. Albumin and globulins, as anions, are not equivalent since their isoelectric points are different and the distribution and number of charges on different proteins vary. The values will also vary with pH. The formula used for calculating their anion equivalent are

for albumin, mEq/liter = g/100 ml Alb $\times$ 1.25(pH -5.16)
for globulins, mEq/liter = g/100 ml Glob $\times$ 0.77(pH -4.89)

For a pH of 7.40, an albumin level of 4.0 g/100 ml, and a globulin level of 3 g/100 ml these become

for albumin, mEq/liter = $4 \times 1.25(7.40 - 5.16) = 11.2$
for globulins, mEq/liter = $3 \times 0.77(7.40 - 4.89) = 5.8$

or a total of 17.0 mEq/liter of anion equivalent.

4.5 PHOSPHATE BUFFER

The concentration of phosphate in human plasma is of the order of 3.5 mg/100 ml as phosphorus. Phosphorus has an atomic weight of 31, and thus there are approximately 1.1 mmol of phosphorus and of the

PO_4 radical per liter. At the pH of the plasma (7.4), it can be seen from Figure 2.1 that substantially all the phosphate is in the form of HPO_4^{2-} and $H_2PO_4^-$. From the equation

$$pH = 7.4 = 7.2 + \log \frac{[HPO_4^{2-}]}{[H_2PO_4^-]}$$

and Table 2.1 listing the pK values for phosphoric acid, one can calculate the ratio of $[HPO_4^{2-}]$ to $[H_2PO_4^-]$. This turns out to be 1.6. Dividing the 1.1 mmol/liter of phosphate in the plasma into this ratio, we calculate that 0.68 mmol is in the form of Na_2HPO_4 and 0.42 mmol is in the form of NaH_2PO_4. In calculating the equivalence of phosphate to sodium, we note that 0.68 mmol is equivalent to 1.36 mEq of Na and 0.42 mmol is equivalent to 0.42 mEq of Na^+. Adding the two, we get 1.78, or approximately 1.8 mEq. For purposes of balancing cations against anions in plasma, the phosphate concentration is therefore considered as 1.8 mEq/liter (see Figure 5.1).[11]

Considering its relatively low concentration, phosphate contributes little to the buffering properties of the blood. However, its concentration in the urine is approximately tenfold that in the blood. The urine is at a pH of ∼5. At this pH, most of the phosphate is in the form of NaH_2PO_4, maintaining electroneutrality with much less sodium ion. This is one mechanism for conserving sodium ion as discussed in the chapter on the kidney. Phosphate is therefore important in maintaining constant pH of the plasma on a long-range basis through exchange of hydrogen ion for sodium ion as the phosphate is excreted. In this way it *acts to conserve base*, which is the major problem in maintaining constant pH in the human, as discussed above. Note that "conserving base" in the human is mainly equivalent to conserving sodium ions. This will be discussed further in the chapter on acidosis.

4.6   *CHLORIDE SHIFT*

When venous blood is oxygenated, chloride ion shifts from the erythrocyte to the plasma, and an increase in chloride concentration in the plasma of approximately 4% will be observed. If the blood is reduced, the chloride ion will move into the erythrocyte; this phenomenon is called the "chloride shift." The explanation for the chloride shift follows from the studies of Donnan.

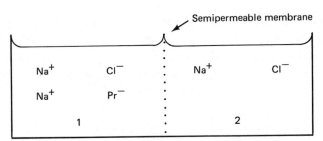

FIGURE 4.3.   Two solutions separated by a semipermeable membrane.

According to the Gibbs–Donnan law,[12] the ratio of the concentration of a negative ion (anion) in the cell to the same ion in the plasma is equal to the inverse of the ratio of the positive ion (cation) in the cell to the same ion in the plasma.

The chloride shift follows from this law. Let us assume that we have two solutions separated by a semipermeable membrane which permits salts and water to diffuse, but not protein molecules, as shown in Figure 4.3. Let us also assume that initially the concentration of NaCl is the same on both sides of the membrane. We now add some protein as the sodium salt to one side as shown. It is apparent that water will move to the compartment on the left. The concentrations of Na^+ and Cl^- are not balanced on both sides of the membrane. At equilibrium, the osmotic pressure must be balanced on both sides of the membrane, and there must be electrical neutrality.

When the concentration of a salt is not the same on both sides of a membrane, we can place a platinum electrode on each side and measure a voltage and draw current from the cell. This type of cell is called a concentration cell. The energy we obtain as the system moves to equilibrium is called the *free energy* obtained from the system. Considering each ion separately, the free energy change $(-\Delta F)$ is given by the formula $-\Delta F = RT \ln (C_1/C_2)$, where the negative sign indicates that energy leaves the system, R is the gas constant, T is the absolute temperature, and $\ln(C_1/C_2)$ is the natural logarithm of the ratio of the concentrations on both sides of the membrane.

At equilibrium, the free energy change of a system is equal to zero, since no work can be obtained from such a system.

Applying this equation to the erythrocyte in plasma, the erythrocyte contents make up one chamber, the cell wall is the membrane, and the plasma is the other chamber. Note that the protein concentration within the erythrocyte is about five times that in the plasma, simulating the model of the figure. We can now represent the system at equilibrium by adding the free energy change for each ion on both sides of the membrane and setting the sum equal to zero. For simplicity, considering only sodium and chloride ions, we get

$$RT \ln \frac{[\text{Na}^+]_{\text{cell}}}{[\text{Na}^+]_{\text{plasma}}} + RT \ln \frac{[\text{Cl}^-]_{\text{cell}}}{[\text{Cl}^-]_{\text{plasma}}} = 0 \qquad (4.3)$$

Transposing and cancelling out the RT factor, and inverting one fraction to eliminate the negative sign, we obtain

$$\ln \frac{[\text{Na}^+]_{\text{cell}}}{[\text{Na}^+]_{\text{plasma}}} = - \ln \frac{[\text{Cl}^-]_{\text{cell}}}{[\text{Cl}^-]_{\text{plasma}}} = \ln \frac{[\text{Cl}^-]_{\text{plasma}}}{[\text{Cl}^-]_{\text{cell}}} \qquad (4.4)$$

Taking the antilog of both sides of the equation, we obtain

$$\frac{[\text{Na}^+]_{\text{cell}}}{[\text{Na}^+]_{\text{plasma}}} = \frac{[\text{Cl}^-]_{\text{plasma}}}{[\text{Cl}^-]_{\text{cell}}} \qquad (4.5)$$

If instead of sodium ion we consider a bivalent ion like calcium or magnesium, we need to take the square root of the concentration of the ions in the free energy equation. For a trivalent ion like aluminum, we take the cube root, etc. It is noteworthy that at equilibrium, *the ratio of the ion concentrations of the positive ions in the two compartments is equal to the inverse of the corresponding ratio of the negative ions,* this ratio being a constant, if the Donnan law is followed. By measuring the concentrations of the different ions on both sides of the membrane, we can readily see whether the equilibrium law is followed or whether some mechanism occurs to prevent equilibrium from being established. The general law can now be written as follows:

$$\frac{(\text{cation}_1)^{1/z}}{(\text{cation}_2)^{1/z}} = \frac{(\text{anion}_2)^{1/z}}{(\text{anion}_1)^{1/z}} = \text{const} \qquad (4.6)$$

where 1 and 2 represent the two compartments and z the charge on the ion.

If no protein were present, the concentrations of the ions on both sides of the membrane would necessarily be the same. *Therefore the difference in concentration of the nondiffusible protein on both sides of the membrane and its charge determine the constant.*

In the example given, the ratio would change when hemoglobin goes to oxyhemoglobin, since the base binding properties (charge) on hemoglobin are not the same as on oxyhemoglobin. The general equation for the monovalent ions, which takes into account the differences in base binding properties of the different proteins, is given by the equation[13]

$$r = \frac{[\text{plasma cation}]}{[\text{cell cation}]} = \frac{[\text{cell anion}]}{[\text{plasma anion}]}$$

$$= 1 - \frac{[BP]_c + [Hb]_c - [BP]_p}{2([B]_p - [BP]_p)} \tag{4.7}$$

where $[BP]_c$ is the base binding power of proteins in the erythrocyte, $[Hb]_c$ is the concentration of Hb in the cells, $[BP]_p$ is the base binding power of the plasma proteins, and $[B]_p$ is the concentration of the total base in plasma. Since other factors remain essentially constant, it is apparent that $[BP]_c$ is the major factor in determining the value of r.

The Donnan relationship as applied to the erythrocyte can be illustrated as follows:

$$r = \frac{[H^+]_{\text{plasma}}}{[H^+]_{\text{cell}}} = \frac{[Cl^-]_{\text{cell}}}{[Cl^-]_{\text{plasma}}} = \frac{[HCO_3^-]_{\text{cell}}}{[HCO_3^-]_{\text{plasma}}} \tag{4.8}$$

As determined experimentally in oxygenated whole blood for chloride ion, the Donnan ratio is approximately 0.5. For example, erythrocyte chloride concentration is 55 mEq/liter and plasma chloride concentration is 103 mEq/liter. Plasma carbon dioxide content is 26 mmol/liter and erythrocyte CO_2 content is 14 mEq/liter.

The apparent degree of ionization (activity coefficient) of bicarbonate ion is less than that for chloride ion, and these ratios differ somewhat. The HCO_3^- fraction should be multiplied by a factor to correct for this. The factor is of the order of 0.8–0.9 when expressing concentrations in molality (see below). The Gibbs–Donnan law is difficult to apply to so complicated a system as the erythrocyte, partly because of the fact that the membrane is not inert and partly because of active metabolism going on in the erythrocyte and in the cell wall.[13a] Calculations from this law are only rough approximations to indicate general tendencies.

If a particular ion obeys the law expressed in equation (4.6), then this suggests that the ion crosses the membrane by *passive transfer*, subject only to the physical laws of diffusion. For example, the concentration of uric acid within the erythrocyte is approximately one-half that in the plasma. This supports the idea that most urate passes the cell wall by passive transfer. On the other hand, from Table 4.1, one can readily see that this is not the case for sodium and potassium. For sodium the plasma/cell ratio is 140/16 and for potassium it is 4.6/95.

For calcium and magnesium, which are bivalent ions, one substitutes the square root of the concentration in the equation for r:

$$r = \frac{[Ca]^{\frac{1}{2}}_{plasma}}{[Ca]^{\frac{1}{2}}_{cell}} = \frac{[Mg]^{\frac{1}{2}}_{plasma}}{[Mg]^{\frac{1}{2}}_{cell}}$$

As for sodium and potassium, it is apparent that the Donnan law does not apply to these ions.

It was this type of reasoning which led to the postulate of "active transport" mediated by some energy-utilizing systems.[14,15] This gave rise to the discovery of the Na–K and Ca–Mg pumps, which will be discussed later.

In calculating osmotic pressures, concentrations are given in mol/1000 g of water (molality). If one considers that the erythrocyte contains approximately 34% protein as compared to plasma proteins of 7%, it is apparent that a significant difference would exist in the ratio when expressed as molality, that is moles/kg water. For example, when this correction is made, one obtains a ratio of approximately 0.7. From this ratio one can calculate the intracellular pH. For example,

TABLE 4.1

Approximate Electrolyte Composition of the Erythrocyte in Oxygenated Blood at pH 7.4 Compared to the Plasma[a]

Ion	Erythrocyte		Plasma	
	mEq/liter	mEq/kg H_2O	mEq/liter	mEq/kg H_2O
Sodium	16	22	140	155
Potassium	95	131	4.6	5.1
Calcium	0.5	0.7	5.0	5.6
Magnesium	4.6	6.4	1.9	2.1
Chloride	55	76	103	114
Bicarbonate	11	15	26	29
Other anions[b]	50	69	22.5	24.8

[a] Water content of the erythrocyte is approximately 72%. Plasma water content is approximately 88%.
[b] Difference between total base and the sum of chloride and bicarbonate.

using the experimentally determined ratio of 0.5, for whole blood,

$$r = \frac{[H^+]_{plasma}}{[H^+]_{cell}} = \frac{10^{-7.4}}{[H^+]_{cell}} = 0.5$$

so that

$$[H^+]_{cell} = 2 \times 10^{-7.4} = 10^{0.3} \times 10^{-7.4} = 10^{-7.1}$$

Thus the pH within the erythrocyte should be of the order of 7.1. If the value of 0.7 is used, one obtains a pH of 7.25. Experimentally, the pH is approximately 7.2. In any case, the pH within the erythrocyte is more acid than that of the plasma.

4.7   ISOHYDRIC SHIFT

The effect of the change from hemoglobin to oxyhemoglobin at the lungs on the blood pH and ion distribution can now be presented in the light of the above. This requires a movement of CO_2 out of the erythrocytes and into the plasma with a readjustment of ion distribution between cells and plasma. This is accomplished without significant change in pH within the cell or the plasma and for this reason is called the "isohydric shift."[16]

At the lungs, the erythrocytes are exposed to a p_{O_2} of approximately 100 mmHg. From Figure 3.1 one can readily see that at this p_{O_2} value 95% of the hemoglobin is converted to oxyhemoglobin. This change converts a relatively basic substance to an acid. *Hydrogen ions are thus generated:*

$$O_2 + Hb \rightarrow HbO_2 + H^+$$

The liberation of hydrogen ion sets in motion a series of reactions designed to excrete CO_2 while maintaining constant pH in the plasma. The hydrogen ion combines with the bicarbonate ion within the cell (to form H_2CO_3) and also hydrolyzes the carbamino compounds within the cell. Under the catalytic effect of carbonic anhydrase, H_2CO_3 is decomposed to H_2O and CO_2:

$$Hb \cdot NH \cdot COOH \xrightarrow{H^+} H_2CO_3 + Hb$$

$$H_2CO_3 \xrightarrow[\text{anhydrase}]{\text{carbonic}} CO_2 + H_2O$$

The CO_2 now diffuses out of the cell into the plasma and finally out through the alveoli to the air. Recombination of H_2O and CO_2 in plasma is slow because of the absence of carbonic anhydrase and the presence of carbonic anhydrase inhibitors in the plasma.

The drop in $HCO_3{}^-$ concentration in the erythrocyte causes a high concentration gradient to exist between the plasma and the erythrocyte for this ion. Bicarbonate ion then moves into the cell, where it combines with the generated hydrogen ion to form carbonic acid and then CO_2 and water and returns to the plasma. Thus a continuous flow of CO_2 leaves the cell at the lungs, a substantial part of which originated from the plasma.

Approximately 20% of the potassium in the cell is complexed with the hemoglobin in the erythrocyte in venous blood.[17] Potassium, unlike sodium, readily forms complexes (chelates) with numerous substances. Combined in this manner, with reduced hemoglobin, the potassium is not ionized and acts therefore as a potential reservoir for neutralization of acids. When hemoglobin is oxidized, 35% of the bound potassium dissociates, combining with bicarbonate ion. The potassium bicarbonate so formed buffers the hydrogen ion that has been generated by oxidation of the hemoglobin. Per mole of oxygen, 0.7 mol of potassium ion is liberated to combine with bicarbonate.

This reaction can be formulated as follows, keeping in mind that the potassium tied to reduced hemoglobin does not readily dissociate. $KHCO_3$ is approximately 40% ionized:

$$K_2Hb + O_2 + 0.7HCO_3{}^- \rightarrow K_{1.3}HbO_2 + 0.7KHCO_3$$

By the time the erythrocyte has passed through the lungs, a new Donnan ratio r for bicarbonate has been set up somewhat lower than the ratio originally in venous blood. This can be seen from the distribution of CO_2 in venous blood as compared to arterial blood (Table 4.2). In venous blood, the ratio of bicarbonate in the erythrocyte as compared to the plasma is 0.59. In arterial blood the ratio is 0.57.

Thus one effect of the increase in H^+ concentration in the erythrocyte (denominator) relative to the plasma when the blood is oxygenated is a lowering of the value of the ratio r. In order to maintain equilibrium between the ions, some chloride ion in the erythrocyte must diffuse out to the plasma so that its ratio follows suit. *This phenomenon is called the "chloride shift."* The sequence of events proceeds as follows, reading from

TABLE 4.2

Typical Distribution of CO_2 Found Between the Erythrocytes and the Plasma in Venous and Arterial Blood from the Same Patient

	Arterial		Venous	
	mmol/liter	mmol/kg H_2O	mmol/liter	mmol/kg H_2O
Total CO_2 in whole blood	21.5	—	23.2	—
Total CO_2 in plasma	26.65	—	28.32	—
CO_2 as HCO_3^- in plasma	25.4	27.3	27.0	29.0
Total CO_2 in R.B.C.	14.0	—	15.5	—
CO_2 as HCO_3^- in R.B.C.	10.5	16.2	11.0	16.7
Carbamino CO_2 in R.B.C.	2.4	—	3.7	—
Ratio[a] $[HCO_3^-]_{cell}/[HCO_3^-]_{plasma}$	0.57	—	0.59	—
Hematocrit, % R.B.C. vol	45	—	48	—

[a] Calculated from mmol/kg H_2O or molality.

left to right, initiated by the $Hb \rightarrow HbO_2$ reaction:

$$\frac{[H^+]_{plasma}}{increased\ [H^+]_{cell}} = \frac{[HCO_3]_{cell}}{[HCO_3]_{plasma}} = \frac{[Cl]_{cell}}{[Cl]_{plasma}} = r$$

If venous blood is drawn from a patient and the plasma from that blood is fully oxygenated before it is separated from the cells by exposure to air, the chloride level in the plasma will rise from a level of 103 to 108 mEq/liter in a typical experiment. This "chloride shift" out of the cell when blood is oxygenated requires that blood be taken under anaerobic conditions when intended for chloride analysis of the plasma.

The movement of chloride ion out of the erythrocyte causes a decrease of osmotic pressure within the cell. In order to maintain osmotic equilibrium, some water must move out of the cell, and the cell shrinks, resulting in lower hematocrit values being obtained for arterial blood as compared to venous blood. A hematocrit of 48 in venous blood will be lowered to 45 if the blood is oxygenated (Table 4.2).

When the erythrocyte moves to the tissues it encounters an area of low oxygen tension and high CO_2 tension from the CO_2 generated by cell metabolism. Under these conditions, oxyhemoglobin loses its oxygen and CO_2 diffuses into the plasma and then into the red cell.

This results in a rise in pH in the cell due to the relative basicity of hemoglobin as compared to oxyhemoglobin. The more alkaline pH favors the reaction

$$H_2O + CO_2 \xrightarrow[\text{anhydrase}]{\text{carbonic}} H_2CO_3 \rightarrow H^+ + HCO_3^-$$

$$\underset{\text{hemoglobin}}{R_2NH} + H_2CO_3 \rightarrow \underset{\substack{\text{hemoglobin} \\ \text{carbamate}}}{R_2N-COOH} + H_2O$$

where the hydrogen ion tends to bring the pH back to its original value and $R_2N-COOH$ represents the carbamino compound formed with the histidine residue in the hemoglobin molecule. Some potassium ion recombines with the hemoglobin molecule, reversing the reaction,

$$K_{1.3}HbO_2 + 0.7KHCO_3 \rightarrow K_2Hb + 0.7HCO_3^- + O_2$$

Chloride ion moves into the cell to satisfy the Donnan conditions since the ratio has increased due to a rise of $[H^+]$ in the cells. The increase in bicarbonate concentration in the plasma is greater than the increase in bicarbonate concentration in the cell (see Table 4.2). In order to satisfy the Donnan conditions, some chloride ion from the plasma moves into the cell, reestablishing the equilibrium condition (chloride shift). At this point the osmotic pressure within the cell has risen and water must move into the cell to reestablish osmotic balance. The hematocrit rises as the cells swell. This can be illustrated schematically as follows:

$$\frac{[H^+]_{\text{plasma}}}{\text{decreased } [H^+]_{\text{cell}}} = \frac{[HCO_3^-]_{\text{cell}}}{[HCO_3^-]_{\text{plasma}}} = \frac{[Cl]_{\text{cell}}}{[Cl]_{\text{plasma}}}$$

The changes that take place in erythrocyte and plasma composition at the lungs and at the tissues are summarized in Table 4.3.

4.8   ORGANIC ACIDS

In addition to the effect of hemoglobin, protein, and the bicarbonate and phosphate buffer systems, some consideration needs to be given to the organic acids normally present in blood. These organic acids derive from both the anaerobic and oxidative action on carbohydrates, hydrolysis of proteins, and exogenous and endogenous fatty

TABLE 4.3

Summary of Steps Taking Place at the Lungs and the Tissues Designed to Maintain Constant pH and Ion and Osmotic Balance Between Erythrocyte and Plasma[a]

At lungs	At tissues
1. $Hb \rightarrow HbO_2 + H^+ (-OH^-)$	1. $HbO_2 \rightarrow Hb - H^+ (+OH^-)$
2. $H^+ + HCO_3^- \xrightarrow{\text{carbonic anhydrase}} H_2O + \uparrow CO_2$ Reaction 2 compensates for rise in [H^+] by removing it. Plasma bicarbonate moves from plasma into erythrocyte to correct for concentration gradient. Result is decreased HCO_3^- in both erythrocyte and plasma. The ratio r between erythrocyte and plasma is decreased (decreased r for HCO_3^-)	2. $H_2O + CO_2 \xrightarrow{\text{carbonic anhydrase}} HCO_3^- + H^+$ Reaction 2 balances drop in [H^+] due to reaction 1. Movement of CO_2 is from tissue to plasma to erythrocyte and back to plasma as HCO_3^-. This results in increased HCO_3^- in both erythrocyte and plasma. The ratio r between erythrocyte and plasma is increased (increased r for HCO_3^-)
3. $K_2Hb + O_2 + 0.7HCO_3^- \rightarrow K_{1.3}HbO_2 + 0.7KHCO_3$	3. $K_{1.3}HbO_2 + 0.7KHCO_3 \rightarrow K_2Hb + 0.7HCO_3^- + O_2$
4. $Cl^-(\text{cell}) \rightarrow Cl^-(\text{plasma})$ (decreased r for Cl^-)	4. $Cl^-(\text{plasma}) \rightarrow Cl(\text{cell})$ (increased r for Cl^-)
5. $H_2O(\text{cell}) \rightarrow H_2O(\text{plasma})$ (to restore osmotic balance)	5. $H_2O(\text{plasma}) \rightarrow H_2O(\text{cell})$ (to restore osmotic balance)

[a] The numbers indicate the sequence of events.

TABLE 4.4

Organic Acids in a Normal, Fasting Adult

Acid	Range, mg/100 ml	Mean, mg/100 ml	Mean, mEq/liter
Acetoacetic	0.5–1.5	0.9	0.09
Citric	1.7–2.8	2.2	0.35
β-Hydroxybutyric	0.6–3	1.5	0.14
α-Ketoglutaric	0.2–1.0	0.6	0.1
Lactic	5–15	9.0	1.0
Pyruvic	0.5–1.5	1.1	0.13
Malic	0.2–0.8	0.5	0.08
Succinic	0.3–0.9	0.5	0.09
Free fatty acids			0.7 (0.45–0.90)
Total		16.3	2.68

acids. They include the amino acids, creatine, citric, lactic, uric, acetoacetic, and hydroxybutyric and the fatty acids of serum. Their concentration range in the fasting state is considered as approximately
1–5 mEq/liter. This can rise as high as 50 mEq/liter in diabetic coma
owing to the high levels of acetoacetic, hydroxybutyric, and lactic acids.
The pK values of these acids are of the order of 3 (Table 2.1), and even
in acidosis (pH 7.1–7.2) they have little buffering action. *Their importance lies in the fact that they replace bicarbonate ion with a weaker base and thus
tend to lower the pH of the blood in disease.*

Table 4.4 lists the values found for the aliphatic organic acids in
a normal, fasting adult.[18,19]

4.9   *CARBONIC ANHYDRASE*

In going from CO_2 to H_2CO_3, we go from a linear structure
($O{=}C{=}O$) to a triangular structure. Because of this change, extensive
electronic rearrangement must take place in the hydration of CO_2.
The rate constant of hydration of CO_2 is 0.037 mol/sec at 25°C.[20]

To give some idea as to the slowness of this reaction, the hydration
of SO_2 (a bent triangular structure) goes 100 million times faster. It is
apparent that unless a catalyst were present to accelerate this reaction,
carbon dioxide exchange would be a problem. For example, it takes 6
sec for blood to go from the arm to the lungs. If CO_2 gas is blown
vigorously through 0.05 M sodium bicarbonate solution in the instrument
of Figure 4.1, it takes 90 sec for the pH to move from 8.6 to 6.5. If a
drop of whole blood is added, this time is reduced to 20 sec.

There are three isoenzymes of carbonic anhydrase in the erythrocyte called *A*, *B*, and *C*.[21] Carbonic anhydrase *B* is present in largest
concentration, but carbonic anhydrase *C* has the highest specific
activity. All these enzymes have a molecular weight of approximately
30,000 and contain one atom of zinc (at wt 65.37), tightly bound. The
percentage of Zn is theoretically 0.21%, although experimentally one
obtains 0.3% zinc. There is 1400 μg Zn/100 g of packed erythrocytes.[22]
This is equivalent to 666 mg of carbonic anhydrase per 100 g of packed
erythrocytes. Thus this enzyme occurs in relatively high concentration
in the erythrocyte. The carbonic anhydrases comprise single polypeptide chains differing from each other only slightly in amino acid
composition.[23]

The action of the enzyme seems to be associated with its general ability to hydrate carbonyl groups. It will also hydrate aldehydes[24] and hydrolyze certain esters.[25] Its action in the erythrocyte is to catalyze the hydration and dehydration of CO_2:

$$CO_2 + H_2O \rightleftharpoons H_2CO_3$$

Carbonic anhydrase, as it occurs in the erythrocyte, has one of the highest turnover rates and in this respect can be compared with catalase and peroxidase. The enzyme is uniformly active between pH 5.5 and 9. Alkali catalyzes hydration of CO_2 and acid favors the dehydration reaction both with and without enzyme. *At the pH of the erythrocyte, 7.2, the hydration reaction is favored.*

The slow reaction between CO_2 and H_2O enables some CO_2 to escape from the erythrocyte and move into the plasma. As blood circulates by diffusion, within the lungs, this CO_2 enters the alveolar gas and is expelled during normal breathing. At the tissue level, the CO_2 generated in the cells can move through the plasma and into the erythrocyte, where it can be hydrated. Concentration of HCO_3^- in the erythrocyte and plasma is then adjusted in accordance with the equilibrium laws. Thus a flow of CO_2 into the erythrocyte and out as HCO_3^- occurs until the balance dictated by the Gibbs–Donnan law is reached.

In the human infant, a deficiency in erythrocyte carbonic anhydrase is commonly found.[29] This may rarely persist into adult life.[30] In both instances elevated levels of hemoglobin F occur. This may represent a physiologic adjustment; by utilizing the greater affinity of hemoglobin F for O_2 over hemoglobin A, the loss of O_2-carrying power resulting from buildup of CO_2 in the erythrocyte is prevented. Juvenile myeloid leukemia is also associated with an elevated hemoglobin F and a marked decrease in erythrocyte carbonic anhydrase, particularly isoenzymes B and C.[31,32]

Plasma carbonic anhydrase inhibitors act to preserve the mechanism of respiration by inactivating small amounts of carbonic anhydrase that may leak from the erythrocyte into the plasma due to hemolysis.

Certain drugs are powerful carbonic anhydrase[26] inhibitors. Sulfa drugs are strong inhibitors of carbonic anhydrase, and acetazolamide (Diamox, 2-acetylamino-1,3,4-thiadiazole-5 sulfonamide) is used as a diuretic. By inhibiting H_2CO_3 decomposition, it causes a rise in HCO_3^- level, which is excreted with sodium and potassium in the urine.[27]

Excessive use of sulfa drugs results in acidosis, hypokalemia, and hypo-natremia.

Diamox will also effectively correct an alkalosis produced by malfunction of the controls of the hypothalamus resulting in excessive sodium retention.[28] Carbon monoxide, HCN, and metals like copper, gold, mercury, and vanadium are also carbonic anhydrase inhibitors and this partly explains their toxicity when taken internally in relatively large amounts.

4.10 RECAPITULATION

The major mechanism for maintaining constant pH in the blood is the balance between the rate of production of CO_2 and the ventilation at the lungs. The rate of CO_2 production is controlled by the rate of metabolism. The rate of metabolism is in turn controlled by the central nervous system, which adjusts the rate of ventilation by the lungs to maintain constant pH. Surges in pH values at the tissues and lungs are prevented by the hemoglobin–oxyhemoglobin system. The bicarbonate, phosphate, and protein buffers of the blood serve to act as the fine control for the system. Carbonic anhydrase serves to bring about rapid equilibrium for the bicarbonate system within the erythrocyte so as to facilitate elimination of CO_2 at the lungs. The Na^+ and K^+ ions serve to neutralize base electrically in the plasma and erythrocytes, respectively. Their conservation is extremely important if base levels are to be maintained.

This system can only be effective if the available chloride, sodium, and potassium ions remain constant within the prescribed limits, and if the available water in the various compartments is also constant and at a constant pressure. How these factors are controlled will be discussed in the next chapter.

4.11 SELECTED READING—MAINTENANCE OF CONSTANT pH IN THE HUMAN

Siggaard-Andersen, O., *Acid–Base Status of the Blood*, 3rd ed., Williams & Wilkins, Baltimore, Maryland (1974).

Christensen, H. N., *Body Fluid and the Acid Base Balance*, Saunders, Philadelphia, Pennsylvania (1964).

Pappenheimer, J. R., The ionic composition of cerebral extracellular fluid and its relation to control of breathing, in *The Harvey Lectures*, Series 61, Academic Press, New York (1965–1966), pp. 71–94.

Edsall, J. T., The carbonic anhydrases of the erythrocytes, in *The Harvey Lectures*, Series 62, Academic Press, New York (1966–1967), pp. 191–230.

Peters, J. P. and Van Slyke, D. D., *Quantitative Clinical Chemistry*, Vol. I. *Hemoglobin and Oxygen*, Williams & Wilkins, Baltimore, Maryland (1931), pp. 518–652.

Filley, G. F., *Acid–Base and Blood Gas Regulation*, Lea & Febiger, Philadelphia (1971).

4.12 REFERENCES

1. Schwartz, W. B., Brackett, N. C., Jr., and Cohen, J. J., The response of extracellular hydrogen ion concentration to graded degrees of chronic hypercapnia: The physiologic limits of the defense of pH, *J. Clin. Invest.* **44**:291–301 (1965).

2. Needham, C. D., Rogan, M. C., and MacDonald, I., Normal standards for lung volumes, intrapulmonary gas-mixing and maximum breathing capacity, *Thorax* **9**:313–325 (1954).

3. Leusen, I. R., Chemosensitivity of the respiratory center. Influence of changes of H^+ and total buffer concentration in the cerebral ventricles on respiration, *Am. J. Physiol.* **176**:45–51 (1954).

4. Wyman, J., Jr., Analysis of the titration data of oxyhemoglobin of the horse by a thermal method, *J. Biol. Chem.* **127**:1–13 (1939).

4a. Ferguson, J. K. W. and Roughton, F. J. W., Chemical relations and physiological importance of carbamino compounds of CO_2 with Hemoglobin, *J. Physiol.* **83**:87–102 (1934).

4b. Rossi, L. and Roughton, F. J. W., The effect of carbamino-Hb compounds on the buffer power of human blood at 37°C, *J. Physiol.* (proc.) **167**:15p–16p (1963).

5. d'Elseaux, F. C., Blackwood, F. C., Palmer, L. E., and Sloman, K. G., Acid base equilibrium in the normal, *J. Biol. Chem.* **144**:529–535 (1942).

6. Saper, D. G., Levine, D. Z., and Schwartz, W. B., The effects of chronic hypoxemia on electrolyte and acid–base equilibrium. An examination of normocapneic hypoxemia and of the influence of hypoxemia on the adaptation to chronic hypercapnia, *J. Clin. Invest.* **46**:369–377 (1967).

7. Goldwitzer-Meier, K. The buffering action of serum proteins, *Biochem. Z.* **163**:470 (1925).

8. Prasad, A. S., Flink, E. B., and Zinneman, H. H., The base binding property of the serum proteins with respect to magnesium. *J. Lab. Clin. Med.* **54**:357–364 (1959).

9. Edsal, J. T., Reversible combination of serum albumin and other plasma proteins with small molecules or ions: Factors affecting stability to heat, in

Advances in Protein Chemistry, Vol. III Academic Press, New York (1947), pp. 463–473.

10. Van Slyke, D. D., Hastings, A. B., Hiller, A., and Sendroy, J., Jr., Studies of gas and electrolyte equilibria in blood. XIV. The amounts of alkali bound by serum albumin and globulin, *J. Biol. Chem.* **79**:769–780 (1928).

11. Gamble, J. L., *Chemical Anatomy, Physiology and Pathology of Extracelluar Fluid*, 6th ed., Harvard Univ. Press, Cambridge, Massachusetts (1954).

12. Donnan, F. G., *Z. Elektrochem.* **17**:572 (1911).

13. Van Slyke, D. D., Wu, H., and McLean, F. C., Studies of gas and electrolyte equilibria in blood. Factors controlling electrolyte and water distribution in blood, *J. Biol. Chem.* **56**:765–849 (1923).

13a. Kintner, E. P., Chemical structure of erythrocytes with emphasis on Donnan equilibrium, *Ann. Clin. Lab. Sci.* **2**:326–334 (1972).

14. Maizels, M. and Paterson, J. L. H., Base binding in erythrocytes, *Biochem. J.* **31**:1642–1656 (1937).

15. Farmer, S. N. and Maizels, M., Organic anions of human erythrocytes, *Biochem. J.* **33**:280–289 (1939).

16. Henderson, L. J., *Blood: A Study in General Physiology*, Yale Univ. Press, New Haven, Connecticut (1928).

17. Van Slyke, D. D., Hastings, A. B., Heidelberger, M., and Neill, J. M., Studies of gas and electrolyte equilibria in blood. III The alkali-binding and buffer values of oxyhemoglobin and reduced hemoglobin, *J. Biol. Chem.* **54**:482–506 (1922).

18. *Handbook of Clinical Laboratory Data*, 2nd ed., Chem. Rubber Co., Cleveland, Ohio (1968).

19. Antonis, A., Clark, M., and Pilkington, T. R. E., A semiautomated fluorimetric method for the enzymatic determination of pyruvate, lactate, acetoacetate and β-hydroxybutyrate levels in plasma, *J. Lab. Clin. Med.* **68**:340–356 (1966).

20. Gibbons, B. H. and Edsall, J. T., Rate of hydration of carbon dioxide and dehydration of carbonic acid at 25°C, *J. Biol. Chem.* **238**:3502–3607 (1963).

21. Armstrong, J. McD., Myers, D. V., Verpoorte, J. A., and Edsall, J. T., Purification and properties of human erythrocyte carbonic anhydrases, *J. Biol. Chem.* **241**:5137–49 (1966).

22. Natelson, S., Leighton, D. R., and Calas, C., Assay for the elements chromium, manganese, iron, cobalt, copper and zinc simultaneously in human serum and sea water by X-ray spectrometry, *Microchem. J.* **6**:539–556 (1962).

23. Laurent, G., Charrel, M. Marriq, C., Garcon, D, and Darrien, Y., Carbonic anhydrases of human erythrocytes. III Amino acid composition, *Bull. Soc. Chim. Biol.* **48**:1125–1136; 1251–1264 (1966).

24. Packer, Y. and Meany, S. E., The catalytic versatility of erythrocyte carbonic anhydrase. II. Kinetic studies of the enzyme-catalyzed hydration of pyridine aldehydes, *Biochem.* **6**:239–246 (1967).

25. Packer, Y. and Stone, J. T., The catalytic versatility of erythrocyte carbonic anhydrase. III. Kinetic studies of the enzyme catalyzed hydrolysis of p nitro phenyl acetate, *Biochem.* **6**:668–678 (1967).

25a. Natelson, S. and Tietz, N., Blood pH measurement with the glass electrode—Study of venous and fingertip blood, *Clin. Chem.* **2**:320–327 (1956).

26. Mann, T. and Keilin, D., Sulfanilamide as a specific inhibitor of carbonic anhydrase, *Nature* **146**:164–5 (1940).

27. Gilman, A., The Mechanism of action of the carbonic anhydrase inhibitors, *Ann. N. Y. Acad. Sci.* **91**:355–362 (1958).

28. Natelson, S., Chronic alkalosis with damage to the central nervous system, *Clin. Chem.* **4**:32–42 (1958).

29. Weatherall, D. J. and McIntyre, P. A., Developmental and acquired variations in erythrocyte carbonic anhydrase, *Brit. J. Heme.* **13**:106–114 (1967).

30. Eng, L. L. and Tarail, R., Carbonic anhydrase deficiency with persistence of foetal haemoglobin: A new syndrome, *Nature* **211**:47–49 (1966).

31. Weatherell, D. J., Edwards, J. A., and Donohue, W. T. A., Haemoglobin and red cell enzyme changes in juvenile myeloid leukemia, *Brit. Med. J.* **1**:679–681 (1968).

32. Weatherell, D. J. and Brown, M. J., Juvenile chronic myeloid leukemia, *Lancet* **1**:526 (1970).

Maintenance of Constant Osmotic Pressure in Body Fluids

5.1 DEFINITION AND UNITS OF OSMOTIC PRESSURE

At 0°C and 760 mmHg pressure (1 atm), 1 mol of an ideal gas will occupy 22.4 liters. If 1 mol of glucose is dissolved in 22.4 liters of water, the osmotic pressure will be 1 atm.[1] The glucose molecules, like the gas molecules, exert their pressure by bombarding the walls of the container. They act as though they were in the gaseous form and the water was merely space. The assumption being made here is that glucose is an "ideal" substance, that is, it does not react with the water molecules. This is not quite the case for glucose but we will make this assumption for the present.

The fundamental gas laws are expressed by the equation $PV = nRT$, where P is the pressure, V is the volume, R is the gas constant, n is the number of moles, and T is the absolute temperature.[2] A similar

---

[1] Osmotic pressure is defined as the pressure required to be placed on a solution separated from water by a membrane, to prevent osmosis from taking place. Osmotic pressure derives from the fact that there are more molecules of water bombarding the membrane on the pure water side than on the side containing a solute.

[2] The equation relating P, V, and T needs to be corrected for the volume occupied by the molecules and their affinity for each other. These equations are discussed in detail in textbooks on physical chemistry.

equation applies to osmotic pressure, and we can write[1]

$$\pi V = nRT$$

The symbol π is used to represent the osmotic pressure. Just as the equation written for the gas laws is an approximate equation because the "ideal" gas does not exist, so this equation is also an approximation because every solute either associates, dissociates, or reacts with the water to some extent. Since n represents the number of moles, we can write

$$\pi = (n/V)RT \quad \text{and} \quad \pi = CRT$$

since the number of moles divided by the volume is the concentration C. Thus we see that the osmotic pressure is proportional to the concentration at constant temperature.[3] We could then express osmotic pressure in atmospheres, pounds per square inch, or millimeters of mercury. None of these is practicable for clinical purposes. Instead we use a unit called the *osmole* (abbreviated osmol). An osmole is the osmotic pressure exerted by 1 mol of an undissociated substance, like glucose, dissolved in 1 kg (liter) of water.

As pointed out above, 1 mol of glucose in 22.4 liters of water would have an osmotic pressure of 1 atm. Then an osmolal solution (1 mol/liter of water) would exert 22.4 atm pressure. This is a concentration out of the range of what we find in biological solutions, and we then resort to a unit 1000 times smaller, or the milliosmole (mosmol). The convenience of this unit can be seen from the fact that a K^+, Na^+, or Cl^- solution of 1 mEq/liter would exert an osmotic pressure of 1 mosmol.

Properties that depend upon the number of particles in solution are called *colligative properties*. Osmotic pressure, freezing point depression, and vapor pressure depression are all examples of colligative properties. Therefore, if a substance like NaCl is dissolved, assuming 100% ionization, one would obtain for a 1 mEq/liter solution, 2 mosmol. For a 100 mEq/liter solution, of the order of the concentration in serum, we should get 200 mosmol. Actually, the apparent ionization (activity) of a 0.1 M solution of NaCl is 78% and we get $0.78 \times 200 = 146$

---

[3] The implication here is that the solute is in the gaseous state. In going from a solid to the gaseous state, one needs to supply the heat of fusion and the heat of vaporization. When a substance that does not react with the solvent dissolves, the heat taken from its surroundings is approximately equal to the heat of fusion. Technically speaking, then, the solute goes from the solid to the liquid state when it dissolves.

mosmol for the ions and 22 mosmol for undissociated NaCl, for a total of $146 + 28 = 174$ mosmol.

It is to be noted that in discussing osmotic pressure we used grams per liter of water, or molality, assuming a density of one for water. A solution of 180 g of glucose in 1000 ml of water would have a larger volume than 58 g of NaCl in 1000 ml of water, yet they would have the same molality. For dilute solutions, molality and molarity (grams made up to one liter with water) are approximately the same, and in clinical chemistry when we talk of osmoles per liter, we usually mean grams made up to a liter. Thus, in place of *osmolality* we utilize *osmolarity*.[2]

Since osmotic pressure is difficult to measure, we usually use some other colligative property and translate its findings in terms of osmotic pressure. Most commonly, modern instruments use freezing point depression.[3] A mole of glucose dissolved in one liter of water will lower the freezing point 1.86°C. Any solution producing this freezing point depression would be considered to contain 1 osmol or 1000 mosmol. The average freezing point of blood serum is -0.56°C. This is equivalent to $(0.56/1.86) \times 1000 = 301$ mosmol.

Recently, a rapid reading osmometer has been introduced based on vapor pressure lowering by salts in solution. A small drop of serum is placed on a thermistor. The faster the evaporation rate, the colder the drop. The thermistor senses the temperature and translates the results to osmolality of the solution (Wescor, Inc., Logan, Utah 84321).

The lowering of vapor pressure is directly proportional to the mole fraction of the solute. This is apparent from the fact that the number of molecules of water reaching the surface per unit time is decreased in proportion to the number of nonvolatile ions present. At 37°C the vapor pressure of water is 47.07 mmHg. The molecular weight of water is 18. Per kilogram of water we have therefore $1000/18 = 55.56$ mol. One mole (1000 mosmol) of an ion per kg of water would reduce the vapor pressure by $1/55.56 \times 47.07 = 0.846$ mmHg. For a human serum of 300 mosmol/liter, the vapor pressure lowering is $0.846 \times 0.3 = 0.254$ mmHg. The aqueous vapor tension of this serum is then $47.07 - 0.254 = 46.816$ mmHg.

A 5.4% glucose solution is said to be *isotonic* with blood because it has the same osmotic pressure. *Hypotonic* solutions are solutions with a lower osmotic pressure than blood and the word *hypertonic* is applied to solutions with an osmotic pressure higher than blood.

5.2 BODY WATER

We can consider the body as comprising a series of compartments separated from each other by membranes.[4,5] In the steady state, the compartments are in osmotic balance with each other, that is, they are *iso-osmotic*. The compartments comprise the *intracellular* compartment, the *intravascular* compartment, and the *interstitial* space occupied between the cells and the blood vessels. One can subdivide the intravascular compartment between the plasma and the erythrocytes. If one lumps the erythrocytes with the other cells of the body, one can then divide the body into only two compartments, namely the *extracellular* and *intracellular* spaces.

The intracellular fluid comprises approximately 45% of the body weight, while the extracellular fluid represents approximately 20% of the body weight. The remainder is due to the weight of minerals, including the minerals of bones and teeth, protein, fat, carbohydrates, and nucleic acids. The approximate water content of various tissues is listed in Table 5.1.[6]

The total body water is usually estimated by administering a substance that penetrates all the membranes of the body and measuring its dilution. Substances used have been deuterium oxide (D_2O),

TABLE 5.1

Approximate Water Content of Various Body Tissues and Fluids

Tissue	Water, g/100 g tissue
Adipose tissue	6–10
Blood plasma	87
Blood erythrocytes	60–65
Bone (extremities and skull)	14–22
Brain	78–84
Cerebrospinal fluid	99
Connective tissue	60
Liver	70–75
Muscle	75–78
Thymus	81
Thyroid	79
Skin	72
Tooth enamel	3.7
Total body water (fat free)	73

TABLE 5.2

Distribution of Body Water in Humans

Compartment	Mean % of body weight	Range
Total body water		
male	70	60–76
female	60	50–70
infant	78	70–83
Intracellular fluid		
male	45	36–50
female	35	26–40
infant	48	45–50
Extracellular fluid		
male	25	20–30
female	25	20–30
infant	30	25–33
Interstitial fluid		
male	21.5	18–23
female	21.5	18–22
infant	25.9	22–27
Intravascular fluid		
male	4.5	3.4–5.8
female	4.3	3.2–5.0
infant	4.1	3.2–5.0

tritiated water (HTO), and antipyrine derivatives.[7] The extracellular compartment is measured by a substance that penetrates all compartments but does not pass the cell membrane.[8] Such substances as inulin, mannitol, sucrose, thiocyanate, sulfate, bromide, thiosulfate, radioactive Na, radioactive chloride, and glucose have been used.[9–11] The results obtained are summarized in Table 5.2.

The percentage of total body water will vary in the same individual depending upon his state of nutrition. If an individual is overweight, the percentage of body water will decrease due to the fact that deposited fat contains very little water compared to the rest of the body.[12,13] The percentage of water as shown in Table 5.1 for the body, excluding fat, is 73%. If we measure the body water and find it to be a lower value, the difference is due to fat. The formula used is

$$\% \text{ fat in body} = 100 - \frac{\% \text{ body water found}}{0.73}$$

Thus, by using D_2O, we may find 60% by weight of total body water in an adult male. We then calculate $100 - 60/0.73 = 18\%$ fat. By this equation, it can readily be shown that women are generally fatter than men. Body water in women is often of the order of 50% and this calculates out to 32% fat.

Specific gravity is readily measured in humans by volume of water displacement and weight.[14] The mean specific gravity of the male has been measured as 1.068 ± 0.012.[15] For the percentage of fat, the following formula has been proposed:

$$\% \text{ fat} = 100\left(\frac{5.548}{\text{sp gr}} - 5.044\right)$$

Substituting 1.068 in this formula gives 14% as the mean fat content of adult males. Substituting 1.034, the mean specific gravity for women, gives 32% fat.

5.3 MAINTENANCE OF OSMOTIC BALANCE BETWEEN INTRAVASCULAR AND INTERSTITIAL FLUID

The capillary wall that separates the plasma from the interstitial fluid is apparently an inert membrane, being impermeable only to high molecular weight compounds like protein and certain compounds such as Evans blue dye.[16] The interstitial fluid is simply an ultrafiltrate of plasma, the difference in composition of both sides obeying the Gibbs–Donnan law. Here $r = 0.95$, when concentrations are expressed in molality.[17–19]

$$r = \frac{[H^+]_i}{[H^+]_p} = \frac{[Na^+]_i}{[Na^+]_p} = \frac{[K^+]_i}{[K^+]_p}$$

$$= \frac{[Ca^{2+}]_i^{1/2}}{[Ca^{2+}]_p^{1/2}} = \frac{[Mg^{2+}]_i^{1/2}}{[Mg^{2+}]_p^{1/2}} = \frac{[Cl^-]_p}{[Cl^-]_i} = \frac{[HCO_3^-]_p}{[HCO_3^-]_i}$$

where

$$r = \left(\frac{[A^-]_p}{[A^-]_p + [Pr^-]_p}\right)^{1/2}$$

$[A]_p$ represents the concentration of anions in the plasma other than protein, and $[Pr^-]_p$ represents the anion equivalence of the plasma proteins. Thus the concentration of Cl$^-$ and bicarbonate should be greater in the interstitial fluid, and the concentration of the cations should be greater in the plasma. Table 5.3 lists typical values obtained for the anions and cations in the extracellular fluid of the different compartments.

Since $r = 0.95$, the concentration of the cations in the interstitial fluid should be lower than that in the plasma when calculated on the basis of molality (Eq/kg H$_2$O) by a ratio of 0.95/1.0. Only sodium approaches this ratio (Table 5.3).

While K, Ca, and Mg values are lower in the interstitial fluid, they are too low to satisfy the Gibbs–Donnan law. The reason for this is that Ca, Mg, and K all form chelates with protein and other substances, like citrate, in the plasma. As a result, 40% of the calcium is known to be complexed with protein. Thus the ionic concentration of calcium in the plasma water does not exceed $0.4 \times 5.4 = 2.16$ mEq/liter of H$_2$O. However, the Ca is also complexed to phosphate and citrate in the interstitial fluid.

TABLE 5.3

Typical Distribution of Anions and Cations in the Extracellular and Intracellular Fluid in a Normal Adult

	Plasma		Interstitial fluid,	Intracellular fluid
Electrolyte	mEq/liter	mEq/kg H$_2$O	mEq/kg H$_2$O	mEq/kg H$_2$O
Cations:				
Sodium	140	150	146	15
Potassium	4.6	5.0	4.5	150
Calcium	5.0	5.4	3.0	2
Magnesium	2.0	2.1	1.5	27
Total cations	151.6	162.5	155.0	194
Anions:				
Chloride	101.8	109	115	10
Bicarbonate	26	28	29.5	1
Proteinate	16	17.1	0.02	63
Phosphate	1.8	1.9	2.0	90
Organic acids[a]	5	5.4	7.5	10
Sulfate	1	1.1	1	20
Total anions	151.6	162.5	155.0	194

[a] By difference, to make anions and cations balance.

The concentration of the anions in the interstitial fluid should be higher than in the plasma. This is approximately the case for chloride and bicarbonate ion. One can say that the Gibbs–Donnan law is followed and can be used to indicate the general requirements for osmotic balance on both sides of the capillary wall, provided one uses the *activities* of the ions rather than their concentrations. If this is true, then the *interstitial fluid can be considered as an ultrafiltrate of blood.*

5.4 VARIATION IN EXTRACELLULAR FLUID VOLUME IN DEFENSE OF INTRACELLULAR OSMOTIC PRESSURE

One reason for maintaining constant osmotic pressure in the body fluids is to protect the tissue cells from rupturing. Another is to maintain an orderly metabolism in the tissue by maintaining a constant environment. During the day, individuals imbibe water at different intervals and during their daily activity lose water sporadically. With these overall large changes, the osmotic pressure of the body fluids remains constant. This mechanism can be examined using a simple example. One must keep in mind that movements of ions must not disturb the electrical balance between the cations and anions. The balance maintained in the plasma is shown in Figure 5.1.

In Figure 5.2 the distribution of water is shown in the different compartments in the normal as compared to the same individual after dehydration. The individual voids and then drinks 1000 ml of water over a 2-hr period. Urine is then collected until 1500 ml of urine has been obtained (24 hr). During this period the individual fasts. Analysis of the urine shows a sodium level of 75 mEq/liter, or a loss of 113 mEq of Na^+. During that same period, approximately 800 ml of additional water is lost in the sweat (along with some Na^+) and through the lungs, so that the total loss of water is approximately $1500 + 800 = 2300$ ml. The net loss is then $2300 - 1000 = 1300$ ml of H_2O and somewhat more than 113 mEq of Na^+. The reason for this loss is explained as follows. If 1000 ml of water is imbibed and 1000 ml of water is excreted through the urine, the 1000 ml excreted in the urine contains salts. This would result in a decrease in osmotic pressure in the extracellular fluid, which would then become hypotonic. To compensate for this, *additional water is lost to bring the osmotic pressure back up.* Some sodium, from

Acid–Base Composition of Blood Plasma

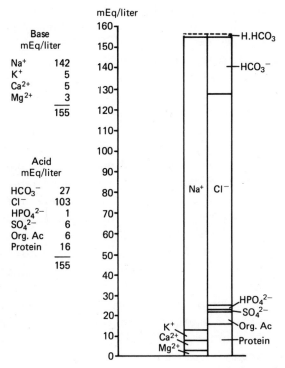

FIGURE 5.1 Graphic representation of the electrical balance between anions and cations in human serum (after Gamble).

BEFORE DEHYDRATION

Plasma	Interstitial		
Na = 140 mEq/liter Pr = 7.0% Hematocrit = 46% Volume = 2.5 liters	8.3 liters	Volume = 28 liters	Overall Volume 38.8 liters
Extracellular Fluid (10.8 liters)		Intracellular Fluid	

AFTER MILD DEHYDRATION

Plasma	Interstitial		
Na = 140 mEq/liter Pr = 8.1 Hematocrit = 50% Volume = 2.1 liters	7.4 liters	Volume = 28 liters	Overall Volume 37.5 liters
Extracellular Fluid (9.5 liters)		Intracellular Fluid	

FIGURE 5.2 Distribution of water, Na^+, and protein in the intracellular fluid in a normal adult weighing 85 kg before and after dehydration.

sodium stores in the bones, is dissolved and moves into the extracellular fluid also.

From Figure 5.2 it is apparent that this is done *at the expense of the extracellular fluid volume and not the intracellular fluid volume.* Thus the hematocrit rises, and so does the protein level, indicating a decreased plasma volume. The osmotic pressure remaining the same in the interstitial fluid results in the volume remaining constant in the cells. The individual will then be mildly dehydrated.

If this process is continued, as occurs with an individual lost in a desert with a limited water supply and no food, the plasma volume and interstitial volume will continue to shrink until a critical point is reached. At this point the kidneys can no longer limit the excretion of sodium ion, and the sodium levels will drop in the plasma and interstitial fluid. The extracellular fluid would then become hypotonic with respect to the cells and water moves into the cells, causing them to swell. *The intracellular fluid volume then increases.* This lowers the already reduced volume of extracellular fluid. If the process is continued, tissue breakdown occurs. *The surviving cells would be engorged with water, but the net content of water in the tissue would be markedly decreased.* Metabolism is interrupted, oliguria or anuria develops, organic acids pour into the extracellular fluid from the bloated cells, and a severe acidosis results, blood pH dropping as low as 6.9.

The above explanation also describes the reason for the well-known fact that drinking water during prolonged exercise, such as while participating in a football game on a hot day, may result in severe salt depletion, with shrinkage of extracellular fluid volume and severe symptoms of dehydration. It is for this reason that fluids given to athletes during prolonged exercise contain some salt to preserve the extracellular fluid volume and osmotic pressure.

5.5   *MECHANISMS FOR MAINTAINING CONSTANT VOLUME AND OSMOTIC PRESSURE*

There are several mechanisms by means of which the brain is alerted to the presence of an increased or decreased volume or osmotic pressure of the extracellular fluid so that its volume may be kept constant. These comprise the "stretch receptors" and "osmoreceptors," which signal the posterior pituitary gland, through the hypothalamus, to release or withhold the antidiuretic hormone (ADH), and to implement other mechanisms discussed below.[20-23]

5.5.1   Stretch Receptors

With change in blood volume, a reflex-mediated change in the levels of the antidiuretic hormone (ADH, vasopressin) can readily be demonstrated in the circulatory system. The secretion of this hormone is sensitive to *stretch receptor zones* in the circulation, such as in the left atrium and the *carotid sinus*. Elevation of left atrial transmural pressure by as little as 2–7 cm of water results in a significant reduction in blood ADH and diuresis.[24] Thus volume changes signal the hypothalamus, which in turn stimulates or inhibits posterior pituitary secretion to maintain a constant volume of the extracellular fluid.[25] These stretch receptors resemble pressure transducers in behavior, generating a signal from pressure changes.

5.5.2   Osmoreceptors

In addition to "stretch receptors," it can be readily demonstrated that "*osmoreceptors*" respond promptly to a change in osmotic pressure in the extracellular fluid.[26,27] The supraoptic nucleus in the hypothalamus contains osmoreceptors that are sensitive to changes in blood osmolarity stimulating or inhibiting vasopressin secretion.[29,29] Vasopressin will promote water reabsorption and thus lower the osmolarity of the blood when it is too high. *The osmoreceptors in the portal vein carrying water from the intestine to the liver respond promptly before the cerebral osmoreceptors are invoked.*[30]

The liver can absorb water and increase its water content by as much as 10%. Drinking water causes a shift of the Na^+, Cl^-, and HCO_3^- from the blood in the splanchnic circulation to the water in the gut. Portal blood osmolarity drops, causing diffusion of water into the liver from the hypotonic blood with a swelling of the liver. The swelling of the liver inhibits vasopressin secretion, which results in sweating and diuresis. Thus the liver also acts as a large reservoir to hold surges of water from the gut while an opportunity is given for the water to be excreted.

5.5.3 *Other Mechanisms for Maintaining Constant Volume and Osmotic Pressure in the Extracellular Fluid*

There are several other mechanisms that come into play when an increase or decrease in extracellular fluid volume or osmotic pressure takes place.[31,32]

1. Increased blood flow to the kidneys results in increased glomerular filtration rate, which results in an increase in sodium and water excretion.[33]

2. It has been suggested that excretion of sodium (natriuresis) in response to blood volume expansion is partly the result of inhibition of release of renin. Effective blood volume seems to be the primary determinant of renin release and thus circulating angiotensin levels. A reciprocal relationship between blood angiotensin levels and effective blood volume levels seems to exist. With increased blood volume, renin secretion drops, with resultant dilation of glomerular arterioles and engorgement of postglomerular vessels, resulting in a constriction of the proximal renal tubule. This would result in a reduction of fractional sodium reabsorption and a natriuresis and diuresis, serving to correct the increase in blood volume.[35] This is discussed in further detail in the chapter dealing with the role of the kidney in fluid balance. Figure 5.3 summarizes the relationship between renin, angiotensin, and aldosterone.[38]

3. An increased secretion of aldosterone will cause sodium retention. A decrease will cause sodium excretion.[34] Aldosterone production, in turn, is modulated by angiotensin levels.

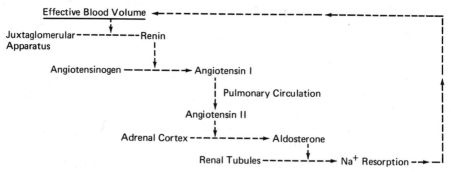

FIGURE 5.3 Schematic diagram of the renin–angiotensin–aldosterone feedback system.

4. The antidiuretic hormone will cause reabsorption of large volumes of water in the distal tubules along with ions such as sodium and chloride. Absence of the antidiuretic hormone from any of several causes will cause a large loss of water along with some sodium chloride.

None of the above mechanisms *completely* explains the phenomena observed with changes in blood volume and osmotic pressure.[36,37] For example, when animals are infused with isotonic saline solution, a marked rise in sodium excretion occurs. This increase is not affected by administration of aldosterone and vasopressin in large quantities. This excretion is also not inhibited by decreasing glomerular filtration rate produced by decreasing blood flow to the kidney by inflating a balloon in the aorta. If albumin is given along with the saline to maintain constant plasma protein levels, excretion of NaCl continues increased as before. If the kidney is denervated, the increase in NaCl excretion still occurs. Several hypotheses have been proposed, none of which completely fits the experimental data. It has been suggested that infusion of saline causes release of some unidentified substance in the blood, which causes decreased proximal tubule reabsorption. The evidence for this hypothesis has been questioned.[31]

If the plasma volume is increased by injection of hyperoncotic solutions of albumin, a decrease in reabsorption of water in the proximal as well as the distal tubules is observed. Increase in sodium excretion is noted but not nearly as much as when a similar decrease in proximal reabsorption is induced by saline. Thus factors other than the known processes are also involved in influencing the kidney in water and electrolyte excretion.

5.6   RECAPITULATION

The objective for maintaining constant osmotic pressure in body fluids is to provide a stable milieu for carrying out the metabolic functions of the cells. Large fluctuations may take place in the extracellular fluid compartments while cell volume and osmotic pressure remain constant. Pressure transducers called "stretch receptors," distributed in the blood vessels, sense changes in pressure and thus volume, and signal the hypothalamus to stimulate or inhibit the posterior pituitary excretion of vasopressin. Osmoreceptors also sense changes in osmotic pressure.

These and other mechanisms serve to ensure that any change of plasma volume or osmotic pressure promptly results in adjustment of excretion by the sweat glands and the kidneys to bring the extracellular fluid volume and osmotic pressure to their original values. *This process is mediated by humoral agents whose secretion is regulated directly and indirectly by the central nervous system.* Changes in osmotic pressure and/or changes in plasma volume are monitored by stretch and osmoreceptors, which signal the hypothalamus of these changes. *In dehydration, constant osmotic pressure takes precedence over constant volume, and extracellular fluid volume shrinks in order to maintain constant osmotic pressure within the cell.*

5.7 SELECTED READING—CONSTANT OSMOTIC PRESSURE AND BODY FLUIDS

Brooks, S. M., *Basic Facts of Body Water and Ions*, 3rd ed., Springer Publishing, New York (1973).

Chapman, G., *Body Fluids and Their Functions*, St. Martin, New York (1967).

Weisberg, H. F., *Water, Electrolyte and Acid Base Balance*, 2nd ed., Williams and Wilkins (1962).

Potts, W. T. and Parry, G., *Osmotic and Ionic Regulators in Animals*, Pergamon, New York (1963).

Muntwyler, E. and Mautz, F. R., Electrolyte and water equilibria in the body, in *Medical Physics*, Glasser, O., Ed. Year Book, Chicago, Illinois (1944), pp. 371–377.

Strauss, M. B., *Body Water in Man, The Acquisition and Maintenance of the Body Fluids*, Little, Brown, and Company, Boston, Massachusetts (1957).

Wolf, A. V., *Thirst, Physiology of the Urge to Drink and Problems of Water Lack*, C. C. Thomas, Springfield, Illinois (1958).

Brozek, J., *Human Body Composition*, Pergamon, New York (1945).

5.8 REFERENCES

1. Van't Hoff, J. H., The role of osmotic pressure in the analogy between solutions and gases, *Z. Physik. Chem.* **1**:481–508 (1887).
2. Dormandy, T. L., Osmometry, *Lancet* **1**:267–270 (1967).
3. Bowman, R. L., Trantham, H. V., and Caulfield, P. A., An instrument and method for rapid dependable determination of freezing point depression, *J. Lab. Clin. Med.* **43**:310–315 (1954).
4. Berger, E. Y., Dunning, M. F., Brodie, B. B., and Steele, J. M., Body Water compartments in man, *Fed. Proc.* **8**:10 (1949).

5. Levitt, M. F. and Gaudino, M., Measurement of body water compartments, *Am. J. Med.* **9**:208–215 (1950).
6. Edelman, I. S. and Leibman, J., Anatomy of body water and electrolytes *Am. J. Med.* **27**:256–277 (1959).
7. Deane, H., Intracellular water in man, *J. Clin. Invest.* **30**:1469–1470 (1951).
8. Check, D. B., Extracellular volume, its structure and measurement and the influence of age and disease, *J. Pediatrics* **58**:103–125 (1961).
9. Fellers, F. X., Barnett, H. L., Hare, K., and McNamara, H., Change in thiocyanate and sodium-24 spaces during growth, *Pediatrics* **3**:622–629 (1949).
10. Berson, S. A. and Yalow, R. S., Critique of extracellular space measurements with small ions, Na^{24} and Br^{82} spaces, *Science* **121**:34–36 (1955).
11. Tomaszewski, L., A new principle for the determination of the extracellular fluid, *Clin. Chim. Acta* **16**:417–427 (1967).
12. Lesser, G. T., Perl, W., and Steele, J. M., Determination of total body fat by absorption of an inert gas, measurements and results in normal human subjects, *J. Clin. Invest.* **39**:1791–1806 (1960).
13. Osserman, E. F., Pitts, G. C., Welham, W. C., and Behnke, A. R., In vivo measurement of body fat and body water in a group of normal men, *J. Appl. Physiol.* **2**:633–639 (1950).
14. Werdein, E. J. and Kyle, L. H., Estimation of the constancy of density of the fat-free body, *J. Clin. Invest.* **39**:626–629 (1960).
15. Behnke, A. R., Feen, B. G., and Welham, W. C., The specific gravity of healthy men; body weight ÷ volume as index of obesity, *J.A.M.A.* **188**:495–498 (1942).
16. Von Porat, B., Blood volume determinations with the Evans Blue Dye, *Acta Med. Scand. Suppl.* p. 256 (1951).
17. Hastings, A. B., Salvesen, H. A., and Van Slyke, D. D., Studies of gas and electrolyte equilibria in blood; distribution of electrolytes between transudates and serum, *J. Gen. Physiol.* **8**:701–711 (1927).
18. Van Slyke, D. D., *Factors Affecting the Distribution of Electrolytes, Water and Gases in the Animal Body*, Monograph on Experimental Biology, Lippincott, Philadelphia, Pennsylvania, (1926).
19. Greene, C. H. and Power, M. H., Distribution of electrolytes between serum and *in vivo* dialysate, *J. Biol. Chem.* **91**:183–202 (1931).
20. Henry, J. P., Gauer, O. H., and Reeves, J. L., Evidence of the atrial location of receptors influencing urine flow, *Circ. Res.* **4**:85–90 (1956).
21. Gilmore, J. P., Contribution of cardiac nerves to the control of body salt and water, *Fed. Proc.* **27**:1156–1159 (1968).
22. Share, L. and Levy, M. H., Cardiovascular receptors and blood titer of antidiuretic hormone, *Am. J. Physiol.* **203**:425–428 (1962).
23. Share, L., Vasopressin, its bioassay and the physiological control of its release, *Am. J. Med.* **42**:701–712 (1967).
24. Arndt, J. O., Reineck, H. and Gauer, O. H., Renal excretory function and hemodynamics on stretching of the left atrium in anesthetized dogs, *Arch. Ges. Physiol.* **277**:1–15 (1963).
25. Perlmutt, J. H., Contribution of carotid and vagal reflex mechanisms. Sym-

posium on neural control of body salt and water, *Fed. Proc.* **27**:1149–1155 (1968).

26. Gaunt, R. and Birnie, J. H., *Hormones and Body Water*, C. C. Thomas, Springfield, Illinois (1951).

27. Chambers, G. H., Melville, E. V., Hare, R. S., and Hare, K., Regulation of release of pituitrin by changes in osmostic pressure of plasma, *Am. J. Physiol.* **144**:311–320 (1945).

28. Verney, E. B., Antidiuretic hormone and the factors which determine its release, *Proc. Roy. Soc. (Lond.)* **135**:25–106 (1948).

29. Bornstein, P., Brandt, I., and Epstein, F. H., Selective failure of osmotic receptors in diabetes insipidus, *Clin. Res.* **9**:199 (1961).

30. Haberich, F. J., Osmoreception in the portal circulation, *Fed. Proc.* **26**:1137– (1968).

31. Berliner, R. W., Intrarenal mechanisms in the control of sodium secretion, *Fed. Proc.* **27**:1127–1131 (1968).

32. Gauer, O. H., Osmocontrol versus volume control, *Fed. Proc.* **27**:1132–1136 (1968).

33. Earley, L. E. and Friedler, R. M., The effects of combined renal vasodilation and pressor agents on renal hemodynamics and the tubular reabsorption of sodium, *J. Clin. Invest.* **45**:542–51 (1966).

34. August, J. T., Nelson, D. H. and Thorn, G. W., Aldosterone. *New Eng. J. Med.* **259**:917–923 (1958).

35. Hodge, R. L., Lowe, R. D., and Vane, J. R., The effects of alteration of blood volume on the concentration of circulating angiotensin in anaesthetized dogs, *J. Physiol.* **185**:613–26 (1966).

36. Sunsten, J. W. and Sawyer, C. H., Electroencephalographic evidence of osmosensitive elements in olfactory bulb of dog brain, *Proc. Soc. Exp. Biol. Med.* **101**:524–27 (1959).

37. Corvian, M. R. and Antunes–Rodriguez, J., Specific alterations in sodium chloride intake after hypothalamic lesions in the rat, *Am. J. Physiol.* **205**:922–926 (1963).

38. Haber, E., Recent developments in pathophysiologic studies of the renin-angiotensin system, *New Eng. J. Med.* **280**:148–155 (1969).

6

Maintenance of Constant Ion Concentration in Body Fluids: Sodium, Potassium, and Chloride

6.1 *INTRODUCTION*

In the discussions in the preceding chapters, the following facts have been pointed out:

1. Stretch receptors and osmotic receptors signal the central nervous system to set in motion neural and humoral mechanisms to maintain constant volume and constant pressure in the extracellular fluid spaces.
2. Bicarbonate and pH levels are maintained constant. Change in these levels results in stimulation of the central nervous system, setting in motion mechanisms for their readjustment.
3. Blood pressure and metabolic rate, which in turn affect body temperature, are also under central nervous system control.
4. Function of individual organs, like the liver, kidney, and sweat glands, is also directly and indirectly controlled by the central nervous system tied to the composition and volume of the body fluids.

All of these observations point to the obvious conclusion that the central nervous system is the "control room" of the servomechanism that maintains homeostasis in the human. It is to be expected, therefore, that ionic concentration (activity) of such ions as Na, Cl, K, Ca, Mg, and phosphate is also under control of the central nervous system. This is emphasized by the fact that *plasma levels of these ions are kept constant within narrow limits despite wide fluctuations in their dietary intake.*

6.2   *SODIUM AND CHLORIDE*

The sodium and chloride ions in the body fluids are responsible for the major portion of the osmotic pressure of the blood. As such, they must be maintained constant if tissues are to remain intact. *Injury to the hypothalamic portion of the brain disturbs this mechanism, resulting in drastic changes.*

There are two mechanisms by means of which these ions, once absorbed, are maintained at constant level. These include:

1. A steady state set up among the various compartments of the body such as bone, tissue, and extracellular fluid. This includes the regulation of the permeability of the tissues to these ions and the movement of these ions into and out of these various compartments.
2. The regulation of the excretion of these ions by the kidney and sweat glands. In addition to the material already presented, this will be discussed in greater detail in the chapter on the role played by kidney function in maintaining the steady state in the human.

The interplay of the various mechanisms designed to maintain constant osmotic pressure, and thus salt levels, and constant volume is outlined in Figure 6.1. From this figure it can be seen that a complicated mechanism exists for maintaining constant sodium levels in the blood plasma, which includes local factors, but overall is controlled by the central nervous system. Since the anion most commonly available for maintaining electrical neutrality with sodium ion, and which can readily pass the cellular membranes, is chloride ion, regulation of sodium levels indirectly controls chloride levels also. However, it must be pointed out at this point that sodium retention and excretion are not always linked to chloride movement (see Chapter 11).

In Figure 6.1 the various mechanisms for maintaining constant Na^+ level in the plasma and extracellular fluid are integrated to show how they work toward a common end. These can be listed as follows:

1. Renal blood flow affects glomerular filtration rate. Increased

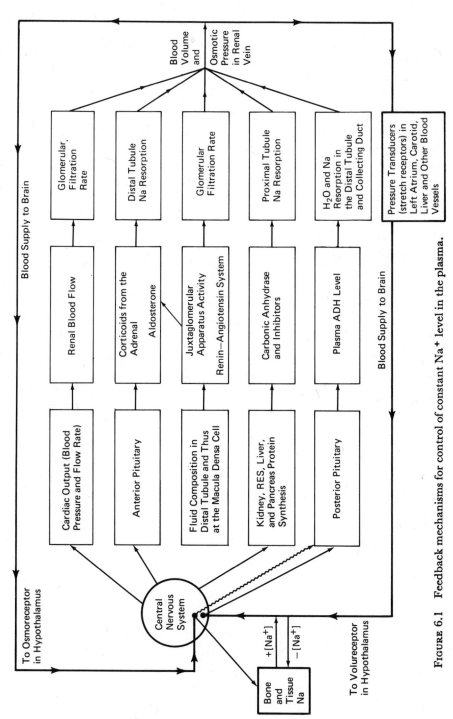

FIGURE 6.1   Feedback mechanisms for control of constant Na⁺ level in the plasma.

renal blood flow to the glomeruli will result in increased Na^+ and Cl^- excretion.[1] Decreased blood flow to the glomeruli will result in Na^+ and Cl^- retention and edema. This occurs in certain patients with reduced cardiac output.[2]

2. The action of carbonic anhydrase on sodium excretion is controlled by the activity of the enzyme in the blood and kidney and the concentration of inhibitors in the plasma.[3-5] Inhibition of this system results in increased sodium reabsorption in the tubules. Thus the level of activity of this system is an important factor in control of rate of sodium excretion. A similar effect has been demonstrated on secretion from the pancreas[6] and sweat glands.[7] Rate of synthesis of this enzyme and thus its activity are controlled by the central nervous system (see also Chapter 11).

3. Aldosterone acts on the distal tubules and also affects sodium reabsorption.[8] The regulation of aldosterone secretion is primarily mediated by the renin–angiotensin system and secondarily by ACTH, plasma sodium, and potassium concentrations.[9,10] In primary hyper-aldosteronism, sodium will be retained and hypertension will result. In exchange for sodium, potassium will often be excreted in this con-dition and low plasma potassium levels may be found in these patients.[11] (For details see Chapter 11.)

4. Corticoid synthesis and thus indirectly the aldosterone level are controlled by the action of the anterior pituitary secretion on the adrenal; this, in turn, is controlled by secretion from the hypothala-mus.[12,13] In addition to the effect of aldosterone, other steroids whose plasma level is controlled by the anterior pituitary can also be effective in salt and water retention for specific purposes. For example, during the menstrual cycle the estrogen–progesterone mechanism also causes salt and water retention before menstruation and diuresis if fertilization has not taken place.[14]

5. The juxtaglomerular apparatus, located at the afferent tubule of the glomerulus (see Section 11.6) at its impingement upon the distal tubule, senses changes in blood flow in the afferent arteriole and changes in the composition of the contents of the distal tubule. Based upon this composition, it controls the secretion of renin, a proteolytic enzyme from the juxtaglomerular apparatus.[15] Renin activates the angiotensinogen of the plasma, converting it to angiotensin. This polypeptide is a potent stimulus to aldosterone secretion. This also controls the diameter of the glomerular capillary and renal tubule and regulates renal blood flow and thus glomerular filtration rate and salt and water excretion.[16] In

certain renal diseases excessive amounts of renin secreted into the plasma result in salt retention and hypertension.[17]

6. The posterior pituitary secretes the antidiuretic hormone (ADH), vasopressin, which controls the reabsorption of water at the distal tubules of the kidney. Secretion of this hormone is responsive to changes in extracellular fluid volume.

7. Bone acts as a major store for sodium and chloride. Various possible mechanisms for regulation of the flow of Na^+ and Cl^- to and from the bone will be discussed below.

All of these factors determine the salt content and volume of the urine excreted. These, in turn, determine the content and flow rate in the renal vein returning the blood that has been processed. Stretch receptors, located in the left atrium, liver, carotid bodies, and elsewhere, sense minute changes in pressure since, in effect, they are *sensitive pressure transducers.* The signal is translated in the hypothalamus to stimulate or inhibit the posterior pituitary to secrete antidiuretic hormone (vasopressin). The level of this hormone in the plasma then adjusts mainly the reabsorption of water, but also salts in the distal tubules of the kidney. At the same time the center in the brain sensitive to osmotic pressure also acts to adjust the balance between salt and water in the plasma to maintain constant osmotic pressure.

Variation in the osmotic pressure of plasma will change blood volume by causing water movement into or out of the vascular spaces. This will modify antidiuretic hormone secretion from the posterior pituitary and thus modulate water reabsorption in the distal tubules. Since this action is indirect, the response to changes in osmotic pressure is slower than that to change in volume.[18]

Thus from Figure 6.1 one can see that while local mechanisms exist each of which can become faulty, resulting in disease, alternative mechanisms exist and the overall mechanisms are integrated at the central nervous system. In this manner, constant salt and volume levels and thus osmotic pressures are maintained in the extracellular and, as a result, intracellular fluids.

6.3   ELECTROLYTE BALANCE BETWEEN BONE AND EXTRACELLULAR FLUIDS

In Figure 6.1 a probable mechanism is proposed for the action of the osmoreceptors in the brain in controlling the movement of ions to and from the bones. The evidence for this concept will now be reviewed.

1. If an animal is subject to intestinal lavage with hypotonic fluids, sodium will move out of the blood stream and into the lavage fluids. The plasma sodium levels will remain constant for some time in spite of the large amount of sodium being removed in the lavage fluids.[19-21] Eventually, salt levels in the plasma will drop only after large quantities of salt have been removed. The source for this can readily be found when bone biopsy studies are performed. The bones contain 43% of the body sodium. Twenty-eight percent of this is rapidly exchangeable. From Table 6.1 it can be seen that approximately 490 mEq is available from bone stores alone, while an additional 455 mEq is normally present in plasma. Also readily available from bone stores is 350 mEq of chloride. Additional NaCl (approximately 500 mEq) is also available from the cartilage and connective tissue.

If salt is now fed to these animals and the lavage discontinued, the Na and Cl levels in the bone will eventually return to their original values *after* the extracellular fluid levels have been restored.

Thus a dynamic balance exists between body sodium stores and plasma levels. In Figure 6.1 it is suggested that this is one function of the osmoreceptors.

2. *If the hypothalamus is invaded surgically, a readjustment of osmotic pressure takes place. Plasma, sodium, and chloride levels rise to new levels without the patient receiving salt from the outside.*[22,23] In premature infants who are experiencing repeated bouts of apnea, a similar observation has been made. A rapid surge of plasma, Na, and Cl levels is noted. These

TABLE 6.1

Distribution of Na, K, and Cl Among Various Compartments of the Body in a 70-kg Man

		Plasma	Interstitial fluid[a]	Connective tissue, cartilage and transcellular	Bone	Intracelluar	Total
Na	mEq	455	1246	511	1750	98	4060
	mEq/kg	6.5	17.8	7.3	25	1.4	58
K	mEq	14	35	49	287	3381	3766
	mEq/kg	0.2	0.5	0.7	4.1	48.3	53.8
Cl	mEq	322	861	497	350	287	2317
	mEq/kg	4.6	12.3	7.1	5.0	4.1	33.1

[a] Includes lymph, cerebrospinal fluid, etc.

will persist for some time and eventually return to normal levels after normal respiration is restored.[24]

3. Some patients with chronic hypernatremia and hyperchloremia exhibit this condition due to brain tumor invading the hypothalamic area,[25,25a] head injury,[26] drug intoxication,[27] encephalitis,[28] brain surgery,[29] anoxia, or some other disease process.[30,31] Some individuals born with congenital brain anomaly also exhibit this syndrome.[24,32] In the latter cases, the osmoreceptor *is offset to a higher point of osmotic pressure at birth*. None of the mechanisms other than possible damage to the osmoreceptor seem to apply. These patients do not exhibit abnormal thirst, ACTH insufficiency, abnormal ADH blood levels, abnormal cardiac function, renal disease, or dehydration.[33–37] Blood pH is normal and no evidence of increased bicarbonate levels can be demonstrated. Thus, defects in their carbonic anhydrase system apparently do not exist. The terms "neurogenic hypernatremia" and "high salt syndrome" have been proposed as names for this condition. If the erythrocytes of these patients are incubated with normal saline, they will swell, since they have been equilibrated with an hyperosmotic solution in the plasma.

From points 1–3, one can draw the tentative conclusion that the osmoreceptor acts by rapidly moving sodium and chloride into the extracellular fluid when required, independent of the other mechanisms that involve the kidney and sweat glands. This is a rapid mechanism. The only source of sodium large enough for this purpose is in the bones and cartilage. For this reason, in Figure 6.1 the action of the osmoreceptor is tentatively given as acting on the bone and tissue stores.

6.4   THE "HIGH SALT SYNDROME"

In the congenital "high salt syndrome," Na and Cl levels are elevated and remain so indefinitely. Development of the infant is retarded, probably due to congenital cerebral cortical atrophy.[24] In those cases where the condition is induced by head injury or a disease process, plasma, Na, and Cl levels will remain elevated in spite of hypotonic fluid therapy until the brain lesion heals or the patient expires. Autopsy in these cases always shows damage to the hypothalamic area of the brain.[21,38]

The "high salt syndrome" was first noted by one of the authors (S.N.) in collaboration with Dr. Benjamin Kramer in 1935 in a six-month-old child with diffuse cerebral cortical atrophy. Maintenance on a salt-free diet for nine days resulted in serum Na and Cl levels remaining at 170 and 125 mEq/liter, respectively. Extracellular fluid volume continued to shrink and the infant expired on the ninth day in circulatory collapse. Autopsy confirmed the brain anomaly. Kidneys and adrenals were normal. In that same year, the condition was produced experimentally.[33] For some time many confused this condition with severe dehydration.[36,37] The fact that this was a distinct syndrome was only finally clarified in 1955.[26]

A typical case is shown in Table 6.2, in a three-month-old infant who was admitted in acidosis. Note that at the 145 mEq/liter serum level, the infant is suffering from *hyponatremia*. Corrective fluid therapy brought him back to his "normal" sodium level and normal blood pH. This strongly suggests than an osmoreceptor is defective in controlling the electrolyte concentrations at "normal" levels.

Dehydration is evident on admission from his elevated protein and urea levels. Normal serum levels for protein in infants are of the order of 5.5 mg/100 ml. On hydration, the child was found to be anemic and blood was administered. The hematocrit value of 37% also actually indicated dehydration for this infant. Incubation of the erythrocytes of this child with normal saline resulted in swelling and hemolysis of the cells.

Table 6.3 shows a case admitted with head injury and in coma. When the high salt levels were noted, the patient was maintained on 5% glucose in water. After 48 hr, the patient went into circulatory collapse with a markedly reduced blood volume. Expansion of extracellular fluid volume was accomplished with 2% saline. The patient was allowed 10 g of salt daily thereafter. This patient was in coma for 54 days, subsisting on I.V. and tube feeding. In spite of apparently adequate

TABLE 6.2

Chronic Hypernatremia and Hyperchloremia in an Infant with Diffuse Cortical Atrophy of the Brain: Blood and Serum Levels of Pertinent Constituents[26]

Hospital day	Na, mEq/liter	Cl, mEq/liter	K, mEq/liter	Protein, g/100 cm³	Hematocrit value, %	Blood pH	Urea, mg/100 ml
1	145	115	4.6	6.9	37	7.26	24.0
2	165	135	4.0	6.4	35	7.41	16.8
6	158	128	4.6	5.0	34	7.39	12.2
10	158	126	4.1	5.8	35	7.37	13.1

TABLE 6.3

Blood and Plasma Levels of Pertinent Constituents in a 67-Year-Old Man Following Head Injury, Illustrating Hypernatremia and Hyperchloremia Which Did Not Respond to Nonsaline Fluids but Spontaneously Returned to Normal on Fluids Containing the Normal Daily Salt Requirement[26]

Hospital day	Urea N, mg/100 ml	Protein, g/100 cm³	Na, mEq/liter	K, mEq/liter	Cl, mEq/liter	Hematocrit value, %
14	24.0	5.9	168	4.5	125	46
16	37.4	6.3	173	3.6	130	52
17	27.2	5.6	173	3.3	140	43
19	19.5	—	171	3.3	133	43
20	35.6	5.1	168	3.2	128	40
21	—	4.1	159	3.3	118	—
22	34.6	5.3	165	3.6	123	—
23	25.1	5.0	165	3.3	115	—
26	17.8	5.3	145	2.7	109	—
27	14.6	5.0	143	2.8	101	35
28	12.0	4.3	146	2.8	100	—
33	18.2	5.5	143	3.3	95	41
43	9.8	6.3	140	4.3	105	40

amounts of KCl administered, potassium levels of the serum remained low. From the 26th day, Na and Cl levels dropped spontaneously and returned to normal. Potassium levels also adjusted themselves to the normal range on the 43rd day when the patient began to show signs of recovered consciousness. Full recovery was noted on the 54th day and the patient eventually returned to work.

From Table 6.3 it appears that the mechanism for maintenance of constant Na and Cl levels has been displaced so that these elements are maintained at constant levels that are higher than that of the normal. It is also to be noted that the mechanism for maintenance of the potassium level is not the same, since this was corrected spontaneously ten days later without increase in daily KCl intake.

6.5 THE "LOW SALT SYNDROME"

An abnormal condition occurs in the human which is apparently the opposite of the high salt syndrome. This is called the "*low salt syndrome*," "encephalogenic hyponatremia," or "cerebral salt wasting syndrome."[38–40] In recent years it has been shown that in many of these cases there appears to be an inappropriate secretion of ADH and hence a positive water balance. These cases are referred to as the "inappropriate ADH syndrome."[40] This condition is often associated with pulmonary abnormalities.

In this condition, levels of sodium and chloride remain constant at a level of the order of 120 and 80 mEq/liter, respectively. If saline is administered to these patients, *the extracellular fluid volume increases and generalized edema follows without an increase in serum salt levels.* This phenomenon is distinct from the low salt levels observed in patients with cardiac disease as a result of withholding salt and administering diuretics. The latter patient's Na and Cl levels will return to approximately normal levels when salt is given. Additional salt may, however, cause edema in these patients.

The patient with cardiac disease and the low salt syndrome is more prone to develop edema than the patient who does not show this phenomenon. Examination of electrocardiograms of patients with cardiac abnormalities cannot predict with any degree of certainty which will show the low salt syndrome and marked tendency to develop edema.

Newborn infants will often show lower serum sodium and chloride levels than adults. Sodium is at 135 mEq/liter and chloride level is 95 mEq/liter.[41] These observations can be summarized as follows, listing typical values observed in the various conditions:

	Na, mEq/liter	Cl, mEq/liter
Normal	140	103
High salt syndrome	170	125
Low salt syndrome	120	80
Infant	135	95

From these data and the examples given above, one may draw the conclusion that the levels of Na^+ and Cl^- are controlled by an osmoreceptor located in the hypothalamus. In infants, this "osmoreceptor" is set at a different point of balance than in the adult. Injury to the brain can either raise or lower the setting of this "osmoreceptor."

6.6 *DISPARITY BETWEEN SODIUM AND CHLORIDE EXCRETION*

Under normal circumstances, sodium and chloride will be excreted in approximately equimolar amounts in the urine. This is probably partly due to the fact that Na and Cl are taken in equimolar amounts as NaCl in the diet. To maintain electroneutrality, chloride will generally move with sodium through membranes.[42] As has been pointed out earlier, chloride levels on both sides of a membrane in the human erythrocyte and in the muscle cells obey the Donnan law,[47a] indicating that these membranes are porous to chloride ions. This is not true for sodium ion. *It is therefore not unexpected that conditions may exist where disparity between Na and Cl excretion can occur.*

In certain patients with brain damage from various causes such as a tumor or brain abscesses in the hypothalamic area, head injury, and chronic hypoxia to the brain, a disproportionate amount of sodium with respect to chloride is excreted in the urine, resulting in sodium retention and alkalosis.[43,44] Not only does this disproportionate secretion occur in the urine, but it also occurs from other secreting organs such as the stomach. This can be seen in Table 6.4.

Administration of acetazolamide (Diamox, a carbonic anhydrase

TABLE 6.4

Composition of Fluids Excreted by Certain Patients in Chronic Alkalosis Associated with Brain Injury, Illustrating Disproportion in Sodium and Chloride Excretion[43]

Case	Fluid	Na, mEq/ liter	Cl, mEq/ liter	K, mEq/ liter	Fluid pH	Blood pH	Volume, ml (24 hr)
1. Multiple acute brain abscesses, including hypothalamic area	Urine	6.5	96	15.2	5.1	7.70	2050
2. Cerebral thrombosis	Urine	13.5	100	34.1	5.5	7.52	1850
3. Cerebral hemangiosarcoma, extending to the hypothalamus, demonstrated at surgery	Urine	15.0	140	28.2	5.3	7.68	2800
4. Head injury due to fall	Urine	12.3	96	30.0	4.7	7.60	1750
5. Chronic hypoxia due to cardiac anomaly	Gastric drainage	18.0	140	32.0	1.5	7.70	125
6. Brain tumor invading hypothalamus with diabetes insipidus	Urine	11.7	78.7	—	—	—	—

TABLE 6.5

Effect of Carbonic Anhydrase Inhibitor on Blood pH and Sodium-to-Chloride Ratio in Urine in Patient with Brain Damage[43]

	Urine, mEq/liter			Serum, mEq/liter				
	Na	K	Cl	Na	K	Cl	Total CO_2	pH
Before Diamox	12	35	54	143	4.5	96	30.2	7.65
After Diamox	64	72	44	136	3.8	94	24.5	7.46

inhibitor) to these patients results in increased sodium and potassium excretion with a fall in chloride excretion and correction of blood pH (Table 6.5). From Table 6.5 one can readily see that inhibition of carbonic anhydrase stimulates sodium and potassium excretion without concomitant chloride excretion. This occurs by linkage of the sodium to the bicarbonate ion, whose level in turn is affected by carbonic anhydrase activity. $NaHCO_3$ and $KHCO_3$ are poorly ionized, so that one can consider that these ions form complexes with HCO_3^-.

From the above, one can also draw the conclusion that chloride excretion is not necessarily tied to sodium excretion and that the mechanism for maintenance of constant chloride levels in the serum is independent, to a degree, of the mechanism for maintenance of constant Na^+ levels. Evidence has been presented which suggests a chloride ion pump mechanism operative in the loop of Henle, independent of sodium ion secretion (see Chapter 11, on the kidney).

Mercury forms a complex with chloride ion and ^-SH sites of enzymes. This may relate to the fact that administering mercurials to the edematous patient causes excretion of chloride. To maintain electroneutrality, sodium ion and some potassium are also excreted simultaneously. Thus, just as the carbonic anhydrase system acts primarily on the cations, the mercurials act primarily on the chloride ion. Thus a separate mechanism may act on chloride ion directly, in the kidney. This is discussed further in Chapter 11.

Up to this point very little has been said about active transport of sodium across membranes. Instead, stress has been laid on the mechanisms of passive transport, such as simple diffusion. Active transport of sodium ion is coupled with active transport of potassium ion in the human and this will be discussed in the next section.

6.7 POTASSIUM

The serum potassium level is maintained, in the normal, at a fairly constant level of 4.6 mEq/liter $\pm$ 10%. Intake of potassium in the normal adult will range from 65 to 125 mEq/24 hr. Normally, 80–90% of this is excreted in the urine. The remainder is excreted in the sweat and stool.[45,46]

Potassium is found in highest concentration in the cells. Of a total body potassium content of 54 mEq/kg of body weight, 48 is intracellular, 0.2 intravascular, 0.5 interstitial, 0.2 in cartilage and connective tissue, 4.1 in the bones, and 0.5 in the cell walls (Table 6.1). On this basis, a 75-kg man would contain a total of ~ 4 Eq of potassium, or 156 g, of which only 0.6 g would be in the plasma.[45] Thus a massive reserve of potassium exists in the body to maintain plasma levels. Potassium is the major cation within the cell. It also carries the charge in nerve conduction. It also affects the activity of the Na–K-dependent ATPase in the cells [47] and with calcium and magnesium, controls the rate and force of contraction of the heart and thus the cardiac output.

Almost all natural foods are made of cells. On a reasonable diet, more than adequate amounts of potassium are supplied daily. This imposes the problem of excreting excess potassium. This is in contrast to sodium ion, which the body must conserve, since in many areas of the world, salt is in short supply. It is for this latter reason that animals will travel miles to reach salt licks.

The body is adapted to efficient potassium excretion. Not only is it excreted in the glomeruli of the kidneys, but it is also actively secreted by the tubules (see Chapter 11). Even when no potassium is taken into the body, as in fasting, 40–50 mEq is excreted daily in the urine.

The marked disparity between sodium and potassium concentrations in cells as compared to extracellular fluid, in contradiction to the laws of diffusion, forces the conclusion that energy must be continuously consumed to maintain this differential.[47a]

As pointed out earlier, the free energy that could be given off by a system consisting of a membrane and an ion of different concentrations on the sides of the membrane is $RT \ln(C_1/C_2)$. The ratio C_1/C_2 for potassium in the erythrocyte, as compared to the plasma, is approximately 20/1. The natural logarithm of 20 is approximately 3.0, body temperature in absolute units is $310°K$, and R has the value of 2 cal/mol. Thus we require $2 \times 310 \times 3 = 1860$ cal/mol to create this concentration differential. To this must be added the energy required to maintain the sodium differential. A source of energy must be continuously available to maintain this imbalance. In support of this concept, it is noted that if the tissue stops its metabolism, the potassium rapidly diffuses out of the cells. Blood stored at room temperature will show a marked drop in glucose levels and eventually a rise of potassium levels in the plasma, even without hemolysis. This is accompanied by a gradual decrease of sodium in the plasma. This indicates that only as long as glucose is available as a source of energy can the potassium and sodium concentration differential be maintained. For the above reasons a *sodium–potassium pump* was postulated which continuously pumps potassium into the cell and pumps sodium out of the cell.

The concept of a dynamic system where Na^+ and K^+ are continuously moving into and out of the cells is supported by the fact that if radioactive sodium or potassium is added to the plasma, a steady state is set up rapidly, indicating that the ions are in rapid exchange between the cell and the plasma. Theoretically, with the potential gradient set up for K^+ due to the difference in concentration on both sides of the erythrocyte membrane (20:1), the rate of outflow should be 1.6 mEq/kg of cells, and inflow should be 0.015 mEq/kg.[1] The rate observed in the radioactive studies[48] is too low by a factor of two. It has been proposed that the Na–K pump is reversible and that the reverse reaction of the pump is responsible for this effect.

Prior to 1960, various theories were propounded in order to explain the nature of the sodium–potassium pump. Although the explanation of the data was in error, these theories were based on sound experimental observations. For this reason, they will be presented here.

1. K^+ is less hydrated than the Na^+ and is therefore a smaller ion. The "sieve" theory was therefore proposed, suggesting that the cell wall, being more permeable to the K^+ ion, permitted it to penetrate the cell while retarding the passage of the sodium ion.[49]

---

[1] The formula for the flux of ions due to a differential of ion concentration on both sides of a membrane is derived from the Nernst equation, $E = (RT/ZF) \ln(a_0/a_1)$. Put in exponential form this becomes $M_1/M_0 = (a_0/a_1)e^{ZEF/RT}$, where M_1 and M_0 are fluxes (mol/sec) into and out of the cell, a_1 and a_0 are ion activities inside and outside the cell, Z is the charge (valence) of the ion (for K^+ and Na^+, $Z = 1$; for Ca^{2+}, $Z = 2$), T is the absolute temperature, and F is the Faraday constant. R is the gas constant.

2. It was proposed that an unidentified metabolite that can move out of the cell as the Na complex, but not into the cell in this form, carries the Na^+ out, exchanges the Na^+ for K^+, and then returns into the cell. This has been called the "ferry boat" hypothesis.[50]

3. It was suggested that the energy required to maintain the K^+–Na^+ differential on both sides of the cell membrane is used to maintain a fixed charge distribution in the cell contents, which favors attraction of the positive ion. Since K^+ forms a complex with organic compounds more readily than Na^+ and since the K^+ is smaller, it passes into the cell preferentially over Na^+. This is the "fixed charge" hypothesis.[51] Data in support of this concept as a partial explanation of the disparity in K^+ levels on both sides of the erythrocyte membrane come from some relatively recent studies.

   With nuclear magnetic resonance techniques, it has been shown that most of the potassium in the cell is undissociated. This would suggest that, while the total potassium concentration within the cell is high, the concentration of potassium ion could be lower than that in the plasma. Thus the cell could accumulate high concentrations of potassium by simple diffusion, the K^+ activity being kept low within the cell by reaction with protein.[52] This phenomenon must also be considered when studying the mechanism of ion transport.

4. Positively charged ions cannot pass the cell wall effectively because of an opposing charge on the membrane. Potassium forms a complex with glucose phosphate, thus wiping out its charge. The potassium glucose phosphate passes the membrane. Once in the cell, the glucose is metabolized and the potassium is trapped, being partly neutralized by Cl^- and HCO_3^-, the remainder forming a complex with hemoglobin. Less Na^+ moves in, since its salt with glucose phosphate is more highly ionized.[53] The concept probably originates with the observation that administration of insulin causes glucose to move into the cells along with potassium and phosphate. Thus there is a simultaneous drop of plasma glucose and phosphate as K^+ moves into the cell.[54]

It is a fact that K^+ ions form chelates with numerous compounds with which sodium does not. For this reason numerous methods for potassium determination exist, since many of these chelate compounds are insoluble and precipitate. Determination of sodium, on the other hand, was a problem before the advent of the flame photometer because of the difficulty in finding compounds with which it would form a complex.

A major advance to an understanding of the nature of the process that results in this large differential in cellular K^+ concentration as compared to the plasma came about with the discovery in 1960 of a Na–K-dependent ATPase, inhibited by cardiac glycosides, present in the walls of the erythrocyte.[55a] This enzyme, under the proper conditions, serves ultimately to hydrolyze the ATP to form ADP and inorganic phosphate.

It is possible to remove hemoglobin without rupturing the erythrocytes by allowing them to swell in successively hypotonic solutions. After all the hemoglobin, other protein, and electrolytes have passed the swelled membrane of the wall, the cells can then be reconstituted by placing them back in a series of solutions culminating in an isotonic solution. The unruptured cell ghosts can then be used to study transport across the cell wall. The composition both on the inside and outside can then be controlled.

Studies with such models revealed the fact that the cardiac glycoside-sensitive, Na–K-dependent ATPase was *stimulated by potassium on the outside of the cell wall and by sodium ion on the inside of the cell wall.* Both sides of the wall had to be activated simultaneously for normal behavior of the Na–K pump.[56] This is illustrated schematically in Figure 6.2.

Addition of ouabain, scillaren, or strophanthin to the solution on the outside of the normal erythrocyte results in rapid leakage of the potassium out of the cell. In the cell ghost models, the reaction is usually followed by adding ATP as a substrate and measuring the rate of generation of phosphate ion in the reaction $ATP \rightleftharpoons ADP + phosphate$.[56] In the absence of Na^+ inside the cell or potassium ion outside the cell, no reaction takes place attributable to the ouabain-sensitive ATPase.

The pump operates normally by pumping K^+ into the cell and Na^+ out of the cell. If only sodium ion is placed outside the cell and only potassium ion inside, no phosphate is generated, but the reaction is reversed.[57] In this case, phosphate combines with ADP to form ATP

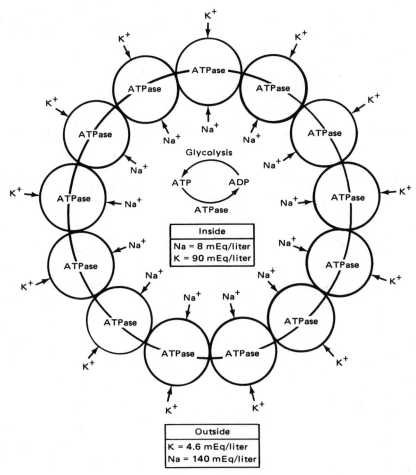

FIGURE 6.2   Schematic representation of the action of cardiac glycoside-inhibited, Na–K-dependent ATPase.

to some extent. This indicates that the pump is reversible, but biased toward the pumping of Na$^+$ out of the cell.[57] A similar mechanism probably operates in the loop of Henle.[58]

If the sodium ion concentration is increased outside the cell in the presence of K$^+$ outside the cell, the generation of phosphate from ATP is rapidly decreased.[59] This is interpreted as indicating that sodium ion can compete for sites on the enzyme with potassium ion and prevent its action. This inhibitory effect is more rapid than that observed with ouabain.

It is of interest to note that the half-maximum activity of the ATPase enzyme is achieved with levels of 5 mEq/liter of potassium, which is approximately the normal level in plasma. At 12 mEq/liter, maximal activity is observed. Below 1.6 mEq/liter, the activity of the ATPase is almost nil. *This is of clinical significance in that such low levels of potassium do occur in patients* and will result in cardiac irregularities and eventually in cardiac arrest. On the other hand, in the absence of pump action, increased leakage of K out of the erythrocyte occurs at this level.[60] The erythrocyte then serves as a reservoir to maintain the plasma potassium levels.

A sodium–potassium-dependent ATPase system occurs throughout the body, including neurological tissue.[61] Of significance is the effect of the cardiac glycosides on the heart. Thus one effect of digitalizing a patient is to regulate the ratio of potassium and sodium inside and outside the cell.

In the model experiments with erythrocyte ghosts, ATP is added as the substrate. In the intact erythrocyte the ATP is generated by glycolysis, each glucose molecule generating two molecules of ATP. The reaction ATP $\rightarrow$ ADP releases 10,000 cal/mol. The energy requirement to maintain a ratio of 20/1 for potassium between the inside and outside of the erythrocyte is at most 1860 cal/mol as discussed above. The value is substantially less, since, as has been pointed out, a significant amount of K^+ is complexed with Hb. The concentration gradient is therefore readily maintained by the energy available for glycolysis.

Studies of the movement of glucose into the erythrocyte indicate that the cardiac glycosides will act as inhibitors. This is partly due to the decreased utilization of glucose within the cell when the Na–K-dependent ATPase is blocked.[62]

Discovery of an Na^+–K^+-dependent ATPase set in motion various theories as to its action. Since primitive cells can keep sodium out against a high concentration gradient, as in sea water, many have proposed that the major action is the extrusion of sodium. The sodium combines with an ATPase complex. The molecule rotates, exposing the Na^+ to the outside. Splitting of the ATP releases the sodium, potassium now combining with the ATPase. On rotation again, ATP, in combining with the ATPase, displaces the potassium.[62] This complex reacts with sodium to repeat the process. This was called the *revolving door hypothesis.* The hydrolysis of the ATP to form ATP and inorganic phosphate supplied the energy for the rotation. Present thinking is that it is not rotation but a change in conformation of the Na–K-dependent ATPase that shifts the ions into and out of the cell.

Variations of this idea assume that ATPase does not rotate but a carrier moves back and forth, depositing sodium on the outside and returning with potassium to be deposited on the inside. 2,3-Diphosphoglyceric acid has been one substance among many suggested as a possible carrier. The evidence for this is that only when 2,3-diphosphoglyceric acid is present in high concentration in the cell is the cell able to maintain the potassium differential.[63] This type of model revives the "ferry boat" hypothesis. It is more likely that the presence of 2,3 diphosphoglycerate merely indicates normal glycolysis and availability of ATP for the pump.

The exact nature of the mechanisms involved were further clarified when the ATPase was isolated and shown to be a protein of approximately 100,000 molecular weight.[64] The action of ATP was to phosphorylate the protein at the carboxyl group of aspartate or glutamate to form a phospho-ATPase. The Mg^{2+} ion is required for this action. Recent work favors phosphorylation at the aspartate group.[65,66] The active site is the sequence: serine, aspartate (phosphorylated), and lysine. This can be formulated as follows in the form of the potassium complex:

serine   aspartate   lysine
              (phosphorylated)

Phosphorylation of the ATPase is readily reversible and can be written for the erythrocyte (i = inside cell; o = outside cell)

$$K_o^+ + Na_i^+ + ATP + H_2O \xrightleftharpoons[Mg^{2+}]{ATPase} K_i^+ + Na_o^+ + ADP + P(inorg)$$

Rubidium may substitute for potassium. ATP is not unique in its properties and other phosphorylating agents, such as acetyl phosphate, other nucleoside triphosphates, and nitrophenyl phosphate may substitute for ATP in the model experiments.

When ATP adsorbs to its active site on the inside of the cell in the presence of Mg^{2+}, a change in conformation of ATPase occurs, generating a form called the *B* form. Presence of K^+ outside the cell causes the *B* form to change back to the *A* form. Both forms have been separated by ultracentrifugation:

$$\text{ATPase}(A \text{ form}) \underset{K^+(\text{external})}{\overset{ATP + Mg}{\rightleftharpoons}} \text{ATPase}(B \text{ form})$$

The action of ouabain is to attach itself to the B form and prevent it from returning to the A form. The point of attachment is not at the active site where K^+ has reacted. The molecule of ATPase remains paralyzed with the K^+ outside the cell. If the ouabain is removed, then the molecule changes shape, locating the K^+ now inside the cell (the *A* form). The phosphate is hydrolyzed, pinning the K^+ inside the cell. The molecule is phosphorylated again (*A* form), carrying Na ion out of the cell as it changes to the *B* form. This resembles the "revolving door" or "ferry boat" hypothesis in principle.

Compounds other than ouabain can also act as inhibitors of the Na–K ATPase system. Fusidic acid is one such example (Figure 6.3). Other compounds that react with –SH groups or inactivate the ATPase will also act as inhibitors. These include maleimide, 1-fluoro-2,4-dinitrobenzene, urea, oligomycin, and guanidine derivatives. The NH_4^+ ion resembles K^+ in its activity and will compete for sites with K^+.

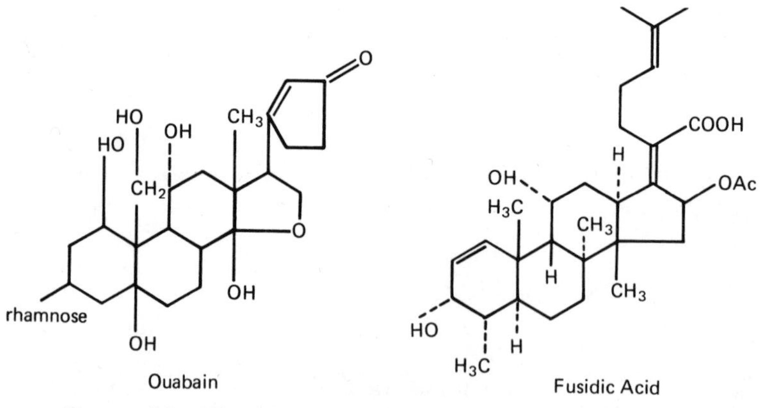

FIGURE 6.3   The ATPase inhibitors ouabain and fusidic acid.

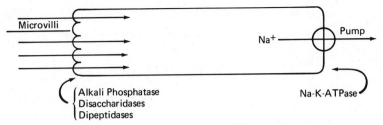

FIGURE 6.4   Transcellular transport of Na^+ in the intestine.

In the villi of the intestinal tract a sodium pump exists based on a Na-dependent ATPase. Absorption of sodium in the gut is done by active transport. This is located mainly in the basolateral plasma membranes of the villi. The apical plasma membranes (microvillous membranes) contain carbohydrates, phosphatases, etc., and serve to actively transport carbohydrates, fats, and proteins. Intermediate products of digestion are further hydrolyzed to terminal products at the microvillous surface and are absorbed along with Na^+. The Na^+ is ejected at the opposite border of the cell.[67-69] This is shown schematically in Figure 6.4.

High and Low K^+ in Erythrocytes. It has been known for some time that there are certain individuals with anomalous K^+ levels in the erythrocyte. These are divided into *low-potassium (LK) and high-potassium (HK) erythrocytes.* As would be expected, this has been shown to be due to genetic variation of the Na–K ATPase of the cell wall. The *HK membranes* are more sensitive to activation and less sensitive to inhibition by K^+ than the normal, and the *LK membranes* are less sensitive to activation. What relationship these observations have to the genetic deficiencies in potassium metabolism, discussed in succeeding pages, needs to be clarified.[69]

6.8   *HYPOKALEMIA AND HYPERKALEMIA*

The total mechanism for maintenance of constant potassium levels in the plasma needs further clarification. However, that such a mechanism exists is demonstrated by the fact that the plasma level of potassium is kept constant at 4.6 mEq/liter and that this level, like that of sodium, varies in different animals and can be changed, with brain injury, to a different constant level without evidence of renal insufficiency.

In contrast to the complex mechanisms of the kidney to conserve sodium, the kidney is effective in actively excreting potassium. The reason for this is that the cation in greatest abundance in the cell, and thus in foods, is the potassium ion. Under normal circumstances this would accumulate unless an efficient mechanism existed in the kidney to excrete the potassium ion. This will be discussed further in the section on the kidney (Chapter 7).

Anomalies in serum potassium levels are associated with a number of disease states. The genetically determined syndromes characterized by recurrent episodes of profound muscular weakness or actual paralysis are now grouped under the heading of *periodic paralysis*. The first such condition to be identified was termed *familial periodic paralysis* when it was discovered that attacks of generalized muscle weakness occur in related individuals, associated with a marked drop in plasma potassium levels. If KCl is not administered promptly, these patients can go on to expire in *hypokalemia* (low potassium level of the plasma).[70] This has since been shown to be only one of a group of genetically determined syndromes associated with anomalies in potassium transport and metabolism and characterized by recurrent episodes of profound muscular weakness or actual paralysis (see Table 6.6).

One *familial* condition is inherited as an autosomal dominant character, with males most severely affected, and is characterized by sudden attacks of progressive flaccid paralysis lasting one to several hours. Usually the serum potassium is reduced at the onset of the attack. An episode of paralysis may be induced by glucose, insulin, corticosteroids, and epinephrine, and may occur spontaneously in the cold or with relaxation after strenuous activity.

TABLE 6.6

Syndromes of Periodic Paralysis

A.　Hypokalemic types
　　1.　Familial
　　2.　Endocrine dysfunction (e.g., thyroid)
　　3.　Renal tubular defects

B.　Hyperkalemic types
　　1.　Myotonia congenita
　　2.　Paramyotonia congenita
　　3.　Adynamia episodica hereditaria
　　4.　Hyperkalemic periodic paralysis

Electron microscopy of muscle tissue extracted during the episode of paralysis identifies vacuolization of muscle fibers, which is reversible. In this condition the potential across the resting muscle cell is not abnormal, and the true defect is not known. It is clear that a measurable fall in the serum potassium does not always occur during a paralytic attack and that there is a poor correlation between the degree of hypokalemia and depth of paralysis. Nevertheless, treatment is usually effective with oral potassium supplements, low carbohydrate diet, and reduction in dietary sodium.

A type of hypokalemic periodic paralysis phenotypically identical to the congenital form may complicate thyrotoxicosis, especially in the oriental male. Correction of the hyperthyroidism results in complete disappearance of the paralytic attacks.[74] In either condition the sudden reduction in serum potassium may be profound, leading to cardiac irregularities and even death.[70] Hypokalemic weakness may also be aggravated by urinary potassium loss in acquired disorders such as renal tubular acidosis and hyperaldosteronism.

Hyperkalemia may impair muscle function and also result in a type of periodic paralysis. The finding of elevated serum potassium levels in association with recurrent attacks of myotonic paralysis which could be induced by the administration of oral potassium salts was pointed out in 1957 by French and Kilpatrick.[71] The current concept is to consider the four syndromes characterized by myotonia and frequently elevated serum potassium levels as varied expressions of the same genetically determined disorder which is transmitted as an autosomal dominant character. In single families with hyperkalemic paralysis, one member will have only myotonia, another myotonia with elevated serum K, and still others with no clinical symptoms but evidence of percussion and electrical myotonia.[72]

The pathology in this condition resides in the muscle cell membrane, where an abnormally low potential can be further lowered by flux of potassium ions from the cell into the extracellular fluid, as occurs during the initial period of weakness, setting off muscle spasm. Although recovery between episodes of paralysis may be complete, occasional individuals develop a proximal muscular dystrophy.

Treatment of this condition with oral thiazide diuretics in order to maintain the serum potassium at a low to low–normal range appears to be quite effective in reducing the frequency and severity of the attacks.[73]

A third type of periodic paralysis, although precipitated by potassium salts, may be distinct from myotonic, hyperkalemic periodic paralysis. In this entity the serum potassium is invariably normal (*normokalemic type*). Treatment with sodium and corticosteroids, detrimental to patients with hyperkalemic paralysis, is beneficial in this disorder.[75]

Most commonly, hypokalemia occurs in inanition and with large losses of body fluids such as during intestinal drainage or following extensive small bowel resections. Excessive diuretic therapy is also a common cause for reduced serum potassium levels. These losses can often be corrected by oral or parenteral administration of potassium salts. Diabetics in coma may initially exhibit hyperkalemia consequent to dehydration resulting from large fluid losses through the kidney, accompanied by diffusion of potassium out of cells. After treatment with insulin and corrective fluids containing sodium ion, however, the potassium moves into the cells. This treatment can result in severe hypokalemia unless subsequently corrected by administration of potassium salts.

In kidney disease, most commonly in acute renal failure, potassium retention and hyperkalemia may reach dangerous levels. The electrocardiogram will show symmetric peaking of the T waves (Figure 6.5).[76,77]

Hyperkalemia can be ameliorated by the use of ion exchange resins orally or by enema to remove this cation. Intravenous solutions containing glucose, insulin, and sodium bicarbonate may be effective more rapidly.

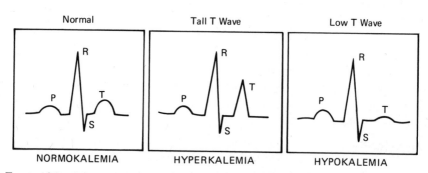

Normal	Tall T Wave	Low T Wave
NORMOKALEMIA	HYPERKALEMIA	HYPOKALEMIA

FIGURE 6.5   Schematic representation of electrocardiogram patterns illustrating the abnormal peaked T waves of hyperkalemia and the depressed T wave in hypokalemia.

In salt-losing nephritis and Milkman's syndrome, large losses of potassium through the kidneys may result in low serum potassium levels.[78-80] In primary aldosteronism, excessive loss of potassium with low serum potassium levels is a common finding. In Addison's disease and Cushing's syndrome hypokalemia and a hypochloremic alkalosis develop due to the loss of these ions in the urine under the influence of the steroid hormones.[81]

Infants with extreme dehydration due to diarrhea may initially exhibit high serum potassium levels. However, large amounts of the body stores have been lost. On hydration, hypokalemia will result unless potassium salts are administered. Premature infants will often show high serum potassium levels with no clinical signs. This is probably due to the fact that potassium is complexed with various substances in the plasma, such as bilirubin, present in high concentrations in the plasma of these infants.

Perhaps the most common cause for hypokalemia in the patient receiving parenteral fluids is water and sodium chloride administration without adequate replacement for potassium lost in the urine and drainage fluids. *These fluids can be collected, their volume measured, and analyzed for Na, Cl, and K levels in order to be able to estimate the amount lost so as to guide replacement therapy.*

6.9 SELECTED READING—MAINTENANCE OF CONSTANT ION CONCENTRATION IN BODY FLUIDS: SODIUM, POTASSIUM, AND CHLORIDE

Cort, J. *Electrolytes, Fluid Dynamics and the Nervous System.* Academic Press, New York (1966).

Black, D. A., *Essentials of Fluid Balance,* 4th ed., Blackwell Davis, Philadelphia, Pennsylvania (1968).

Roberts, K. E., Parker, V. J., and Poppell, J. W., *Electrolyte Changes in Surgery,* Charles C Thomas, Springfield, Illinois (1958).

Shoemaker, W. D. and Walker, W. F., *Fluid–Electrolyte Therapy in Acute Illness,* Year Book, Chicago, Illinois (1970).

Carlson, F. D., *Physiological and Biochemical Aspects of Nervous Integration,* Prentice Hall, Englewood Cliffs, New Jersey (1968).

Surawicz, B., *Potassium and the Heart,* Lea & Febiger, Philadelphia, Pennsylvania (1969).

Berliner, R. W. *et al.*, Physiology society symposium on neural control of body salt and water, *Fed. Proc.* **72**:1127–1159 (1968).

Stein, W. D. *The Movement of Molecules across Cell Membranes.* Academic Press, New York (1967).

Hoken, L. E., Ed., *Metabolic Transport*, Academic Press, New York (1972).

Soffer, A., Ed., *Potassium Therapy*, Charles C Thomas, Springfield, Illinois (1968).

6.10 *REFERENCES*

1. Pitts, R. F. *Physiology of the Kidney and Body Fluids*, 2nd ed., Year Book, Chicago, Illinois (1968).
2. Bajusz, E., Ed., *Electrolytes and Cardiovascular Disease*, Williams and Wilkins, Baltimore, Maryland (1965).
3. Maren, T. H., The relation between enzyme inhibition and physiological response in the carbonic anhydrase system, *J. Pharmacol. Exp. Ther.* **139**:140–153 (1963).
4. Clapp, J. R., Watson, J. F., and Berliner, R. W., Effect of carbonic anhydrase inhibition on proximal tubule bicarbonate reabsorption, *Am. J. Physiol.* **205**:693–696 (1963).
5. Rector, F. C., Jr., Seldin, D. W., Roberts, A. D., Jr., and Smith, J. S., The role of plasma CO_2 tension and carbonic anhydrase activity in the renal reabsorption of bicarbonate, *J. Clin. Invest.* **39**:1706–1721 (1960).
6. Rawls, Jr. J. A., Wistrand, P. J., and Maren, T. H., Affects of acid–base changes and carbonic anhydrase inhibition on pancreatic secretion, *Am. J. Physiol.* **205**:651–657 (1963).
7. Gibbs, G. E., Reimer, K., Kollmorgen, R. L., and Young, P. G., Quantitative microdetermination of enzymes in sweat glands, *Am. J. Dis. Child.* **105**:249–252 (1963).
8. Bledsoe, T., Island, D. P., and Liddle, G. W., Studies of the mechanisms through which sodium depletion increases aldosterone biosynthesis in man, *J. Clin. Invest.* **45**:524–530 (1966).
9. Gross, F., Regulation of aldosterone secretion by the renin-angiotensin system under various conditions, *Acta Endocrinal. Suppl.* **124**:41–64 (1967).
10. Farrell, G., Regulation of aldosterone secretion, *Physiol. Rev.* **38**:709–728 (1958).
11. Conn, J. W., Evolution of primary aldosteronism as a highly specific clinical entity, *J. Am. Med. Assoc.* **172**:1650–1653 (1960).
12. Ganong, W. F., The central nervous system and the release of the adrenocorticotropic hormone, in *Advances in Neuroendocrinology*, Univ. of Illinois Press, Urbana, Illinois (1963), p.92.
13. Dorfman, R. I. and Ungar, F., Metabolism of steroid hormones, Academic Press, New York (1965).
14. Hvidberg, E., Szporny, L. and Langgaard, H., The composition of aedema fluid provoked in mice by oestradiol, *Acta Pharmacol.* **20**:243–252 (1963).

15. Bing, J. and Kazimierczak, J., Localization of renin in the kidney. III. *Acta Pathol. Microbiol. Scand.* **50**:1–11 (1960).

16. Luetscher, J. A., Boyers, D. G., Cuthbertson, J. G., and McMahon, D. F., A model of the human circulation. Regulation by autonomic nervous system and renin-angiotensin system and influence of blood volume on cardiac output and blood pressure. *Circulation Res.* **32**:Suppl. **1**:84–98 (1973).

17. Gunnells, J. C. Jr., Grim, C. E., Robinson, R. R., and Wildermann, N. M., Plasma renin activity in healthy subjects and patients with hypertension, *Arch. Intern. Med.* **119**:232–40 (1967).

18. Pabst, K., Renal excretory function following rapid infusion with solutions containing, and free of, sodium chloride, *Arch. Ges. Physiol.* **273**:315–24 (1961).

19. Bergstrom, W. H. and Wallace, W. M., Bone as a sodium and potassium reservoir, *J. Clin. Invest.* **33**:867–873 (1954); The participation of bone in the total body sodium metabolism in the rat, *J. Clin. Invest.* **34**:997–1004 (1955).

20. Bauer, G. C. H. and Carlsson, A., Rate of bone salt formation in a healing fracture determined in rats by means of radiophosphorus, *Acta Orthopaed. Scand.* **24**:27–34 (1955).

21. Forbes, G. B., Bone sodium and sodium-22 exchange: Relation to water content, *Proc. Soc. Exp. Biol. Med.* **102**:248–50 (1959).

22. MacCarty, C. S. and Cooper, I. S., Neurologic and metabolic effects of bilateral ligation of the anterior cerebral arteries in man, *Proc. Staff. Meet., Mayo Clinic* **26**:185–190 (1951).

23. Cooper, I. S., and MacCarthy, C. S., Unusual electrolyte abnormalities associated with cerebral lesions, *Proc. Staff Meet. Mayo Clinic* **26**:354–358 (1951).

24. Natelson, S., Crawford, W. L., and Munsey, F. A., in *Correlation of Clinical and Chemical Observations in the Immature Infant*, Illinois Dept. of Public Health, Div. of Preventive Medicine Publication (1952), p. 16.

25. Engstrom, W. W. and Liebman, A., Chronic hyperosmolarity of the body fluids with a cerebral lesion causing diabetes insipidus and anterior pituitary insufficiency, *Am. J. Med.* **15**:180–186 (1953).

25a. Vajjajiva, A., Sitprija, V., and Shuangshoti, S., Chronic sustained hypernatremia and hypovolemia in hypothalamic tumor. A physiologic study, *Neurology* **19**:161–166 (1969).

26. Natelson, S. and Alexander, M. O., Marked hypernatremia and hyperchloremia with damage to the central nervous system, *Arch. Intern. Med.* **96**:172–175 (1955).

27. Goodale, W. T. and Kinney, T. D., Sulfadiazine nephrosis with hyperchloremia and encephalopathy, *Ann. Intern. Med.* **31**:1118–1128 (1949).

28. Allott, E. N., Hypernatremia and hyperchloremia in bulbar poliomyelitis, *Lancet* **1**:246–250 (1957).

29. Sweet, W. H., Cotzias, G. C., Seed, J., and Yakovlev, P. I., Gastrointestinal hemorrhages, hyperglycemia, azotemia, hyperchloremia and hypernatremia following lesions of frontal lobe in man, *Ann. Res. Nerv. Ment. Dis. Proc.* **27**:795–822 (1948).

30. Daily, W. J. R. and Victorin, J. L. H., Hyperosmolarity (hypernatremia) with cerebral disease. A report of two cases in children, *Acta Paediat. Scand.* **56**:97–104 (1967).

31. Taylor, W. H., Hypernatraemia in cerebral disorders, *J. Clin. Pathol.* **15**:211–220 (1962).

31a. Christie, S. B. M., and Ross, E. J., Ectopic pinealoma with adipsea and hypernatraemia, *Brit. Med. J.* **2**, 699–670 (1968).

32. Allott, E. N., Sodium and chloride retention without renal disease, *Lancet* **1**:1035–1037 (1939).

33. Pleasure, D. and Goldberg, M., Neurogenic hypernatremia, *Arch. Neurol.* **15**:78–87 (1966).

34. Alvioli, L. V., Earley, L. E., and Kashima, H. K., Chronic and sustained hypernatremia. Absence of thirst, diabetes insipidus and ACTH insufficiency resulting from widespread destruction of the hypothalamus, *Ann. Intern. Med.* **56**:131–140 (1962).

34a. Mahoney, J. H. and Goodman, D., Hypernatremia due to hypodypsia and elevated threshold for vasopressin release. Effects of treatment with hydrochlorothiazide, chlorpropamide and tolbutamide, *New Eng. J. Med.* **279**:1191–1196 (1968).

35. Lewy, F. H. and Gassmann, F. K., Experiments on the hypothalamic nuclei in the regulation of chloride and sugar metabolism, *Am. J. Physiol.* **112**:504–510 (1935).

36. Schoolman, H. M., Dubin, A., and Hoffman, W. S., Clinical syndromes associated with hypernatremia, *Arch. Intern. Med.* **95**:15–23 (1955).

37. Montgomery, R., Clinical conference at the Los Angeles Children's Hospital, Case 2, hyperelectrolytemia and hyperosmolarity in an infant, *J. Pediat.* **42**:742–748 (1953).

38. Goldberg, M. and Handler, J. S., Hyponatremia and renal wasting of sodium in patients with malfunction of the central nervous system, *New Eng. J. Med.* **263**:1037–1043 (1960).

39. Peters, J. P., Welt, L. G., Sims, E. A. H., Orloff, J., and Needham, J. W., Salt wasting syndrome associated with cerebral disease, *Trans. Ass. Am. Physicians* **63**:57–64 (1951).

40. Carter, N. W., Rector, F. C., Jr., and Seldins, D. W., Hyponatremia in cerebral disease resulting from the inappropriate secretion of the antidiuretic hormone, *New Eng. J. Med.* **264**:67–72 (1961).

41. Pincus, J. B., Gittleman, I. F., Saito, M., and Sobel, A. F., A study of plasma values of sodium, potassium, chloride, carbon dioxide tension, carbon dioxide, sugar, urea, and the protein base-binding power, pH and hematocrit on the first day of life, *Pediatrics* **18**:39–49 (1956).

42. Clapp, J. R. and Rector, F. C., The mechanism of renal chloride reabsorption, *Clin. Res.* **9**:56 (1961).

43. Natelson, S., Chronic alkalosis with damage to the central nervous system, *Clin. Chem.* **4**:32–42 (1958).

44. Rowntree, L. G. Boucek, R. J., and Noble, N. L., Anomalous type of salt and water retention with persistent edema, *J. Am. Med. Assoc.* **161**:877–879 (1956).

45. Deane, N. and Smith, H. W., The distribution of sodium and potassium in man, *J. Clin. Invest.* **31**:197–199 (1952).

46. Beilin, L. J., Knight, G. J., Munroe-Faure, A. D., and Anderson, J., The sodium, potassium and water contents of healthy human adults, *J. Clin. Invest.* **45**:1817–1825 (1966).

47. Katz, A. I. and Epstein, F. H., The role of sodium–potassium activated adenosine triphosphatase in the reabsorption of sodium in the kidney, *J. Clin. Invest.* **46**:1999–2011 (1967).

47a. Kintner, E. P., Chemical structure of erythrocytes with emphasis on Donnan equilibrium, *Ann. Clin. Lab. Sci.* **2**:326–334 (1972).

48. Ussing, H. H., The distinction by means of tracers, between active transport and diffusion. The transfer of iodide across the isolated frog skin, *Acta Physiol. Scand.* **19**:43–56 (1949).

49. Boyle, P. J. and Conway, E. J., Potassium accumulation in muscle and associated changes, *J. Physiol.* **100**:1–63 (1941).

50. Ussing, H. H., Transport of ions across cellular membranes, *Physiol. Rev.* **29**:127–155 (1949).

51. Ling, G., Muscle electrolytes, *Am. J. Phys. Med.* **34**:89–101 (1955).

52. Ling, G. N. and Cope, F. W., Potassium ion: Is the bulk of intracelluar K^+ adsorbed? *Science* **163**:1335–1336 (1969).

53. Fenn, W. O., Deposition of potassium and phosphate with glycogen in rat livers, *J. Biol. Chem.* **128**:297–307 (1939).

54. Altman, P. L. and Dittmer, D. S., Eds., *Membrane Transport of Nutrients, Classification of Transport Processes in Metabolism*, Fed. Am. Soc. for Exp. Biology, Bethesda, Maryland (1968).

55. Post, R. L., Merritt, C. R., Kinsolving, C. R., and Albright, C. D., Membrane adenosine triphosphatase as a participant in the active transport of Na and K in the human erythrocyte, *J. Biol. Chem.* **235**:1796–1802 (1960).

55a. Perrone, J. R. and Blostein, R., Asymmetric interaction of inside out and right side out erythrocyte membrane vesicles with ouabain, *Biochim. Biophys. Acta* **291**:680–689 (1973).

56. Whittam, R., Control of membrane permeability to potassium in red cells, *Nature* **219**:610 (1968).

57. Garrahan, P. J. and Glynn, I. M., Driving the sodium pump backwards to form adenosine triphosphate, *Nature* **211**:1414–1415 (1966).

58. Shelburne, J. D. and Trump, B. F., Disorders of cell volume regulation. I. Effects of inhibition of plasma membrane adenosine triphosphatase with ouabain, *Am. J. Pathol.* **53**:1041–1071 (1968).

59. Whittam, R. and Ager, M., Dual effects of sodium ions on erythrocyte membrane adenosine triphosphatase, *Biochim. Biophys. Acta* **65**:383–385 (1962).

60. Glynn, I. M., Relation between ouabain-sensitive potassium efflux and hypothetical dephosphorylation step in transport ATPase system, *J. Gen. Physiol.* **51**:385–388 (1968).

61. Potter, H. A., Charnock, J. S., and Opit, L. J., The separation of sodium and potassium-activated adenosine-triphosphate from a sodium or potassium

inhibited adenosine triphosphatase of cardiac muscle, *Austral. J. Exp. Biol. Med. Sci.* **44**:503–518 (1966).

62. Opit, L. J. and Charnock, J. S., A molecular model for a sodium pump, *Nature* **209**:471–474 (1965).

63. Asukuta, T., Sato, Y., Minikami, S., and Yoshikawa, H., pH dependency of 2,3 diphosphoglycerate content in red blood cells, *Clin. Chim. Acta* **14**:840–841 (1966).

64. Mezumo, N., Nagano, K, Nakao, T., and Tashima, Y., Approximation of molecular weight of $Na^+-K^+-ATPase$, *Biochim. Biophys. Acta* **168**:311–320 (1968).

65. Post, R. L., Kume S., Tobin, T., Orcutt, B., and Sen, A. K., Flexibility of an active center in sodium plus potassium adenosine triphosphatase, *J. Gen. Physiol.* **54**:306$_s$–326$_s$ (1969).

66. Post, R. L., Kume, S., and Rogers, R. N., in *Mechanism in Bioenergetics*, Azzone, G. F., Ed., Academic Press, New York (1973).

67. Barry, R. J. C., Electrical changes in relation to transport, *Brit. Med. Bull.* **23**:266–269, (1967).

68. Fujita, M., Ota, H., Kawai, K., Matsui, H., and Nakao, M., Differential isolation of microvillous and basolateral plasma membranes from intestinal mucosa; mutually exclusive distribution of digestive enzymes and ouabain sensitive ATPase, *Biochim. Biophys. Acta* **274**:336–347 (1972); *J. Physiol.* **227**:377 (1972).

69. Blostein, R., Sodium activated adenosine triphosphatase activity of the erythrocyte membrane, *J. Biol. Chem.* **245**:270–275 (1970).

70. Chen, R. F., Familial periodic paralysis. Report of a case resistant to dextrose and insulin provocation, *Arch. Neurol.* **1**:475–484 (1959).

71. French, E. B. and Kilpatrick, R., A variety of paramyotonia congenita, *J. Neurol. Neurosurg. Psychiat.* **20**:40–46 (1957).

72. Layzer, R. B., Lovelace, R. E., and Rowland, L. P., Hyperkalemic periodic paralysis, *Arch. Neurol.* **16**:455–472 (1967).

73. Gamstorp, I., Hauge, M., Helweg-Larsen, H. F., Mjones, H., and Sagild, U. Adynamia episodica hereditaria, *Am. J. Med.* **23**:385–390 (1957).

74. McFadzean, A. J. S. and Yeung, R. Periodic paralysis complicating thyrotoxicosis in Chinese, *Brit. Med. J.* **1**:451–455 (1967).

75. Poskanzer, D. C. and Kerr, D. N. S., A third type of periodic paralysis with normokalemia and favorable response to sodium chloride, *Am. J. Med.* **31**:328–342 (1961).

75a. Tyler, F. H., Stephens, F. E., Gunn, F. D., and Perkoff, G. T., Studies in disorder of muscles. VII. Clinical manifestations and inheritance of a type of periodic paralysis without hypopotassemia, *J. Clin. Invest.* **30**:492–502 (1951).

76. Braun, H. A., Surawicz, B., and Bellet, S., T waves in hyperpotassemia, their differentiation from simulating T waves in other conditions, *Am. J. Med. Sci.* **230**:147–156 (1955).

76a. Samaha, F. J., Von Eulenberg's paramyotonia, *Trans. Am. Neurol. Assoc.* **89**:87–91 (1964).

77. Bellet, S., Steiger, W. A., Nadler, C. S., and Gazes, P. C., Electrocardiographic patterns in hypopotassemia; observations on 79 patients, *Am. J. Med. Sci.* **219**:542–558 (1950).

78. Evans, B. M. and Milne, M. D., Potassium-losing nephritis presented as a case of periodic paralysis, *Brit. Med. J.* **2**:1067–1071 (1954).

79. Kartal, J. P., Leve, L., Ryder, H. W., and Horowitz, M. G., Renal tubular acidosis with hypokalemia symptoms, *Arch. Intern. Med.* **107**:743–749 (1961).

80. Giebisch, G., Windhager, E. E., and Malnic, G., Renal control of sodium and potassium of body fluids, in *23rd Int. Congr. Physiol. Soc. Lect. Symp.* Tokyo (1965), pp. 167–75

81. Christy, N. P. and Laragh, J. H., Pathogenesis of hypokalemic alkalosis in Cushing's syndrome, *New Eng. J. Med.* **265**:1083–1088 (1961).

Maintenance of Constant Ion Concentration in Body Fluids: Calcium, Magnesium, Phosphate, and Sulfate

7.1 PLASMA CALCIUM

Approximately 99% of the body calcium is located in the skeleton. Thus, along with phosphate, it forms the hard tissues of the body. As such, it acts as a huge reservoir for maintenance of plasma calcium levels.

The body contains approximately one equivalent weight of calcium per kilogram. A 75-kg man would then contain $75 \times 1 \times 20 = 1500$ g of calcium. Of this the plasma contains 5 mEq/liter, or a total of approximately 18 mEq in the 75-kg man at a concentration of 10 mg/100 ml $\pm$ 5%. Interstitial fluid calcium concentration is half that of the plasma. However, interstitial fluid volume is $20/5 = 4$ times the plasma volume and therefore contains more calcium than the plasma. The interstitial fluid in the 75-kg man contains 28 mEq of calcium. It is estimated that the total amount of calcium in the cells is not greater than 40 mEq.[1]

While the total muscle calcium concentration in the guinea pig is of the order of 1 mmol/kg, intracellular concentration of calcium is significantly lower, and the little that is present in the cell is complexed mainly with protein. Direct measurement indicates that muscle calcium ion concentration is less than 10^{-4} mmol/kg.[2] Calcium is conserved by reabsorption, mainly in the proximal tubules of the kidney.[3]

Although the intracellular calcium is at low concentration, it is of great importance in its effect on the activity of the mitochondria and on muscular contraction.[4,5] Along with potassium and magnesium, it exerts an effect on the rate of cardiac output.

Approximately 39% of the calcium in the plasma is in the ionic form, the rest being complexed to protein, citrate, phosphate, and, to a lesser extent, to the various nonprotein nitrogen derivatives.[6] The erythrocytes contain only small amounts of calcium.

Calcium ion acts as a "cementing" material for forming protein conglomerates by forming complexes (called chelates) with these proteins.

A typical chelate is shown in the reaction of calcium ion with the disodium salt of ethylene diamine tetracetic acid (EDTA):

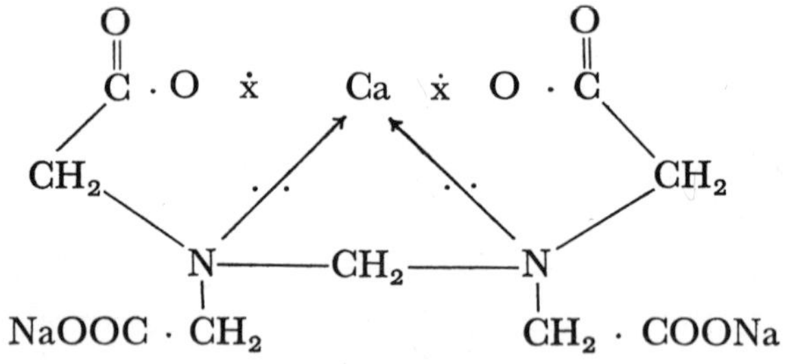

Here calcium has displaced two protons to form two typical polar bonds. In these bonds, one electron derives from the oxygen atom ($\cdot$) and one from the calcium atom (x) to form the electron pair. Nitrogen, which has five electrons in its outer shell, is using only three to form coordinate bonds with three carbon atoms. This permits the other two electrons from each nitrogen to shift toward the calcium atom, forming a stable system with eight electrons around the calcium atom. Since the two electrons derive from one atom only (nitrogen), we call this a semipolar bond. This chelate compound is formed by two semipolar bonds and two polar bonds to form a rigid structure. It is this type of structure which permits calcium ion to bind proteins together, forming polar bonds with carboxy groups, as from aspartate and glutamate, or with certain active hydroxyl groups, such as with tyrosine and the enol form of the polypeptide bonds. The electrons forming the semipolar bonds then may be derived from nitrogen, sulfur, or oxygen.

Chelate formation with the same organic compound such as EDTA yields complexes of varying stability. Thus the lead complex with

EDTA is more stable and dissociates to a lesser extent than Ca-EDTA. The Ca-EDTA complex administered to patients with lead deposited in their bones will therefore exchange the lead for calcium and excrete the soluble Pb-EDTA complex. The reason for this phenomenon is that lead "fits" the structure of EDTA with less distortion than calcium does. Another example is that of 8-hydroxyquinoline, where the lithium complex is much more stable than the sodium or potassium complex. Throughout this text, repeated examples are given where a particular element forms a biologically active chelate, such as Mg for ATP kinase, zinc for carbonic anhydrase and insulin, manganese for arginase, copper for peroxidases, and iron for hemoglobin and catalase. These chelates form with the exclusion of other elements because their complex with the structure is the one most stable.

The complex formed by Ca^{2+} with proteins is relatively easily ruptured by the addition of a chelating agent. For example, $C1$ globulin of the complement system is a complex of three proteins, $C1q$, $C1r$, and $C1s$, with Ca^{2+}. If a solution of EDTA is added to the system, it falls apart, generating the three distinct proteins.

In the clotting system, Ca^{2+} is required for protein binding throughout the system until thrombin is finally generated. If EDTA or citrate is added to this system, clotting will be inhibited due to the sequestering of the Ca^{2+}.

7.2 CALCIUM TRANSPORT

Calcium transport is accomplished across membranes by diffusion (passive transport) and with the Ca–Mg-specific ATPase pump mechanism (active transport). This system exists in all tissues, including even the platelets. The system studied to the greatest extent is calcium transport across the sarcoplasmic reticulum (SR) membranes[7-10] and its intestinal absorption.

The Ca-dependent ATPase is present in large quantities in the human. As much as 70% of the sarcoplasm reticulum is this protein.[11]

When muscle is homogenized with dodecyl sulfate, the Ca^{2+}–Mg^{2+}-ATPase is readily solubilized. Like Na^+–K^+-ATPase, it has a molecular weight of approximately 100,000.[12,13] Active subunits with molecular weights as low as 10,000 have been prepared. Active sites in this enzyme are the same as for the Na^+–K^+-ATPase system at approximately 100 Å apart.[14] The ATPase is phosphorylated by ATP at the carboxyl groups of aspartate, which is suspended between serine and lysine substituents of the polypeptide chain (see Section 6.7). The overall reaction can be written, where Ca_o is the calcium outside the cell wall and Ca_i represents that inside the cell,[15]

$$Ca_o^{2+} + ATP + H_2O \xrightarrow[Mg^{2+}]{ATPase} Ca_i^{2+} + ADP + P(inorg)$$

If ethylene glycol replaces the two sodium atoms in EDTA, then we have a soluble substance for complexing Ca^{2+} which has no Na^+ ion. This is called EGTA. If EGTA is added to the sarcoplasmic reticulum ATPase preparation containing the phosphorylated ATPase, then ADP in the presence of Mg^{2+} and ATP kinase will form ATP, showing the reversibility of the Ca^{2+} pump.

$$ATPase \sim P + ADP\ Mg \xrightarrow{EGTA} ATP + ATPase + Mg^{2+}$$

The sequence of events that takes place when calcium moves into a cell can be shown by the following series of reactions. Phosphorylation of the ATPase takes place when the Ca^{2+} is outside the SR membrane:

ATPase + Mg-ATP(external) + Ca^{2+}(external) $\rightleftharpoons$ ATPase complex

ATPase complex $\rightleftharpoons$ ATPase phosphate Ca + Mg^{2+}

ATPase phosphate Ca $\rightarrow$ $2Ca^{2+}$(inside) + ATPase phosphate

ATPase phosphate + Mg^{2+} $\rightarrow$ Mg ATPase phosphate

Mg ATPase phosphate $\rightleftharpoons$ ATPase + phosphate + Mg^{2+}

Associated with the influx of Ca^{2+} we have a simultaneous efflux of Mg^{2+} and also K^+ ions to maintain electrical neutrality.[16] The binding of 1 mol of Mg^{2+} to the active site is required for the hydrolysis of the phosphorylated ATPase and release of inorganic phosphate.

In the sarcoplasmic reticulum, affinity of the ATPase phosphate for Ca^{2+} and Mg^{2+} is in the ratio of 2.5:1. Therefore a large excess of Mg^{2+} will interfere with Ca^{2+} binding and transfer. *This may explain the interference of Mg^{2+} with mineral deposition in in vitro calcification* experiments. The Sr^{2+} ion may replace Ca^{2+} in the reaction, and Sr^{2+} will deposit in the bones. In the case of ^{90}Sr this represents a potential health hazard of nuclear test blasts.

The above can be summarized as follows:

1. An active pump exists in sarcoplasmic reticulum (SR) for

calcium. Calcium, in moving in or out, exchanges for Mg and K. The total concentration inside (i) remains constant [16a]:

$$[Mg^{2+}]_i + [Ca^{2+}]_i + [K^+]_i = constant$$

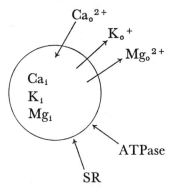

2. This is mediated by a Ca–Mg-dependent ATPase which comprises *70% of the membrane by weight*. The membrane disintegrates and is solubilized by detergents such as Triton X-100 (nonionic) or dodecyl sulfate (ionic). Its molecular weight is approximately 106,000. Alternatively, the inert components are removed with 5 mmol of EDTA (pH 8.6) in which they are soluble, leaving insoluble ATPase.[14]

3. In the presence of ATP, Mg^{2+}, and Ca^{2+} the ATPase is phosphorylated. It is the carboxyl group of aspartate which is phosphorylated. The sequence serine, aspartate, and lysine occurs at the active site.[15]

4. The phosphorylated ATPase changes conformation to move the Ca^{2+} into the cell, the phosphate being hydrolyzed during the process.[16,16a,17]

5. A binding protein, calsequestrin, is present within the cell, which binds the calcium, lowering the gradient against which new Ca$^+$ needs to be pumped.[52a]

7.3   MAINTENANCE OF CONSTANT SERUM Ca^{2+} CONCENTRATION

When accurate methods for Ca assay are performed in human serum, one is impressed by the narrow limits within which the calcium concentration is maintained at 9.9 mg/100 ml ± 5%.[17a] The diurnal range of serum calcium is less than $\pm 3\%$.[18] When serum ionized calcium is measured with the ion exchange electrode, a mean value of 39.2 ± 0.2 mg/liter [0.98 ± 0.01 (2σ) mmol/liter] is obtained.[19]

These levels are controlled by several humoral factors, the most studied being parathormone, calcitonin, and vitamin D. Prostaglandin E and other factors also come into play. These will now be discussed.

One must distinguish between *ultrafilterable calcium* and *ionic calcium*. Ionic calcium measurements are made of the activity of the calcium ion itself. On filtration of serum through a membrane by pressure, usually by centrifugation, all calcium not bound to protein filters through. Thus, in citrated blood, or where EDTA has been added, all the calcium becomes ultrafilterable, since it is all complexed and passes the membrane. In this case, calcium ion concentration is very low.

7.3.1 *Parathormone*

If EDTA is injected intravenously, the calcium ion concentration (not the total calcium level) is lowered, due to chelate formation. On the other hand, serum calcium ion concentrations may be increased by intravenous administration of calcium gluconate. In either case, the serum calcium level, both total and ionic, rapidly returns to normal levels in the intact dog. This is not true if the parathyroid and thyroid glands have been removed.

That the parathyroid glands are somehow involved in the maintenance of serum calcium levels was first demonstrated in 1891 by Gley.[20] He showed that removal of the glands in young animals resulted in tetany and death. That lowered serum calcium concentration was the problem was demonstrated finally in 1908 when microprocedures for calcium assay in serum became available.[21] MacCallum and Voegtlin also demonstrated that extracts of the parathyroid gland would ameliorate the symptoms in these animals. Potent extracts of the parathyroids were not produced until the middle 1920's.[22] In 1959 several investigators[23,24] prepared pure enough preparations for generation of antibodies for immunoassay.[25-27]

The parathyroid hormone exerts a direct effect on bone to promote mobilization of calcium, which then moves into the plasma. It is proposed that parathormone (PTH) acts directly on osteocytes or osteoclasts[27] or on certain mesenchymal cells to stimulate their conversion to osteoclasts. Others question this mechanism.[28] In any case, lysozomal enzymes are released which attack the bone matrix by one of several proposed mechanisms. Citrate and lactate accumulate[29] and have been proposed as being the apatite solubilizing factors. Collagenase activity is also increased locally under the influence of PTH to help bone dissolution.

In both *in vitro* and *in vivo* experiments, PTH acting on bone releases cyclic AMP.[30] Dibutyl-cyclic AMP injected into thyroparathyroidectomized rats mimics the effect of the parathyroid hormone. It is therefore felt that PTH activates a specific adenyl cyclase in bone, which then sets in motion the sequence of events resulting in the release of lysozomal enzymes and dissolution of bone.[31]

Prostaglandin E increases the level of cyclic AMP in bone in tissue culture and stimulates bone reabsorption. If the cells are broken, however, the prostaglandin loses its effect.[32]

The level of circulating PTH in the animal is controlled by the plasma calcium level. If serum calcium level decreases, the parathormone level increases; if the calcium level is increased, the reverse effect takes place.[33] This is illustrated in Figure 7.1, simulating tissue culture experiments performed by several investigators.

The parathyroid and the bone are incubated in two different compartments. A simulated plasma solution is circulated around the para-

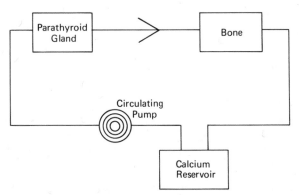

Figure 7.1 Schematic illustration of an experiment to study the effect of circulating calcium levels on parathormone release and calcium dissolution. A solution of 5, 10, or 20 mg/100ml of calcium salt in isotonic tissue culture solution is circulated. The parathyroids and bone are contained in two separate containers. From time to time portions of bone are removed for microscopic examination.

thyroids. This solution, also containing phosphate and other ions simulating the plasma, may then communicate with the container holding the bone, such as a rat tibia. When the perfusing liquid is circulated with a low calcium level (e.g., 5 mg/100 ml), the parathormone is released, causing bone dissolution. If a solution with a high level of calcium is now circulating, deposition of bone will take place and parthormone secretion will be suppressed. *Bone and the parathyroids taken together therefore comprise a servomechanism for maintaining serum Ca^{2+} concentration constant.*

It has long been known that in addition to its effect in mobilizing bone, PTH also acts on the proximal tubules of the kidney in such a way as to increase phosphate excretion in the urine. This comes about by either an increase in phosphate excretion or, more likely, a decrease in phosphate reabsorption in the proximal tubule.[34]

Injection of PTH intramuscularly causes a marked increase (as much as four times) in urinary cyclic AMP excretion.[35] In *in vitro* experiments it has also been shown that PTH causes a marked increase in cyclic AMP formation from rat *renal cortex homogenate.* On the other hand, *vasopressin, whose action is on the distal* tubules, causes a marked increase of cyclic AMP formation in *renal medulla preparations.* Renal cortical tubules (proximal tubules) are unresponsive to vasopressin. The action of vasopressin and that of PTH are independent of each other.[36]

The action of PTH in helping to excrete phosphate disposes of the excess phosphate in the plasma as bone is dissolved with the intent of raising only calcium levels.

For the relationship between parathormone and vitamin D see Section 7.3.3 below.

7.3.2. Calcitonin

In 1962 it was noted that acid extracts of the parathyroid glands would lower serum calcium levels.[37] The name calcitonin was given to this material. Subsequently, an extract made from the thyroid glands (thyrocalcitonin) was shown to have a greater effect.[38] Extracts of the adrenal gland made in the same way also have some serum calcium-lowering effect. These polypeptides were shown to be the same, and only the term *calcitonin* is now used.

It was soon shown that the tissue common to these organs was ultimobranchial tissue present in these glands.[39] The ultimobranchial cells of the thyroid are now called the "C cells" because they secrete calcitonin. By fluorescent antibody techniques, calcitonin-containing granules are located in these cells. These dissolve and move into the circulation when the gland is stimulated by elevated serum calcium levels.[40]

The ultimobranchial glands occur in all vertebrates. They obtain their name from the fact that they derive from the terminal branchial pouch in amphibia. In chickens and turkeys, the gland is formed separate and posterior to the thyroid and the parathyroids.

The principal effect of calcitonin on serum calcium is the opposite of parathormone. It lowers serum calcium and inorganic phosphate concentration in the plasma. The maximum drop occurs in 1–2 hr after injection. Secretion of calcitonin is augmented when serum calcium levels drop.

The principal action of the hormone is on bone. Addition of calcitonin to tissue cultures of bone reduces the number of osteoclasts and increases the number of typical osteoblasts. The effect is to shift the balance to bone formation, as distinct from parathormone, which shifts the balance to bone resorption. This has also been demonstrated *in vivo* in rats that had been parathyroidectomized.[41]

At first it was proposed that the action of calcitonin was to interfere with the effect of parathormone by activation of the phosphodiesterase, which destroys cyclic AMP generated by the parathyroid hormone. However, it was soon found that calcitonin served to stimulate cyclic AMP formation in bone tissue and, in this regard, was synergistic with PTH in raising cyclic AMP levels in tissue-cultured bone.[42] Present thinking is that calcitonin and PTH act on different cells in the bone, producing their opposing effects independently.

7.3.3.　Vitamin D

Egyptian inscriptions written approximately 3000 years ago record the fact that exposure to the sun will "harden" bones and cure rickets. It was in the middle 1920's that Hess and Steenbock demonstrated that irradiated foods would also heal rickets. This was eventually related to the dehydrocholesterol content of the foods.

It was soon established that serum calcium and phosphate levels were kept in balance under the influence of vitamin D. If the diet were deficient in calcium and phosphate, administration of vitamin D would cause dissolution of the shaft of the bone, in order to elevate serum concentration of Ca^{2+} to normal levels and calcify newly growing bone, at the expense of the older bone. *In vitro* experiments also showed that a minimum Ca × P product had to be reached before calcification would take place.[43] If that product was not reached, then calcification would not take place. Blood has a product of greater than 35 (Ca = 10 mg/100 ml, P = 3.5 mg/100 ml) and is supersaturated with respect to calcium phosphate. Hydroxyapatite should then precipitate if a first crystal is formed in the bone matrix to initiate crystallization.

It has also been conclusively established, that, in addition to its direct action on bone, vitamin D stimulates calcium absorption from the small intestine[44,45] and reabsorption of calcium in the proximal tubule of the kidney.[46]

In blood, vitamin D is bound mainly to protein. On electrophoresis, it travels with the α_2 globulin and albumin.[47] On protein precipitation, it comes down with the precipitate. Only by alkaline saponification can it be separated from the protein.

The relationship between the action of the parathyroid hormone and that of vitamin D on bone was explored extensively. Until 1967 the general conclusion was that vitamin D and PTH acted synergistically at the bone site, but had very little relationship at the absorption sites in the gut or the kidney.

It had been shown that intestinal loops from rachitic rats would not effectively transport calcium from the mucosal to the serosal side. If the animal were given vitamin D, this rate of absorption did not change for approximately 9 hr and then rapidly increased to normal.

The 9-hr lag suggested that some metabolic change had to be taking place. It was at first shown that oxidation of the vitamin had taken place in the liver to generate 25-hydroxy vitamin D_3.[48,49] When this substance was administered to rats, the lag phase was reduced to 3 hr before calcium absorption took place.

Further studies showed that the 25-hydroxy vitamin D_3 was further oxidized to generate 1,25-dihydroxy vitamin D_3. This substance reacted promptly. With nephrectomized animals, this second hydroxylation did not take place, and it then became apparent that this second hydroxylation took place in the kidney.

Intestinal transport of Ca is apparently not associated with protein synthesis, since it is not blocked with actinomycin D. On the other hand, this protein synthesis inhibitor does block bone mobilization by 1,25-dihydroxy vitamin D_3.

Further studies demonstrated that oxidation of 25-hydroxy vitamin D_3 to 1,25-dihydroxy vitamin D_3 was under control of parathormone secretion, which in turn was under control of serum calcium levels. Thus, when the calcium level dropped, there was a prompt increase in the 1,25-dihydroxy vitamin D_3 generated. Apparently, parathormone controls the synthesis of 1,25-dihydroxy vitamin D_3 formation. Thus, the "symbiotic effect" of parathormone and vitamin D_3 on calcium reabsorption in the tubule, in the intestine, and in bone mobilization appears to be clarified by these observations.[51,53,53a]

When serum calcium levels return to normal or are elevated, the synthesis of 1,25-dihydroxy vitamin D_3 is inhibited, and the 25-hydroxy vitamin D_3 is converted to 24,25-dihydroxy vitamin D_3. 24,25-Dihydroxy vitamin D_3 may be further hydroxylated in the kidney to form 1,24,25-trihydroxy vitamin D_3. These reactions can be formulated as follows (Figure 7.2).

Vitamin D_3 and 1,25-dihydroxy vitamin D_3 have been shown to stimulate the goblet cells of the intestine to synthesize a low molecular weight (mol wt 25,000) protein that strongly binds calcium ion. This calcium binding protein (CaBP) apparently serves to concentrate the calcium at the surface of the brush borders of the villi so as to facilitate the action of the calcium pump. Lysolecithin and other phospholipids serve to complex this protein and release the calcium for absorption.[50,50a] A similar protein also seems to be present in the proximal tubule of the kidney.[52]

The designation vitamin D was reserved for the natural vitamin. This turns out to be a mixture, in humans, of mainly vitamin D_3 and small amounts of other related antirachitic substances.

On irradiation of ergosterol (24-methyl dehydrocholesterol with a double bond between carbons 23 and 24), a 1:1 complex between vitamin D_2 (below) and lumisterol was obtained. This was called *vitamin D_1*. On removal of the lumisterol, the pure calciferol was called *vitamin D_2*. Subsequently, dehydrocholesterol was prepared and converted to an antirachitic substance. This was called cholecalciferol and assigned the designation *vitamin D_3*. Vitamin D_4 is the substance obtained after reduction of the double bond in the ergosterol side chain

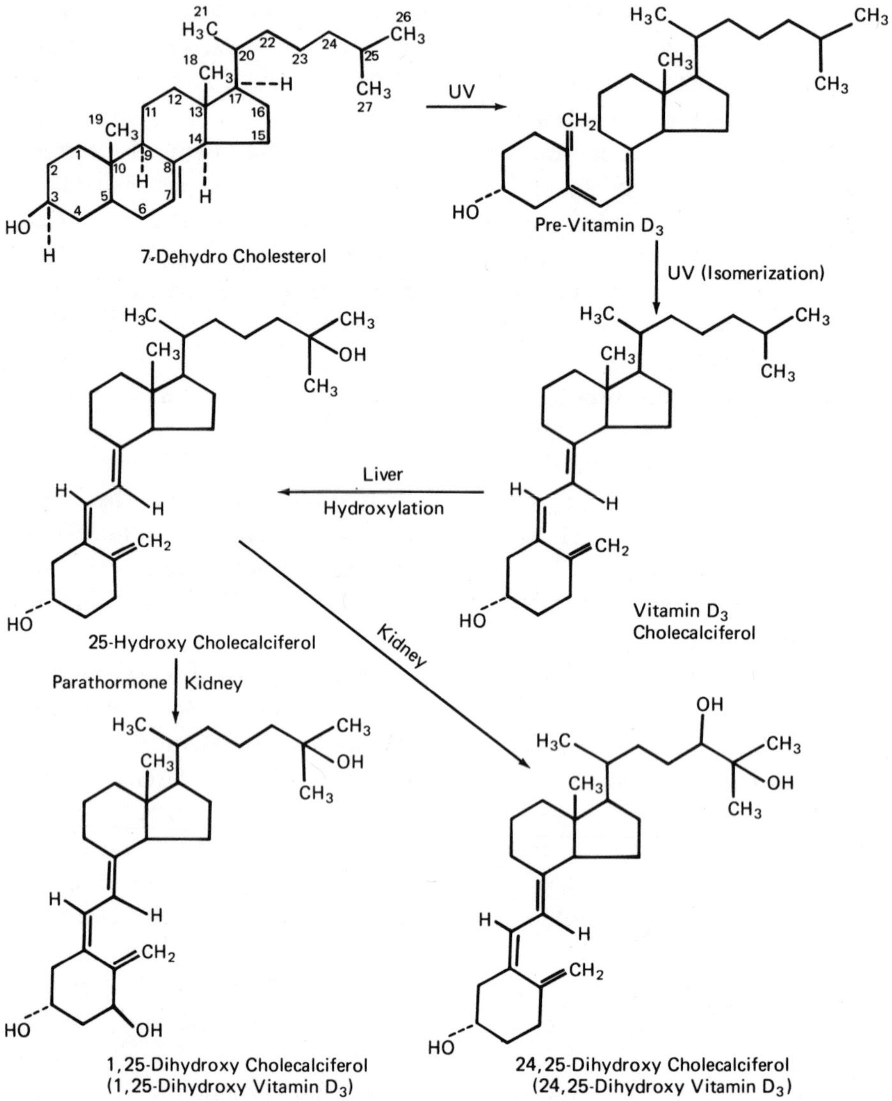

FIGURE 7.2 Schematic representation of the conversion of dehydrocholesterol to 1,25-dihydroxy vitamin D₃.

and irradiation, and is called *dihydrocalciferol*. If the double bond at the methylene groups adjacent to the 19 position of calciferol (vitamin D_2) is reduced, we have *dihydrotachysterol*, also called *AT-10*. This is used to correct hypocalcemia and resembles the parathyroid hormone in its

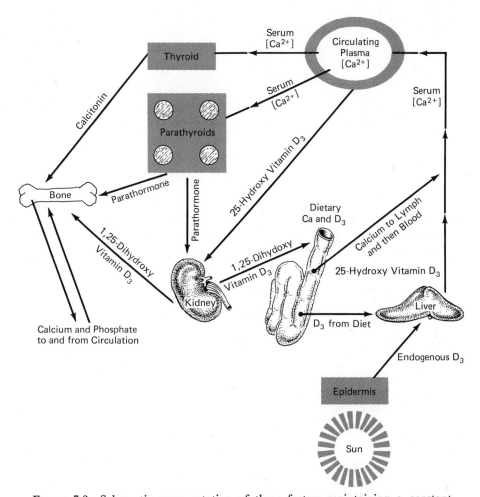

FIGURE 7.3 Schematic representation of those factors maintaining a constant serum calcium level. The thyroid and parathyroids respond to changes in serum Ca^{2+} levels in order to modulate calcitonin and parathormone secretion. Vitamin D_3 absorbed at the gut and from the epidermis is converted at the liver to 25-hydroxy vitamin D_3. This compound moves to the kidney, where under the influence of parathormone, it is hydroxylated to form 1,25-dihydroxy vitamin D_3. This substance stimulates Ca^{2+} reabsorption in the kidney tubules, dissolution of bone, and absorption of calcium from the gut, all of these processes serving to raise serum Ca^{2+} levels. Elevated serum Ca^{2+} levels, in turn, stop synthesis of 1,25-dihydroxy cholecalciferol by shutting off parathormone secretion and thus bone dissolution and absorption from the intestine and kidney tubules. Ca^{2+} is also shunted to the bone under the influence of calcitonin. When 1,25-dihydroxy cholecalciferol is not needed, the 25-hydroxy compound is hydroxylated at the kidney to form 24,25-dihydroxy cholecalciferol. This may be further hydroxylated to 1,24,25-trihydroxy cholecalciferol, which is biologically active in stimulating Ca^{2+} resorption.

action. It is effective in raising serum calcium levels. Dihydrotachysterol is hydroxylated in the 25 position at the liver, which activates it. It does not need further hydroxylation at the kidney, as for vitamin D.[167]

Calcitonin has no effect in stimulating or inhibiting the hydroxylation of vitamin D_3, and its action in this regard is independent of the action of parathormone.[56]

The relationship between the plasma Ca^{2+} concentration, the parathyroid hormone, and calcitonin, and the three active sites of calcium absorption (the intestine, kidney tubules, and bone) is summarized in Figure 7.3. This is in effect a schematic representation of *the servomechanism which is designed to maintain constant serum calcium concentrations.*

7.4   OTHER FACTORS AFFECTING PLASMA CALCIUM LEVELS

7.4.1   *Pituitary Extracts*

In 1951 it was reported that acid extracts of the anterior pituitary gland, injected intramuscularly, would lower rabbit serum calcium levels by approximately 30–50% and produce tetany, convulsions, and death in these animals.[54] Adrenaline, ACTH, and cortisone were without effect in this respect.[55] Maximum depression of serum calcium levels was within 2–3 hr and returned to normal at the end of 5 hr. The curves resembled those subsequently obtained with calcitonin (1962 *et seq.*), except that the serum calcium levels were depressed to much lower levels with the pituitary extracts. Tetany was demonstrated with the electromyograph, which could be relieved by I.M. injection of Ca salts.[57]

A significant finding in these studies was that citrate levels rose to as much as double their initial value and also returned to normal levels at the end of approximately 5 hr. This was remarkable in that it had been commonly observed that calcium and citrate levels generally moved up and down together in an apparent attempt to maintain constant calcium ion levels.[58] One could then say that these pituitary extracts *uncoupled citrate and calcium,* the rise in citrate aggravating the hypocalcemia, since it complexes calcium. This is distinct from the action of calcitonin which causes a drop in serum citrate levels.

The rise in citrate could have been partly an adrenaline effect, since adrenaline causes an elevation of serum citrate levels,[59] while administration of glucose or insulin causes a lowering of serum citrate levels.[60,61] It was soon shown that pituitary extracts, with no ACTH activity (no corticotrophic activity), still could produce the calcium-lowering effect.[62] ACTH with 270 U/mg also had similar effects.[63] On the other hand, TSH, STH, LTH, oxytocin, parathormone, adrenaline, hydrocortisone, ascorbic acid, and corticosterone were without effect. Vasopressin would lower serum calcium levels by approximately 20%. A similar but less dramatic effect was demonstrated in guinea pigs and dogs, but not rats.

When ACTH was administered to normal children (2 U/kg), citrate levels rose on an average to 60% above initial levels, and serum Ca concentration *rose somewhat (12%)*. This was in marked contrast to the behavior of ACTH in rabbits.[55] It might also be pointed out that calcitonin does not lower serum Ca levels in normal humans with normal serum calcium levels.[56]

When ACTH was administered to children who had been admitted to the hospital because they had had at least one convulsive seizure recently, from various causes, some depression of serum Ca levels was noted in seven out of eight cases. Calcium concentration returned to normal after 5 hr. In one case serum calcium concentration dropped below 5 mg/100 ml. At the end of 5 hr, serum Ca level was still not normal, after calcium gluconate was administered.

These observations suggest that in the human and rat a strong antagonistic mechanism exists to protect against lowered serum Ca levels. This is more easily overcome in the rabbit. In some individuals this mechanism may be defective. To explore this idea, Librium and Dilantin were administered after ACTH administration to rabbits, with the idea that they might prevent the rise in citrate and drop in calcium levels. Dilantin was without significant effect. Librium, however, aggravated the effect of the ACTH and prolonged the period during which the serum Ca level was depressed and serum citrate level was elevated. In effect, Librium seemed to block the mechanism that returns the Ca and citrate levels to normal.[64] Librium *per se* had no effect on the serum Ca or citrate levels.

The above observations indicate that the pituitary is involved in some way in maintenance of constant serum calcium levels. It has been pointed out that hypophysectomy results in a fairly rapid atrophy of the parathyroids (42% in eight days).

The pituitary is also involved, long range, in maintaining serum calcium levels by its action on the various target glands. Estrogens and androgens exert a significant influence on calcium metabolism. In birds, hypercalcemia appears just before ovulation. In mammals (rats), administration of estrogens produces increased calcium deposition, resulting in hyperossification of the proximal epiphyseal zone of the tibia and thickening of the trabeculae. Androgens antagonize this effect.

After the menopause, a type of osteoporosis may occur in humans, which responds to estrogen administration. Hyperfunction of the adrenal cortex or thyroid may also result in hypocalcemia and decalcification of the bones. This has been attributed to decreased collagen synthesis with resulting dissolution of the bone matrix.

7.4.2 *Sugars*

In the human, addition of lactose to the diet increases the absorption and retention of calcium. While glucose and sucrose are ineffective in this regard, other sugars and sugar derivatives, such as mannose, raffinose, cellobiose, mannitol, sorbital, and inositol, have similar effects.

It has been demonstrated that glucose can enhance Ca absorption if the duodenum is bypassed. If the glucose is injected directly into the ileum, calcium absorption will be increased by glucose.[65]

There are several reasons for the effects of carbohydrates on calcium absorption. First, calcium salts form complexes with carbohydrates such as lactose.[66] The calcium complexes then diffuse readily across the membrane. $CaCl_2$ is used routinely to adsorb glucose from serum in performing certain tests, such as for lactate estimation. Second, calcium may form salts with phosphorylated sugars, which then pass the membrane more readily. For example, calcium fructose diphosphate will be more readily absorbed into the blood stream than calcium acid phosphate.[67]

A third mechanism proposed is that the sugar supplies energy locally in order to supply energy to the calcium pump and thus enhance its activity. A fourth mechanism suggests that sugar metabolism by bacteria lowers pH in the intestine, increasing calcium solubility.

Regardless of the mechanisms that apply, carbohydrates play a significant role in increasing intestinal absorption of calcium and maintaining serum calcium levels.

7.4.3. *Effect of Amino Acids*

Calcium salts are kept in solution in milk as a complex with the phosphoprotein, casein. It is for this reason that those animals, such as cows, with a higher percentage of bone in their makeup have a much higher casein content in milk as compared to animals, such as pigs and humans, with a lower percentage of bone in their bodies. When the casein is hydrolyzed in the intestine, calcium is absorbed very efficiently. Several explanations have been suggested. It has been proposed that calcium is absorbed as a chelate with low-molecular-weight polypeptide phosphates generated from the casein. Individual amino acids such as lysine[68] and arginine[68a] will enhance calcium absorption in the intestine; glycine is less effective in this regard. Lysine also forms a chelate with calcium ion, which may explain its behavior.

From the above it is apparent that amino acids also serve to aid in calcium absorption so as to maintain serum calcium levels.

7.5 *ABNORMAL SERUM CALCIUM CONCENTRATION*

7.5.1. *Hypercalcemia*

Hyperparathyroidism due to adenoma or hyperplasia (polyostotic fibrous dysplasia) results in elevated serum Ca levels, lowered serum phosphate levels, increased urinary excretion of calcium, and increased volume of 24-hr urine collections and normal or elevated urinary phosphate excretion in the face of a low serum phosphate concentration.[69] Serum alkaline phosphatase levels are elevated, and on fractionation for isoenzymes, the elevation is due mainly to bone-derived alkaline phosphatase.[70] In hyperparathyroidism, serum calcium levels usually range from 11 to 16 mg/100 ml as compared to 10 mg/100 ml in the normal. Values higher than 20 mg/100 ml have also been reported. In some cases, during periods of spontaneous remission, serum calcium levels may return to normal.

Symptoms of hypercalcemia start with atony of the gastrointestinal tract, with nausea, vomiting, and abdominal pain. There is impaired renal tubular reabsorption of water and phosphate, resulting in polyuria with the large 24-hr volume of urine mentioned above accompanied by thirst and chronic dehydration. In the electrocardiogram the $Q-T$ interval is shortened.

Calcium phosphate stones in the ureters and bladder are a frequent finding in hyperparathyroidism. If uremia develops with phosphate retention, then the serum calcium level may be lowered to the normal range. Urinary calcium concentration may be markedly elevated in some of these patients, so that a qualitative test for calcium is sometimes helpful for diagnostic purposes. Quite often, however, urinary calcium concentration is in the normal range.

The patient is placed on a diet containing 50 mg calcium per 24 hr for two days.[71] If the second 24-hr urine specimen contains substantially more than 50 mg of calcium, the test is considered positive. A qualitative test may also be done on the urine. Equal parts of urine and Sulkowitch reagent (2.5 g of ammonium oxalate, 2.5 g of oxalic acid plus 5 ml of glacial acetic acid to 150 ml with H_2O) are added.[72] A heavy precipitate is considered positive. No precipitate at all is also considered positive for abnormally low urine Ca levels.

Hyperplasia of the parathyroids may be a result of some other condition tending to maintain lower serum Ca levels, such as glomerulonephritis, renal rickets, malabsorption of Ca due to sprue, celiac disease, or other causes such as pregnancy and lactation. In all of these conditions, excessive calcium losses or poor absorption results in lowered serum Ca levels. This serves to stimulate the parathyroid glands to increase secretion of parathormone. Serum Ca levels may then be maintained at normal or slightly lowered levels. This condition is called *secondary hyperparathyroidism*. In these cases serum alkaline phosphatase is not as markedly elevated as in primary hyperparathyroidism.[73]

Administration of excessive amounts of vitamin D or dihydrotachysterol will serve to elevate serum calcium concentration. If inadequate amounts of calcium are supplied, demineralization of the shafts of the bones will take place, and the calcium will be deposited in the epiphesis, at the expense of the induced osteoporosis. Serum calcium levels and bone alkaline phosphatase levels will be elevated.

In multiple myeloma and other neoplastic diseases which may produce osteolytic lesions, serum calcium levels may rise to as high as 20 mg/100 ml of serum. Alkaline phosphatase levels are usually normal. Serum protein levels rise to as high as 12 g/100 ml (normal is 7 g/100 ml) due to formation of monoclonal antibodies in huge amounts. These form complexes with Ca, reducing serum ionic Ca levels. Destruction of bone with skeletal lesions ensues and is the cause of the hypercalcemia. The markedly elevated serum protein levels stimulate bone dissolution due to the complex they form with calcium. Generally, decrease in renal function accompanies the disease, with resultant acidosis aggravating the condition.[74]

With most metastatic neoplasms not involving bone, serum calcium level is usually normal, unless there is a marked elevation in serum protein levels, in which case it may rise dramatically. With metastases to the bone, hypercalcemia associated with hyperphosphatemia and increase in bone alkaline phosphatase levels is noted. With primary bone tumors, similar observations are made. Increased serum calcium and phosphate levels result in calcium phosphate deposition in the kidneys and renal calculi formation. This condition is sometimes mistaken for hyperparathyroidism.

A condition which is usually observed below the age of 2 yr is an unexplained hypercalcemia (idiopathic infantile hypercalcemia) with anorexia, vomiting, polyuria, and muscular hypotonia.[75] There are two recognizable forms, one mild, which usually remits spontaneously in a few months, and a severe form accompanied by osteosclerosis and mental and physical retardation. The condition is associated with what has been described as an "elfin" appearance, with low-set ears, epicanthic folds, wide mouth, overhanging upper lip, and, occasionally, strabismus. The hypercalcemia may lead to deposits in the lungs and kidneys. In the severe form, death usually ensues from renal failure.

In hypophosphatasia, low serum alkaline phosphatase occurs with the appearance of ethanolamine phosphate in the urine and low serum phosphate levels.[76] This is often accompanied by a hypercalcemia. In order to maintain equilibrium with the bones, the low phosphate level causes bone dissolution and an elevated Ca level, so as to maintain a constant Ca × P product.

Heavy milk drinkers, especially those on a high milk diet, such as patients with peptic ulcers, may develop hypercalcemia. Alkalosis and renal impairment sometimes follow, which aggravates the hypercalcemia, resulting in renal calcium deposits. This condition has also been mistaken for hyperparathyroidism.

Almost any condition that results in chronic elevation of serum protein levels will result in hypercalcemia. In addition to multiple myeloma and viral hepatitis, leukemia and polycythemia are examples which may illustrate this principle. Conditions that tend to produce a chronic acidosis, such as the respiratory conditions, chronic emphysema, pneumonia, silicosis, and congestive heart failure, will also tend to dissolve bone and elevate serum calcium levels. When the patient is immobilized, the normal stimulus to anabolic processes is removed, with decreased resynthesis of osteoid tissue, accompanied by extensive demineralization and elevated serum calcium levels.[77] In these conditions, serum phosphate levels are normal or elevated, which distinguishes this condition from hyperparathyroidism. Serum alkaline phosphatase levels remain normal or low, since extensive turnover of functioning osteoblasts, where the alkaline phosphatase originates, is not taking place.

7.5.2. *Hypocalcemia*

Hypoparathyroidism occurs as a spontaneous condition (idiopathic). It also ensues after surgical removal of the parathyroids, for any of several causes. It also may follow a disease process, such as from some severe infection, hemorrhagic disease, or, occasionally, in tuberculosis or amyloidosis.

The serum changes are inverse of those seen with hyperparathyroidism. These include hypocalcemia, increased or normal serum phosphate levels, decreased urinary excretion of Ca and P, and usually normal serum alkaline phosphatase levels.[78] Serum calcium levels may drop as low as 5 mg/100 ml, accompanied by symptoms of tetany in the patient.

In these patients the $Q-T$ interval in the electrocardiogram is prolonged. Convulsions are not uncommon in these patients, accompanied by carpopedal spasm. The spasm is painful and persists unless the ionic serum calcium level is raised.

A condition occurs in children which has been called *pseudohypoparathyroidism*, where serum calcium levels run chronically low, of the order of 7 mg/100 ml.[79] Administration of PTH *does not increase serum Ca* levels nor increase phosphate excretion. They are also resistant to administration of dihydrotachysterol and vitamin D. Evidence has been presented to show that these individuals are deficient in the hydroxylating enzymes required to convert vitamin D to 1,25-cholecalciferol. This needs to be explored. Along with this condition, one needs to consider that some cases of "*incurable rickets*" may be of similar etiology as the so-called pseudohypoparathyroidism. The defect appears to be in lack of the ability of the liver to hydroxylate vitamin D_3, since these patients respond to 25-hydroxycholecalciferol.[79a] They do not respond to dihydrotachysterol because this also requires hydroxylation in the liver but not the kidney.[79b]

With vitamin D deficiency, low serum calcium levels are observed, due to conditions that prevent the reabsorption of Ca from the gut and proximal tubules, and the mobilization of bone, as has been discussed before.

If absorption of fatty acids is interfered with, then insoluble calcium soaps appear in the stool. Such a condition is seen with celiac disease, sprue, prolonged obstructive jaundice, pancreatic and bile duct obstruction, and in any of the conditions leading to the malabsorption syndrome. This condition may result in rickets, dwarfism, and osteomalacia, along with hypophosphatemia and hypocalcemia.

In the nephrotic syndrome, serum protein levels are reduced by as much as 30–40%. This would result in a marked increase in calcium ion activity except that total serum calcium levels drop to levels as low as 5 mg/100 ml. Ionic calcium remains the same as in the normal. Urinary excretion of calcium falls to low levels. Other conditions resulting in low plasma protein levels such as malnutrition (e.g., Kwashiakor) and malignancy, also result in hypocalcemia[80] and hypomagnesemia.

Chronic glomerulonephritis, chronic renal failure, and pyelonephritis, with phosphate retention resulting in hyperphosphatemia, each result in hypocalcemia. The reason for this is twofold. First, the Ca × P product needs to remain at a level to be in equilibrium with the bone, and second, the acidosis that usually accompanies the condition results in increased ionization of the calcium, so that a lower total calcium is needed to maintain the required ionic calcium level at approximately 4 mg/100 ml.

With prolonged renal failure with hypocalcemia, a secondary hyperparathyroidism may develop. When it occurs in children during the period of growth, osteomalacia and osteoporosis may result, with impaired skeletal growth. This condition has been referred to as "*renal rickets*" or "*renal dwarfism.*"[81]

Within the first few days of life, infants placed on artificial feeding formulas may show hypocalcemia, severe enough to induce tetany. This has been called *neonatal tetany*. On a breast-milk diet this syndrome does not seem to occur.[82] It has been suggested that the higher concentration of phosphate in cow's milk suppresses the serum Ca level.[82a] Phosphate concentration in breast milk is approximately 30 mg/100 ml, while in cow's milk it is 100 mg/100 ml.

Hypocalcemia in infants, unrelated to formula intake, occurs occasionally and is also referred to as neonatal tetany. This responds to dihydrotachysterol, which is required for only a relatively short time after birth (overall less than 1–2 weeks). At this time serum calcium levels spontaneously return to normal.[83] This has been attributed to the newborn's parathyroids requiring some time to take over the burden of calcium control which was carried before birth by the maternal parathyroid glands.

In acute hemorrhagic pancreatitis substantial decrease in serum calcium levels is observed. It has been suggested that, as has been discussed above, with the reduction of the lipases and esterases in the intestine, huge amounts of calcium are lost in the gut as calcium soaps. This is probably not the complete explanation for this hypocalcemia. An additional cause is due to that fact that large losses of plasma proteins occur, necessarily causing a decrease in the total serum calcium level in order to maintain a constant ionic calcium level.

In alkalosis, induced by administration of alkaline fluids or by hyperventilation, tetany, relieved by administration of calcium gluconate, will ensue. Hypocalcemia is not the problem in these cases. Decreased ionization of calcium due to rise in pH is the immediate cause. Acidemia increases Ca ionization and alkalosis decreases Ca ionization, as shown in the following scheme:

$$Ca(Ac)_2 \rightleftharpoons Ca^{2+} + 2Ac^-$$

$$2H_2O \rightleftharpoons 2OH^- + \boxed{2H^+}$$

$$\Updownarrow$$

$$2HAc$$

Increase in $[H^+]$ (circled) will drive the reaction to formation of acetic acid, thus tending to increase the ionization of the calcium acetate. Decrease of hydrogen ion concentration (increase in pH) will decrease the extent of ionization of the calcium acetate.

The clinical conditions in which significant changes in calcium ion concentration may take place are summarized in Table 7.1.

Certain individuals with normal serum calcium levels will tend to deposit calcium-containing stones in their bladder. In some unexplained way, the chelates that normally keep the calcium in solution seem to be broken. This has been attributed to various factors, including deficiency in mucopolysaccharide excretion. If the diabasic acids glutamine or aspartic acid are administered to these patients, then the level of these amino acids is elevated in the urine, and the tendency to form kidney stones seems to be decreased.[84] This is attributed to the formation of chelates between calcium and the amino acids.

7.5.3   *Magnesium Deficiency*

Magnesium deficiency from any cause will result in impaired absorption of Ca. This is due to the fact that the calcium pump in the gut is magnesium dependent.

From time to time cases are reported where serum calcium levels are low whose primary cause is magnesium deficiency. Typical is the *idiopathic hypomagnesemia*, which results in decreased serum Ca levels due to poor absorption from the gut. Vitamin D will not correct this condition, but oral or intramuscular administration of magnesium will correct the calcium deficiency.[84]

7.6   *MAGNESIUM*

Magnesium is the cofactor required for the utilization of adenosine triphosphate as a source of energy. It is therefore required for the action of numerous enzyme systems. It is required at several steps in the anaerobic mechanism of carbohydrate metabolism, for protein synthesis and nucleic acid synthesis. It is required for contraction of muscular tissue.[85,86]

Along with Na, K, and Ca ions, it serves to regulate neuromuscular irritability. Since the extent of ionization of Mg and the other ions is pH dependent, the activity of these ions is pH dependent. Decreased neuromuscular irritability results with increased activity of Ca, Mg, or

TABLE 7.1

Some Conditions Where Serum Calcium Concentration Is of Diagnostic Interest

Condition	Total calcium	Ionic calcium	Phosphate	Alk. phos.	Total protein	pH	Comments
Primary hyperparathyroidism	Elevated	Elevated	Low	Elevated	Normal	Normal	Serum parathormone level elevated
Excessive vitamin D intake	Elevated	Elevated	Elevated	Elevated	Normal	Normal	Serum parathormone level depressed
Chronic renal disease leading to secondary hyperparathyroidism	Normal to low	Normal	Elevated	Normal to slightly elevated	Moderately low	Lowered	Serum parathormone level elevated
Nephrotic syndrome	Low	Normal	Low	Elevated	Very low	Lowered	Increase in fecal Ca excretion, decrease in urinary Ca
Hypoparathyroidism	Low	Low	Normal to elevated	Normal	Normal	Normal	Serum parathormone level very low
Malnutrition (e.g., Kwashiakor)	Low	Low	Low	Normal to slightly elevated	Very low	Low	Progressive osteomalacia
Rickets	Low	Low	Low	Elevated	Normal to low	Normal	Corrected by vitamin D administration
Renal rickets (vitamin D-resistant rickets)	Low	Low	Elevated	Normal to elevated	Low	Lowered	Responds to 1,25-dihydroxy vitamin D_3 with rise in serum Ca but not to vitamin D_3

Dietary neonatal tetany	Low	Low	Elevated	Elevated	Normal to low	Lowered	Occurs with cow's milk and not breast milk feeding
Hypophosphatasia	Normal to high	Normal to high	Low	Very low	Low	Normal	Ethanolamine phosphate appears in urine; an alk. phosphatase deficiency
Hormonal osteoporosis (e.g., postmenopausal)	Normal	Normal	Slightly elevated	Normal to elevated	Normal to low	Normal	Responds to estrogen administration
Immobilization	Elevated	Slightly elevated	Slightly elevated	Normal to low	Normal to low	Normal	A common cause for osteoporosis in the bedridden patient
Sarcoidosis	Elevated	Elevated	Normal to high	Normal to slightly elevated	Normal to elevated	Slightly lower	Increased globulins, particular the α-globulins and lower albumin serum Ca decreased with therapy
Neoplastic disease of the bone	Normal to markedly elevated	Normal to elevated	Normal	Elevated	Varies widely, usually low	Normal	Elevation of bone isoenzyme of alkaline phosphatase
Multiple myeloma	Elevated	Normal	Normal to slightly elevated	Normal	Elevated	Normal	Associated with monoclonal gammopathy
Malabsorption (sprue, celiac disease, and others)	Low	Low	Low	Elevated to normal	Low	Slightly lowered	Ca recovered in stool as Ca soaps
Magnesium deficiency	Low	Low	Low	Normal to low	Low	Normal	Ca recovered in stool does not respond to vitamin D

H ions. Increased concentration of K or Na ions results in increased muscular irritability. The relationship of neuromuscular irritability to these ions can be represented as follows:

$$\text{neuromuscular irritability} = F\left(\frac{[\text{Na}^+], [\text{K}^+]}{[\text{Ca}^{2+}], [\text{Mg}^{2+}], [\text{H}^+]}\right)$$

In the adult, approximately 300 mg of magnesium is ingested daily on a normal diet. Only 40% of this is absorbed and excreted in the urine. The remainder is found in the stool. In a normal diet, magnesium deficiency is rare, since all natural foods are rich in magnesium. Ingestion of magnesium increases not only the amount of magnesium absorbed, but also the amount of calcium absorbed. The requirement for magnesium in order for the calcium pump to operate has been discussed above. Thus magnesium is also involved in the rate at which calcium is absorbed from the intestine.

Serum magnesium levels are at 1.8 ± 0.4 mEq/liter (2 ± 0.2 mg/100 ml) for both males and females.[87] The average 75-kg man contains a total of approximately 2250 mEq of magnesium (187.5 mEq/kg). This corresponds to 27 g. Of this, 58% is in the skeleton (1305 mEq), and the remainder is distributed as 7 mEq in the plasma, 668 mEq intracellular, 8.4 mEq in the interstitial fluid, and 262 mEq in cartilage, transcellular, and connective tissue. In the cell, the magnesium is concentrated mainly in the mitochondria, where major energy changes are taking place.[88]

Inanition related to excessive losses from the gastrointestinal tract, such as chronic diarrhea, after hemodialysis (for severe uremia), chronic renal disease, hepatic cirrhosis, chronic pancreatitis, ulcerative colitis, hyperaldosteronism, toxemia of pregnancy, and hyperparathyroidism has been reported as the cause for *low plasma magnesium levels*. Magnesium moves with potassium into the cells after insulin administration. Treatment of diabetics in coma will therefore often result in low plasma magnesium levels.

The symptoms of magnesium deficiency simulate those observed with hypocalcemic tetany.[89,90] Severe neuromuscular irritability, such as hyper-reflexia, tremor of the extremities, carpopedal spasms, positive Chvostek and Trousseau signs, tachycardia, and hypertension are observed. Such patients are markedly susceptible to auditory, mechanical, and visual stimuli. Magnesium deficiency, like potassium deficiency, causes a marked reduction in protein synthesis.[91]

Elevated plasma magnesium levels are often observed in oliguric states, dehydration, and diabetic coma.[92] Generally, excessive oral magnesium intake, such as with Epsom salts administered for constipation, does not result in elevated plasma magnesium levels. However, if plasma magnesium levels are low, they can readily be raised by oral administration of magnesium. With a tumor invading the hypothalamus, chronic high magnesium levels have been observed. Plasma magnesium levels are therefore apparently regulated by homeostatic mechanisms ultimately controlled by the central nervous system. The symptoms of elevated serum magnesium levels are noted as depression of cardiac activity and depression of neuromuscular activity.

The ionic activity of the magnesium in the plasma is dependent on the pH and plasma protein, phosphate, and citrate concentration.[93] Normally, 1.32 $\pm$ 0.1 mEq/liter is the ionic activity for magnesium in human plasma at normal pH.[94]

Serum alkaline phosphatase is a magnesium-dependent enzyme, and magnesium deficiency results in lowered serum alkaline phosphatase levels. Phosphoglucomutase activity requires Mg as a cofactor. Mucolytic enzyme activity is also magnesium dependent.[95]

Normally, *95% of the magnesium that is filtered through the glomerulus is reabsorbed in the tubule.* With decreased glomerular filtration rate due to any of numerous causes, greater amounts of magnesium are retained, resulting in elevated serum levels above the 1.9–2.2 mg/100 ml normally observed.

As pointed out above, it has been known for some time that magnesium is important in absorption of calcium from the gut and in calcium metabolism. The relatively recent discovery of the role of magnesium in the movement of calcium across the various membranes in the living organism has clarified to some extent the relationship between the two elements. Extending these studies, it has been shown that magnesium deficiency results in enlargement of the thyroid C cells, which secrete calcitonin, with decrease in cellular function.[96,97] It is apparent that, with chronic low serum calcium levels, there is little need for calcitonin secretion, and loss of function of the C cells may be a protective mechanism.

Magnesium deficiency results in the drift of calcium out of the bones and in the absence of a pump mechanism, abnormal calcification takes place in various organs, such as in the aorta [98] and the kidney.[99] This condition responds to administration of magnesium salts.

Magnesium is present in cow's milk to the extent of 11.5–13 mg/ 100 ml. For this reason magnesium deficiency in infants is rare where cow's milk is readily available.[100] However, magnesium deficiency does occur in older children, associated with malabsorption of various types, with prolonged parenteral infusions of solutions containing no magnesium and in rickets. In addition, *transient idiopathic hypomagnesia* is not uncommon, but responds readily to magnesium administration.[101]

The serum of newborn children contains 1.94 ± 0.27 mg of magnesium, which rises to 2.12 ± 0.27 mg at the end of 30 days and remains at that level. In cord blood, the magnesium level is 1.89 mg/ 100 ml ± 0.27. It is of interest to note that infants show higher magnesium levels on breast milk than on cow's milk.[102] Breast milk contains approximately one-third the magnesium of cow's milk (approximately 4 mg/100 ml as compared to 12 mg/100 ml for cow's milk). This should be compared to the hypocalcemia and tetany observed in newborns on a cow's-milk diet with high phosphate concentration.[103] In some manner, high phosphate concentration in the diet suppresses both calcium and magnesium absorption. This may be due to depression of the magnesium ion concentration due to combination with phosphate, reducing the Mg^{2+} activity and suppressing active transport of both calcium and magnesium.

Serum magnesium concentrations are very low with cirrhosis of the liver due to alcoholism and other causes.[104] The caloric requirement of the alcoholic is met in large part by the whiskey he drinks. As a result, his intake of magnesium with foods is decreased to the point where serum magnesium levels are severely lowered. It is proposed that the symptoms of chronic alcoholism are partly due to magnesium deficiency.[104]

The importance of magnesium in the clotting mechanism became apparent when Mg^{2+} as well as Ca^{2+} was shown to be involved in the effect of ADP on platelet aggregation.[105] It needs to be emphasized that the ATP $\rightleftharpoons$ ADP reaction requires Mg as a cofactor.

From the above observations, it is apparent that *Mg^{2+} and Ca^{2+} are intimately tied together in their biological function, and deficiency of either one has a marked effect on the metabolism of the other*. It is for this reason that increasing attention is being paid to losses of magnesium during fluid therapy and hemodialysis in order to make adequate replacement of this ion.

7.7 MECHANISMS CONTROLLING SERUM PHOSPHATE LEVELS

Phosphate is the major anion found within the cells. Phosphate concentration within the cell is 50 times that in the plasma water and 100 times that in the interstitial fluid. Assuming that at pH 7.4 the equivalence of phosphate is 1.8 (see discussion of buffers in Chapter 2), then phosphate is 100 mEq/liter in intracellular fluid, 1 mEq/liter in the interstitial fluid, and 2.1 mEq/liter in plasma water.

The total phosphorus content in the adult is 5.4 g/kg of fat-free body weight. For 70 kg of body weight (fat-free), this calculates to 37.8 g in the body. Approximately 85% of this is in the bone, and a major part of the remainder is within the cells. Table 7.2 lists the distribution of the phosphorus found in the plasma and erythrocyte.[106,107]

Phosphate is involved in transport of metabolites, such as glucose and lipids, across membranes. It is also utilized in the form of purine and pyrimidine pyrophosphates, such as ATP, GTP, and UTP, for storage and transfer of energy from one site to another. It is for this reason that circulating phosphate levels need to be controlled within reasonably constant limits.

Serum inorganic phosphorus levels are maintained at 2–4 mg/

TABLE 7.2

Concentration of the Major Fractions of Phosphate in Plasma and Erythrocyte[106,107]

Fraction	Plasma or serum phosphorus, mg/100 ml of P		Erythrocyte phosphorus, mg/100 ml of P	
	Range	Mean	Range	Mean
Total	9.7–13.3	11.7	39–62	52.1
Inorganic	2.7–4.3	3.5	1.0–3.3	2.4
Organic	7–9	8.2	38.5–58.7	49.7
ATP	0.1–0.7	0.16	4.2–15.1	10.6
Diphosphoglycerate	0.1–0.4	0.03	19.0–40.4	29.2
Hexosephosphate	0.05–0.22	0.04	3.5–10.7	7.5
Phospholipid	6.4–12.0	9.2	7.5–13.0	11.9
Nucleic acid	0.44–0.65	0.54	5.1–7.1	6.2

100 ml in the adult and 4–8 mg/100 ml in the newborn. This latter value decreases gradually during the first few years of life and is reduced to the adult level only after rapid growth has ceased, or at approximately 15–17 yr of age.[108,109]

The plasma phosphate level is maintained constant by a complex series of reactions taking place in the living organism (Figure 7.4). Phosphate is taken in with the food in the form of organic and inorganic phosphate. Most of the organic phosphate is hydrolyzed and converted to inorganic phosphate. Absorption from the gut is by three different routes, simple diffusion (passive transport), facilitated transport, and active transport. This is true for the passage of phosphate through the various membranes of the body, *including the mitochondria.*

In *passive transport,* the concentration on one side of the membrane is higher than on the other side, resulting in simple diffusion. This can be demonstrated with radioactive tracers for phosphate.[110] In *facilitated transport,* the same conditions as for passive transport are present, namely that one is moving from a high to lower concentration on the other side of the membrane. However, some chemical reaction occurs, which speeds up the process. For example, the phosphate may serve to phosphorylate glucose or glycerol and then pass the membrane more rapidly as the phosphate ester. Generally, passive diffusion is too slow and some accelerating process is usually present in the tissues.

Active transport requires movement of phosphate uphill to a region of higher concentration, requiring the expenditure of substantial amounts of energy.[111–113] This is particularly true in moving phosphate from the interstitial fluid (less than 1 mmol/liter) to 30 times that concentration within the cell. This energy is supplied by metabolism of carbohydrates and fats. The effect of parathormone in stimulating reabsorption of phosphate in the proximal tubule is *prima facie* evidence that transport of phosphate is active and "energy dependent."

Active transport is usually associated with some polypeptide on the side of the membrane where the higher concentration occurs. This results in the anion or cation reacting with this polypeptide to remove it (in effect) from the medium. This cuts down the concentration gradient and facilitates the active transport mechanism. Examples have already been shown with potassium (combines with Hb) and calcium (combines with calsequestrin). This is also true for various steroids such as the estrogens. Thus active transport is usually associated with some facilitating process and should really be called *facilitated active transport.*

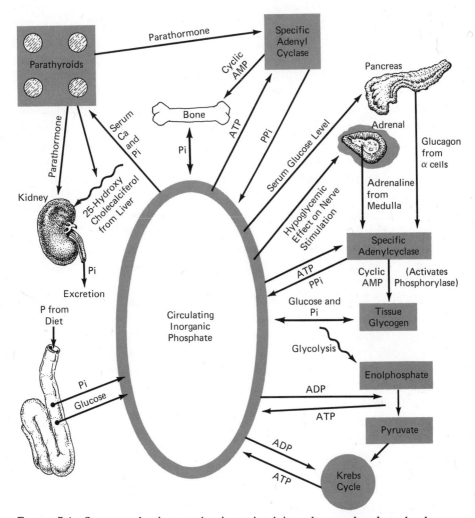

FIGURE 7.4 Some mechanisms active in maintaining plasma phosphate levels. The parathyroids maintain plasma phosphate levels by controlling flow of phosphate to and from the bones and phosphate excretion from the kidneys. The flow of phosphate to and from the tissues is associated with movement in inorganic phosphate, ATP, and ADP as indicated in the figure. Phosphate absorption from the intestines is mainly by facilitated transport along with glucose. ATP is supplied to the various targets of the hormones to be converted to cyclic AMP so as to activate certain enzyme systems.

Pure active transport does occur experimentally, as with erythrocyte ghosts, but probably rarely *in vivo*.

Figure 7.4 summarizes the mechanisms involved in the maintenance of constant serum phosphate levels. After the phosphate has been absorbed from the intestine it circulates and establishes dynamic equilibrium with bone phosphate.[111] Serum calcium ion level modulates the parathyroid output, and when it is low, secretion of parathormone is stimulated and bone dissolution takes place, causing phosphate to move into the circulation. At the same time reabsorption of phosphate at the proximal tubule of the kidney is inhibited, serving to prevent excessive increase of serum phosphate.

From the figure, one can see that when glucose moves into the blood stream from the gut, it combines with the phosphate to facilitate its absorption. It was with intestinal loops that Cori first demonstrated active transport of glucose and phosphate[114] and discovered phosphorylase.

The glucose that finds its way to the circulation is transported to the tissues under the influence of insulin and is then phosphorylated to glucose 6-phosphate with hexokinase and ATP. It then rearranges to glucose 1-phosphate and transfers with uridine triphosphate (UTP) to form uridine diphosphoglucose, which then forms glycogen. Postprandially, *large amounts of phosphate move into the tissues and plasma phosphate levels are decreased.*

The metabolism of glucose gives rise to substantial amounts of ATP, which by conversion to cyclic AMP, serves to mediate the action of adrenaline and glucagon. These hormones then activate tissue phosphorylase by activating tissue adenyl cyclase so as to stimulate conversion of glycogen to glucose 1-phosphate. Some of this is metabolized, but a substantial amount is hydrolyzed to glucose and phosphate, which then returns to the blood stream. *Thus a continuous flow of phosphate into and out of the blood stream to engage in metabolism and generation of energy is taking place.*

Cyclic AMP is formed at the various sites of action of the glands of the hormone system, the particular site recognizing the stimulus that activates its adenyl cyclase system to form cyclic AMP.[115] This then activates some enzyme system, such as the action of adrenaline and glucagon in activating phosphorylase. Parathormone also acts by stimulating cyclic AMP formation at the bone site[116] so as to activate certain enzymes, the products of whose action dissolve the deposited minerals.

If serum phosphate is elevated, then parathyroid hormone secretion is increased. This has been demonstrated by direct PTH serum measurements.[117] This effect can be abolished by calcium infusion. Thus parathormone secretion is also responsive to the serum phosphate level as well as the serum calcium concentration. This phenomenon has also been demonstrated indirectly, using the tubular reabsorption of phosphate (TRP) test for measuring reabsorption of phosphate in the tubules and thus parathyroid function.[117]

Under the influence of the parathyroid hormone, reabsorption of phosphate is inhibited in the proximal tubules. By measuring serum phosphate and creatinine, and urine phosphate and creatinine, the percentage of phosphate reabsorbed by the tubules, as compared to that filtered through the glomeruli, can be calculated. This TRP test [117–119] is run in the same manner as for the creatinine clearance, collecting urine specimens at 1 and 2 hr. A single blood specimen is taken between the two hourly intervals as in the creatinine clearance test.

Let us assume that the patient shows a creatinine clearance of 100 ml/min. Let us also assume that his serum phosphorus level is 4 mg/100 ml. He therefore filters 4 mg of phosphorus per minute.

Now let us assume that his urine phosphate level is 30 mg/100 ml and that the urine volume for 1 hr is 120 ml. Then the urine volume per minute is 2 ml. Since there is 30 mg of phosphorus in 100 ml of urine, there is 0.6 mg of phosphorus in 2 ml. But 4 mg/min entered the tubules from the glomerulus; therefore $4 - 0.6 = 3.4$ mg was reabsorbed in the tubules. To convert to percentage reabsorbed, we use $3.4/4.0 \times 100 = 85\%$, which is lower than the normal range.

By simple algebra, a formula can be derived for calculating the phosphate reabsorption rapidly. The surface area factors cancel out, as do the times and the urine volumes. This is because we are really comparing a creatinine clearance to a phosphate clearance calculated in the same way and measuring a ratio. The formula then becomes

$$\text{fraction of P absorbed} = 1 - \left(\frac{\text{urine P}}{\text{serum P}} \times \frac{\text{serum creatinine}}{\text{urine creatinine}} \right)$$

Let us now use this formula in the above problem. The urine creatinine is measured to be 50 mg/100 ml. Serum creatinine is 1.0 mg/100 ml. Urine phosphorus is 30 mg/100 ml and serum phosphorus is 4 mg/100 ml. Then the fraction of P absorbed is $1 - (30/4 \times 1/50) = 0.85$. Multiplying by 100 to find the percentage reabsorbed, we report

85% reabsorbed as before. Normally, 92–94% of the phosphate filtered through the tubules is reabsorbed.

The test has little significance with severe tubular disease, or in a condition such as diabetes insipidus. The ratio of creatinine in the urine to that in the serum is usually at least 50:1. If this ratio is substantially lower, then the lowered phosphate reabsorption may be due to tubular damage.

7.8 *ABNORMAL SERUM PHOSPHATE LEVELS*

7.8.1 *Hyperphosphatemia*

Most commonly, elevated serum phosphate levels are encountered in association with kidney dysfunction and uremia. Retention of phosphate tends to contribute, along with other causes, to the acidosis associated with uremia. Phosphate levels somewhat parallel creatinine levels more closely than they do urea levels. A phosphate level of more than 8 mg/100 ml may indicate significant loss of renal function. Levels twice this value are common with severe kidney damage.

Serum phosphorus values are markedly elevated in renal insufficiency due to many causes, such as nephrosclerosis, congenital polycystic kidney, pyelonephritis, and glomerulonephritis. With chronic glomerulonephritis, elevation of serum phosphate concentration tends to suppress serum calcium and magnesium levels, leading eventually to a chronic increase in secretion of parathormone and hyperplasia of the parathyroid.

Administration of large doses of vitamin D will also result in elevated serum phosphorus levels, but not as high as that observed with renal dysfunction. During a period of active bone repair, such as the healing of a fracture, some elevation of serum phosphorus levels is usually observed.

With decreased secretion of PTH from the parathyroid gland (hypoparathyroidism), there is decreased excretion of phosphate in the urine and decreased movement of phosphate from the bone (Figure 7.4). Serum phosphorus concentration is generally high.

7.8.2 *Hypophosphatemia*

Low serum phosphate levels were commonly seen with symptoms of rickets in children before the 1930's and the advent of vitamin Denriched milk. The calcium–phosphorus product is approximately 60 in children, since mean calcium levels are 10 mg/100 ml and mean phosphorus levels in children are approximately 6 mg/100 ml. With rickets, due to malnutrition in humans, this product drops to less than 30, due to a lowering of both calcium and phosphate levels. If vitamin D is administered, both levels are increased and healing of the rickets begins with levels above 40. Serum alkaline phosphatase concentration, originating from bone, is elevated in these cases.[121]

Where vitamin D milk is generally available, low serum phosphate levels are rare in children.[122] However, with low alkaline phosphatase levels (*hypophosphatasia*), low serum phosphate values are found. This condition may occur associated with the appearance of ethanolamine phosphate or serine phosphate in the urine, or in the absence of the ethanolamine phosphate.[123] In either case, serum phosphorus concentration is 1.5–2.5 mg/100 ml, as compared to the normal value of 5–8 mg/100 ml. Alkaline phosphatase is at approximately 2 Bodansky units as compared to 6–12 units in the normal infant. The condition where lowered alkaline phosphatase values and low serum phosphorus concentrations occur in the absence of ethanolamine phosphate has been referred to as "*benign hypophosphatasia.*" In these cases, symptoms are ameliorated by oral administration of phosphate.[124]

With hyperparathyroidism, reabsorption of phosphate in the tubules decreases. Bone dissolution results in an increase of serum Ca levels and suppression of serum phosphate levels to values often of the order of 2 mg/100 ml. In many of these cases, renal involvement in the disease process may ensue, and decreased serum phosphate values may not be observed.[120] Thus, if the condition is associated with renal disease, which is not uncommon, then serum phosphorus levels may be normal or even somewhat elevated, associated with a rise in nonprotein nitrogen.[125]

With various conditions leading to *steatorrhea*, such as sprue, celiac disease, and nontropical sprue, calcium is lost in the stool as calcium soaps. It has been pointed out earlier that this also results in decreased phosphate absorption and decreased phosphate in the serum, simulating rickets.

One of the findings in the *Fanconi syndrome* is low serum phosphorus values. This is due to excretion of abnormal amounts of phosphate in the urine, along with certain amino acids, glucose, Na, K, and Ca. The problem is a genetic defect in the reabsorptive capacity of the proximal tubules. A similar syndrome can also be acquired with dysproteinemia and with certain drugs such as tetracycline. The blood picture, as far as Ca and phosphate are concerned, resembles that of rickets, and osteomalacia is often associated with the disease.

Fanconi's syndrome is one of a group of conditions, including cystinosis, Lowe's syndrome, various states of renal insufficiency and other conditions, including those of hereditary idiopathic origin, referred to as *proximal tubular renal acidosis*.[125] In these conditions, along with the loss of sodium and hyperchloremic acidemia, large losses of phosphate to the urine occur because of defects in reabsorption at the proximal tubule.

Distal renal tubular acidosis is a condition which results in a defect in the sodium-for-hydrogen exchange mechanism in the distal renal tubule. Urine pH is then elevated and may reach the alkaline side of the pH scale. Normally, urine pH is approximately 5. As a result, not only do excessive losses of Na and K occur, but also calcium and magnesium are lost. A generalized acidosis, associated with lowered blood CO_2 levels, and elevated chloride ensues. The condition is often associated with skeletal demineralization and nephrocalcinosis. Since phosphate reabsorption is defective in this condition, low serum phosphate levels are observed.[125b]

By infusing phosphate so as to obtain a constant elevated phosphate level, one can measure the effectiveness of the kidney in excreting phosphate and, indirectly, parathyroid activity. The measurement is called the theoretical renal phosphorus threshold (TRPT). This has been proposed as a useful test for measuring the kidney's ability to excrete phosphate,[126,127] and an improvement over the tubular reabsorption test (TRP).

7.9   *PLASMA INORGANIC SULFATE*

Inorganic sulfate circulating in the plasma is of major importance in maintaining normal function in the human. Some of its applications are as follows:

1. Formation of cartilage, particularly in the form of sulfated polysaccharides such as chondroitin sulfate.[128,129]

2. For solubilizing and thus detoxifying phenolic compounds and alcohols both endogenous and exogenous. For example, many of the steroids, such as estrogens and androgens, are excreted as the soluble sulfate esters. If borneol is taken orally, it appears in the urine as bornyl

sulfate. By converting unwanted metabolites to sulfate esters, they lose their physiological function, and may then be excreted in the urine.[130] Indoxyl and skatole are also excreted as their sulfate esters.

3. The "*sulfatides*" are sulfate esters of polysaccharides attached to ceramide so as to form an integral part of the myelin sheath. These are discussed in the section on the *Sphingolipidoses* (Volume 2). These are formed by sulfation, using sulfate originally circulating in the plasma.[131]

4. Several important substances originate from plasma sulfate, such as taurine, which is an integral part of the taurocholic acids; these serve to solubilize the bile acids and thus to perform their function.[132] Sulfate also activates cystathionine synthetase from liver.[133]

5. Sulfate is necessary in order to form lubricants and substances which maintain the viscosity of body fluids, such as the hyaluronic acid of the synovial fluid and saliva.[124]

In spite of its wide application, the amount of inorganic sulfate circulating at any time is only 0.323 ± 0.08 mmol/liter of plasma or 0.6–1.2 mg of sulfate, as inorganic sulfur, per 100 ml.[135–137,140] Erythrocyte inorganic sulfate is reported as less than 0.1 mmol/liter. However, there is no support for this value, obtained many years ago, in subsequent literature, and this needs to be reexamined. In spinal fluid the concentration is approximately the same as in plasma.[138] This result is also open to question.

Some sulfate occurs esterified with circulating carbohydrate or other substances having a hydroxyl group, such as alcohols and phenols.[139] If this sulfur is added to that which occurs as free inorganic sulfate, one finds from 0.35–0.5 mmol/liter of plasma (1.1–1.6 mg of sulfate as sulfur per 100 ml of plasma).

Total sulfur in plasma, including that in the protein, ranges from 25 to 35 with a mean of 30 mmol/liter. Compare this with a value of approximately 1 mmol/liter for sulfur not precipitated by trichloroacetic acid.[140] Most of this sulfur is in the plasma protein in the form of cysteine, cystine, and methionine. It is this mass of sulfur which serves as a reservoir for the generation of inorganic sulfate for conjugation with the various polysaccharides. Table 7.3 shows the distribution of sulfur in the electrophoretic fractions of human serum.[141,142]

The sulfur in the body derives from the protein intake in food along with smaller amounts of esterified (also called ethereal or conjugated) sulfate and inorganic sulfate. The ratio of nitrogen to sulfur in the diet,

TABLE 7.3

Sulfur and Protein Distribution in the Protein Fractions of Human Serum (26 Controls[141,142])

	Protein fractions					Total serum concentration
	Albumin	Alpha$_1$	Alpha$_2$	Beta	Gamma	
Sulfur						
Mean (% of total)	67.2	2.6	7.5	9.7	13.0	
S.D. (±)	5.76	0.60	2.02	1.90	3.36	
Mean mg/100 ml	63.0	2.4	7.0	9.1	12.2	93.7 mg/100 ml ± 4.60 (29.6 mM)
Protein						
Mean (% of total)	55.5	5.5	9.3	11.5	18.0	
S.D. (±)	2.55	0.08	0.99	1.69	2.26	
Mean g/100 ml	3.83	0.38	0.64	0.80	1.24	6.89 g/100 ml ± 0.38
g sulfur/100 g protein	1.64	0.63	1.10	1.14	1.08	1.36

when expressed in the weight of the elements, is approximately 12:1, and that in the urine is also 12:1. This is a result of the fact that *unless the individual is in a period of rapid growth, a balance exists between the intake and outgo of sulfur.* Both the sulfur and nitrogen in the diet derive from protein. From Table 7.3 it can be seen that the percentage of sulfur in the plasma proteins is 1.36. The percentage of nitrogen is 16.25. The ratio of the two is 16.25/1.36 = 11.95. Experimentally, a value of 12.14 was found for this ratio in the urine of healthy adults.[143]

In the urine, the total output of sulfur varies with intake. On a normal diet output of sulfur ranges from 20 to 45 mmol/24 hr (0.6–1.4 g per 24 hr as sulfur). Of this, most is in the form of inorganic sulfate. Esterified sulfate (ethereal sulfate) is from 2 to 4 mmol/24 hr. The unoxidized sulfur (neutral sulfur) ranges from 2 to 6 mmol/24 hr. The neutral sulfur is mainly in the form of amino acids, such as cystine, taurocholic acid, urochrome, and various –SH compounds. Thiosulfate and thiocyanates have been lumped with the neutral fraction of sulfur in the urine. Table 7.4 lists the distribution of sulfur in human urine.

In starvation or malnutrition an imbalance becomes apparent, more sulfur appearing in the urine than in the diet. This is accompanied by general disintegration of the hard tissues of the body, due to the destruction of the mucopolysaccharide sulfates.[144]

TABLE 7.4

Excretion of Sulfur per 24 hr in Urine of Healthy Adults

Fraction	Range, mmol/24 hr	Mode, mmol/24 hr
Inorganic sulfate	16–35	22.4
Sulfate as ester (etheral)	2–4	3.1
Neutral sulfur	2–6	3.4
Total sulfur	20–45	28.9

Of interest is the urine cystine level, which ranges from 8 to 86 mg/24 hr (mean value 26.6).[145,146] With the disease process known as *cystinuria*, one finds values as high as 400–1000 mg excreted of this amino acid. These patients will often deposit cystine in the bladder (cystine stones).[147]

As pointed out earlier, the sulfur in the urine derives originally

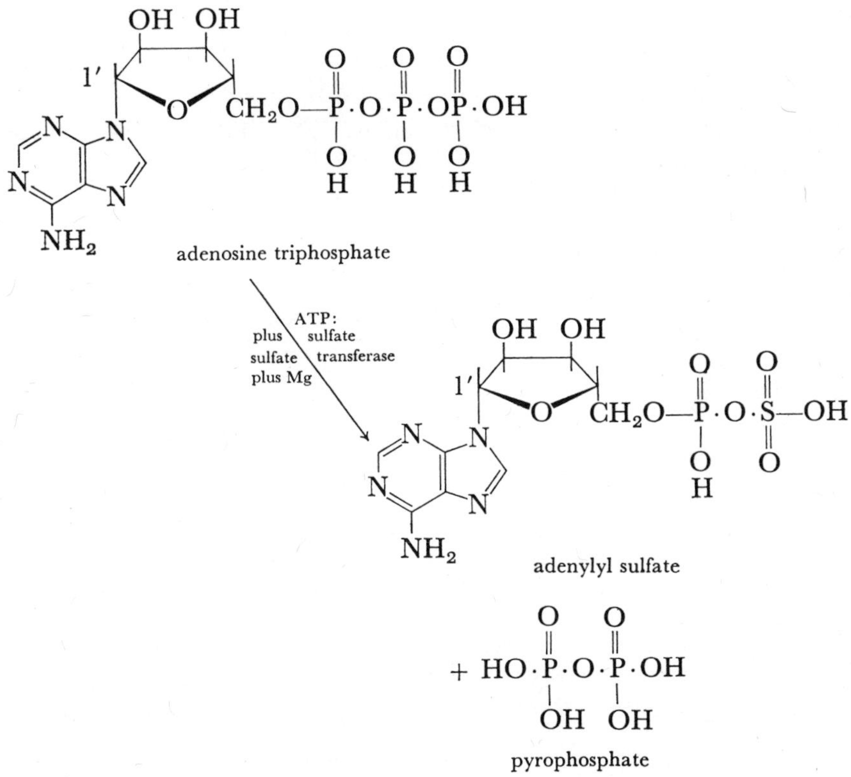

adenosine triphosphate

ATP:
plus sulfate
sulfate transferase
plus Mg

adenylyl sulfate

pyrophosphate

from protein. This is absorbed as the amino acids cysteine and methionine. Thus the ingested sulfur is in the reduced state, and this needs to be oxidized. Reduced sulfur, in the form of inorganic sulfide from the amino acids, reacts with heme in myoglobin, hemoglobin, and the cytochromes to form *sulfhemoglobin*.[148] In the presence of oxygen, this oxidizes to form sulfate. This finds its way into the blood stream as inorganic sulfate, which may be excreted in the urine or utilized for various functions. Any reduced inorganic sulfur that finds its way into the blood stream reacts promptly with hemoglobin to form sulfhemoglobin and, under normal conditions, is eventually oxidized at the lungs. Sulfite is formed intermediate to sulfate and also uses the heme–oxygen system for oxidation.[149]

Sulfate needs to be converted to "active sulfate" before it can form sulfate derivatives with the various organic compounds of biological significance. For this purpose ATP in the presence of ATP:sulfate adenylyl transferase (EC 2·7·7·4) and Mg reacts with sulfate to form adenylyl sulfate and pyrophosphate.

Adenylyl sulfate is then phosphorylated by a second molecule of ATP at the 3′ position of the ribose moiety to form a compound analogous to NADP (TPN). The enzyme that brings this about is a liver transferase, adenylylsulfate kinase or ATP:adenylyl sulfate 3′-phosphotransferase (EC 2·7·1·25),[150,151]

adenylyl sulfate

3′-phosphoadenylylsulfate (PAPS)

The 3′-phosphoadenylylsulfate (PAPS) is the compound that serves to transfer the sulfate radical to various compounds. The enzymes required, PAPS-transferases or sulfotransferases (EC $2 \cdot 8 \cdot 2$) are present in the liver, as are the other enzymes in this series mentioned above. For example, chondroitin sulfate is formed from PAPS using one of these enzymes, (EC $2 \cdot 8 \cdot 2 \cdot c$)[152]:

$$\text{polysaccharide} \cdot \text{OH} + \text{PAPS} \xrightarrow[\text{transferase}]{\text{sulfo-}}$$

$$\xrightarrow[\text{transferase}]{\text{sulfo-}} \text{PAP} + \text{polysaccharide}\text{—}\text{O}\text{—}\text{SO}_3\text{H}$$

Taurine is formed by sulfation with PAPS[153] and not by oxidation of the cysteine –SH group as originally thought. Methionine or serine supplies the carbon atoms of taurine but phosphoadenylylsulfate (PAPS) supplies the sulfate radical. Cysteic acid is an intermediate,

$$\text{PAPS} + \begin{array}{c} \text{CH}_2 \cdot \text{OH} \\ | \\ \text{CH} \cdot \text{NH}_2 \\ | \\ \text{COOH} \end{array} \xrightarrow[\substack{\text{followed by} \\ \text{sulfonation} \\ \text{with transferase}}]{\text{dehydration}} \begin{array}{c} \text{CH}_2\text{SO}_3\text{H} \\ | \\ \text{CHNH}_2 \\ | \\ \text{COOH} \end{array} \xrightarrow{-\text{CO}_2} \begin{array}{c} \text{CH}_2 \cdot \text{SO}_3\text{H} \\ | \\ \text{CH}_2 \cdot \text{NH}_2 \end{array}$$

$$\qquad\qquad\text{serine} \qquad\qquad\qquad\qquad \text{cysteic acid} \qquad\qquad \text{taurine}$$

The formation of the important sulfatides that are components of the myelin sheath are also formed from PAPS by sulfation of ceramide.[154,155]

$$\text{ceramide} \xrightarrow[\text{sulfotransferase}]{\text{PAPS}} \text{sulfatide}$$

7.9.1 Sulfate Transport

Sulfate will pass the membranes of living organisms by both passive and active transport.[156,157] Absorption from the intestines is mediated by an active mechanism which is also associated with transport of sodium ions.[158,159]

The plasma contains a factor that activates the penetration by sulfate and facilitates sulfation of the chondroitin of cartilage. This factor is called the "sulfation factor (SF)." The sulfation factor of plasma[160] originates from the growth hormone. For this reason it is also called *somatomedin*, after somatotropin. Its plasma level is markedly elevated in acromegaly.[161] Concentration of the sulfation factor results in a product that is similar in its characteristics to the growth hormone in its insulin-like action. It is proposed that the effect of human growth hormone on sulfation and also protein synthesis is mediated through somatomedin.[168,169]

Somatomedin is present in extracts of muscle, liver, and brain, as well as in the blood stream. It has been suggested that it is produced in the liver under the influence of the growth hormone.[161] If a liver is perfused with growth hormone, sulfation factor is formed. The question of somatomedin and its relationships to the growth hormone is being explored.

7.9.2 *Sulfatase Activity*

There are three major arylsulfatases $(EC 3 \cdot 1 \cdot 6 \cdot 1)$ of importance in the human. These are designated *A*, *B*, and *C*, depending upon their optimum pH. Optimum pH for *A* is 5.0, for *B* is 6.2, and for *C* is 8.2.

For the acid sulfatase *A* and *B*, catechol sulfate may act as substrate.[162,163] For the alkaline sulfatase, nitrophenyl sulfate is often used as substrate. Deficiency of these sulfatases results in the accumulation of polysaccharide sulfates and these may be recovered in large quantities in the urine. An example is *metachromatic leucodystrophy* (see under this heading, Volume 2).

In addition to these sulfatases, other sulfatases specific for the steroids are found in animal tissues, such as the testes.[164] Phenolphthalein sulfate is commonly used for differentiating among various types of microorganisms by their differences in sulfatase activity.

7.9.3 *Sulfate and Molybdenum*

Several reports seem to indicate that there is a relationship between ceruloplasmin, copper, molybdenum, and sulfate metabolism. This has been extensively investigated in animals by several investigators.[165] This may be related to the fact that molybdenum is an integral part of the sulfite oxidase enzyme. This enzyme contains one atom of molybdenum per heme residue.[166]

7.10   RECAPITULATION

Figure 7.5 summarizes some of the major pathways for sulfate metabolism in the human. Sulfur is taken in the diet mainly as the amino acids cysteine and methionine. It is oxidized to sulfate by the cytochromes, with sulfheme and then sulfite as intermediates, using molybdenum as a cofactor.

Metabolism of food generates ATP, which is converted to PAPS by the various transferases. The pituitary secretes the growth hormone, somatotropin (HGH), which is converted to sulfation factor (SF) in the liver and other tissues. The sulfation factor of the plasma, called

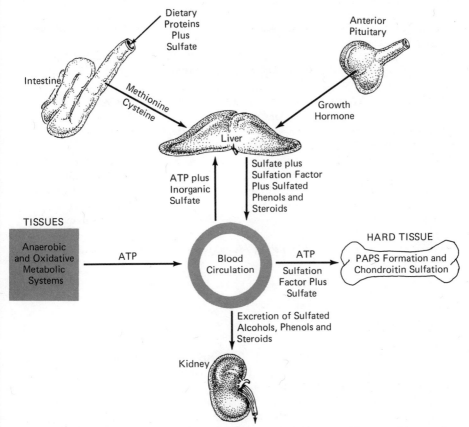

FIGURE 7.5   Proposed mechanism for conversion of dietary sulfur to sulfate and sulfation of chondroitin to form chondroitin sulfate. Certain substances are solubilized and detoxified by sulfation and excretion by the kidney.

somatomedin, then causes movement of sulfate into the tissues and sulfates chondroitin to form chondroitin sulfate. At the liver, the PAPS sulfates various substances such as the estrogens, other phenols, and hydroxy compounds for purposes of detoxification. They are then excreted as their sulfates in the urine.

7.11   SELECTED READING—MAINTENANCE OF CONSTANT ION CONCENTRATION IN BODY FLUIDS: CALCIUM, MAGNESIUM, PHOSPHATE, AND SULFATE

Malm, O. J., *Calcium Requirement and Adaptation in Adult Males*, Universitet, Boston, Massachusetts (1958).

Van Watzer, J. R., *Phosphorus and Its Compounds. Technology, Biological Functions and Applications*, Vol. II, Wiley, New York (1961).

Benesch, R., Ed., *Sulfur in Proteins*, Academic Press, New York (1959).

Bolis, L., Keynes, R. D., and Wilbrandt, W., *Roles of Membranes in Secretory Processes*, North-Holland, Amsterdam (1972).

Nakao, M. and Packer, L., *Organization of Energy-Transducing Membranes*, University Park Press, Baltimore, Maryland (1972).

McLean, F. C., *Calcium and phosphorus metabolism in man and animals with special reference to pregnancy and lactation*, Ann. N.Y. Acad. Sci. **64**:279–462 (1956).

Comar, C. L. and Bronne, F., *Mineral Metabolism*, Vol. III, *Calcium Physiology*, Academic Press, New York (1969).

Jackson, W. P., *Calcium Metabolism and Bone Disease*, Williams and Wilkins, Baltimore, Maryland (1967).

Fourman, P. and Royer, P., *Calcium, Metabolism and the Bone*, 2nd Ed., Davis Co. (1968).

7.12   REFERENCES

1. Eichelberger, L. and McLean, F. C., Distribution of calcium and magnesium between cells and extracellular fluids of skeletal muscle and liver in dogs, *J. Biol. Chem.* **142**:467–476 (1942).

2. Winegrad, S. and Shanes, A. M., Calcium flux and contractility in guinea pig atria, *J. Gen. Physiol.* **45**:371 (1962).

3. Grollman, A. P., Walker, W. G., Harrison, H. C., and Harrison, H. E., Site of reabsorption of citrate and calcium in the renal tubule of the dog, *Am. J. Physiol.* **205**:697–701 (1963).

4. Carafoli, E., Gamble, R. L., and Lehninger, A. L., Rebounds and oscillations in respiration linked movements of Ca^{++} and H^+ in rat liver mitochondria, *J. Biol. Chem.* **241**:2644–2652 (1966).

5. Chance, B. and Johnson, M. L., Hydrogen ion concentration changes in mitochondrial membranes, *J. Biol. Chem.* **241**:4588–4599 (1966).

6. Dale, M. E. and Kellerman, G. M., The binding of calcium by the plasma proteins in hyperparathyroidism, *Clin. Sci.* **32**:433–442 (1967).

7. Yamada, S. and Tonomura, Y., Phosphorylation of the Ca^{2+}–Mg^{2+} dependent ATPase of the sarcoplasmic reticulum coupled with cation translocations, *J. Biochem.* (*Tokyo*) **71**:1101–1104 (1972).

8. Weber, A., Herz, R., and Reiss, I., Study of the kinetics of calcium transport by isolated fragmented sarcoplasmic reticulum, *Biochem. Z.* **345**:329–369 (1966).

9. Ebashi, S. and Lipmann, F., Adenosine triphosphate-linked concentration of calcium ions in a particulate fraction of rabbit muscle, *J. Cell. Biol.* **14**:389–400 (1962).

10. Skou, J. C., The influence of some cations on an adenosine triphosphatase from peripheral nerves, *Biochim. Biophys. Acta* **23**:394–401 (1957); The enzymatic basis for the active transport of sodium and potassium, *Protoplasma* **63**:303–308 (1967).

11. Meissner, G. and Fleischer, S., Characterization of sarcoplasmic reticulum from skeletal muscle, *Biochim. Biophys. Acta* **241**:356–378 (1971).

12. Eletr, S. and Inesi, G., Phospholipid orientation in sarcoplasmic membranes: Spin-label ESR and proton NMR studies, *Biochim. Biophys. Acta* **282**:174–179 (1972).

13. Martonosi, A. and Halpin, R. A., Sarcoplasmic reticulum. X. The protein composition of sarcoplasmic reticulum membranes, *Arch. Biochem. Biophys.* **144**:66–77 (1971).

14. Yu, B. P. and Masoro, E. J., Isolation and characterization of the major protein component of sarcotubular membranes, *Biochemistry* **9**:2909–2917 (1970).

15. Yamada, S. and Tonomura, Y., Reaction mechanism of the Ca^{2+} dependent ATPase of sarcoplasmic reticulum from skeletal muscle, *J. Biochem.* (*Tokyo*) **72**:417–425 (1972).

16. Nakamura, H., Hori, H., and Mitsui, T., Conformational change in sarcoplasmic reticulum induced by ATP in the presence of magnesium ion and calcium ion, *J. Biochem.* (Tokyo) **72**:635–646 (1972).

16a. de Meis, L. and de Mello, M. C. F., Substrate regulation of membrane phosphorylation and Ca^{+2} transport in the sarcoplasmic reticulum, *J. Biol. Chem.* **248**:3691–3701 (1973).

17. Kanazawa, T., Yamada, S., Yamamoto, T., and Tonomura, Y., Reaction mechanism of the Ca^{++} dependent ATPase of sarcoplasmic reticulum. V. Vectoral requirements for Ca and Mg, *J. Biochem.* (*Tokyo*) **70**:95–123 (1971).

17a. Natelson, S., Richelson, M. R., Sheid, B., and Bender, S. L., X-ray spectroscopy in the clinical laboratory. I. Calcium and potassium, *Clin. Chem.* **5**:519–531 (1959).

18. Carruthers, B. M., Copp, D. H., and McIntosh, H. W., Diurnal variation in urinary excretion of calcium and phosphate and its relation to blood levels, *J. Lab. Clin. Med.* **63**:959–968 (1964).

19. Subryan, V. L., Popovtzer, M. M., Parks, S. D., and Reeve, E. B., Measurement of serum ionized calcium with the ion-exchange electrodes, *Clin. Chem.* **18**:1459–1462 (1972).

20. Gley, E., Note on the function of the thyroid gland in the rabbit and dog, *Compt. Rend. Soc. Biol.* **43**:843 (1891).

21. MacCallum, W. G. and Voegtlin, C., On the relation of the parathyroid to calcium metabolism and the nature of tetany, *Bull. Johns Hopkins Hosp.* **19**:91–92 (1908); ibid: *J. Exp. Med.* **11**:118 (1909).

22. Collip, J. B., The extraction of a parathyroid hormone which will prevent or control parathyroid tetany and which regulates the level of blood calcium, *J. Biol. Chem.* **63**:395–438 (1925).

23. Aurbach, G. D., Isolation of parathyroid hormone after extraction with phenol, *J. Biol. Chem.* **234**:3179–3181 (1959).

24. Rasmussen, H. and Craig, L., Purification of parathyroid hormone by use of countercurrent distribution, *J. Am. Chem. Soc.* **81**:5003 (1959).

25. Berson, S. A. and Yalow, R. S., Parathyroid hormone in plasma in adenomatous hyperparathyroidism, uremia, and bronchogenic carcinoma, *Science* **154**:907–909 (1966).

26. Arnaud, C. D., Taso, H. S., and Littledike, T., Radioimmunoassay of human parathyroid hormone in serum, *J. Clin. Invest.* **50**:21–34 (1971).

27. Bélanger, L. F. and Robichon, J., Parathormone induced osteolysis in dogs, *J. Bone Joint Surg.* **46A**:1008–1012 (1964).

28. Doty, S. B. and Talmage, R. V., Stimulation of parathyroid secretion in the absence of the adrenal, thyroid, or pituitary gland, *Gen. Comp. Endocrinol.* **4**:545–549 (1964).

29. Walker, D. G., Lapiere, C. M., and Gross, J. A., A collagenolytic factor in rat bone promoted by parathyroid extract, *Biochem. Biophys. Res. Commun.* **15**:397–402 (1964).

30. Chase, L. R. and Aurbach, G. D., The effect of parathyroid hormone on the concentration of adenosine 3′,5′-monophosphate in skeletal tissue *in vitro*, *J. Biol. Chem.* **245**:1520–1526 (1970).

31. Melson, G. L., Chase, L. R., and Aurbach, G. D., Parathyroid hormone-sensitive adenyl cyclase in isolated renal tubules, *Endocrinology* **86**:511–518 (1970).

32. Klein, D. C. and Raisz, L. G., Role of adenosine-3′,5′-monophosphate in the hormonal regulation of bone resorption: studies in the cultured fetal bone, *Endocrinology* **89**:818–826 (1971).

33. Copp, D. H. and Davidson, A. G. F., Direct humoral control of parathyroid function in the dog, *Proc. Soc. Exp. Biol. Med.* **107**: 342–344 (1961).

34. Egawa, J. and Neumann, W. F., Effects of parathyroid hormone on phosphate turnover in bone and kidney, *Endocrinology* **72**:370–376 (1963); Effect of parathyroid extract on the metabolism of radioactive phosphate in kidney, **74**:90–101 (1964).

35. Chase, L. R. and Aurbach, G. D., Parathyroid function and the renal excretion of 3′,5′ adenylic acid, *Proc. Nat. Acad. Sci. U.S.* **58**:518–525 (1967).

36. Chase, L. R. and Aurbach, G. D., Renal adenyl cyclase: anatomically separate sites for parathyroid hormone and vasopressin, *Science* **159**:545–548 (1968).

37. Copp, D. H., Calcitonin—a new hormone from the parathyroid which lowers blood calcium, *Oral Surg. Oral Med. Oral Pathol.* **16**:872–877 (1962).

38. Hirsch, P. F., Thyrocalcitonin inhibition of bone resorption induced by parathyroid extract in thyroparathyroidectomized rats, *Endocrinology* **80**:539–541 (1967).

39. Pearse, A. G. E. and Carvalheira, A. F., Cytochemical evidence for an ultimobronchial origin of rodent thyroid *C.* cells, *Nature* **214**:929–931 (1967).

40. Matsuzawa, T. and Kurosumi, K., Morphological changes in the parafollicular cells of the rat thyroid glands after administration of calcium shown by electron microscopy, *Nature* **213**:927–928 (1967).

41. Foster, G. V., Doyle, F. H., Bordier, P., Matrajt, H., and Tun-Chot, S., Roentgenologic and histologic changes in bone produced by thyrocalcitonin, *Am. J. Med.* **43**:691–695 (1967).

42. Murad, F., Brewer, H. B., Jr., and Vaughan, M., Effect of thyrocalcitonin on adenosine 3′,5′ cyclic phosphate formation by rat kidney and bone, *Proc. Nat. Acad. Sci. U.S.* **65**:446–453 (1970).

43. Sobel, A. E., Kramer, B., *et al.*, Composition of bones and teeth in relationship to the composition of blood and diet, *J. Biol. Chem.* **158**:475–489 (1945); **159**:159–171 (1945); **176**:1103–1121 (1948); **179**:205–210 (1949).

44. Harris, F., Hoffenberg, R., and Black, E., Calcium kinetics in vitamin D deficiency rickets. *S. Afr. Med. J.* **38**:938 (1964).

45. Schachter, E., Finkelstein, J. D., and Kowarski, S., Metabolism of vitamin D; I. Preparation of radioactive vitamin D and its intestinal absorption in the rat, *J. Clin. Invest.* **43**:787–796 (1964).

46. Gran, F. C., The retention of parenterally injected calcium in rachitic dogs, *Acta Physiol. Scand.* **50**:132–139 (1960).

47. Chen, P. S., Jr. and Lane, K., Serum protein binding of Vitamin D₃. *Arch. Biochem. Biophys.* **112**:70–75 (1965).

48. Morii, H., Lund, J., Neville, P. F., and DeLuca, H. F., Biological activity of a vitamin D metabolite, *Arch. Biochem. Biophys.* **120**:508–512 (1967).

49. Blunt, J. W., DeLuca, H. F., and Schnoes, H. K., 25 hydroxy cholecalciferol, A biologically active metabolite of vitamin D₃, *Biochemistry* **7**:3317–3322 (1968).

50. Corradino, R. A. and Wasserman, R. H., Actinomycin D inhibition of Vitamin D₃-induced calcium binding protein (CaBP) formation in chick duodenal mucosa, *Arch. Biochem. Biophys.* **126**:957–960 (1968).

51. Garabedian, M., Holick, M. F., Deluca, H. F., and Boyle, I. T., Control of 25 hydroxycholecalciferol metabolism by parathyroid glands, *Proc. Nat. Acad. Sci. U.S.* **69**:1673–1676 (1972).

52. Morrissey, R. L. and Rath, Damon F., Purification of human renal calcium binding protein from necropsy specimens, *Proc. Soc. Exp. Biol. Med.* **145**:699–703 (1974).

52a. MacLennan, D. H., Isolation of a second form of calsequestrin, *J. Biol. Chem.* **249**:980–984 (1974).

53. Tanaka, Y. and Deluca, H. F., Bone mineral mobilization activity of 1,25 dihydroxycalciferol, a metabolite of Vitamin D, *Arch Biochem. Biophys.* **146**:574–578 (1971).

53a. DeLuca, F., Vitamin D: The vitamin and the hormone, *Fed. Proc.* **33**:2211–2219 (1974).

54. Pincus, J. B., Natelson, S., and Lugovoy, J. K., Effect of epinephrine, ACTH and cortisone on citrate, calcium glucose and phosphate levels in rabbits, *Proc. Soc. Exp. Biol. Med.* **78**:24–27 (1951).

55. Natelson, S., Rannazzisi, G., and Pincus, J. B., Dynamic control of calcium, phosphate, citrate, and glucose levels in blood serum, *Clin. Chem.* **9**:31–62 (1963).

56. Singer, F. R., Woodhouse, N. J. Y., Parkinson, D. K., and Joplin, G. F., Some acute effects of administered porcine calcitonin in Man, *Clin. Sci.* **37**:181–190 (1969).

57. Pincus, J. H., Feldman, R. G., Rannazzisi, G., and Natelson, S., Electromyographic studies with pituitary extracts which lower serum calcium and raise serum citrate levels, *Endocrinology* **76**:783–786 (1965).

58. Wexler, J. B., Pincus, J. B., Natelson, S., and Lugovoy, J. K., Fate of citrate in erythroblastotic infants treated with exchange transfusion, *J. Clin. Invest.* **28**:478, 481 (1959).

59. Pincus, J. B., Natelson, S., and Lugovoy, J. K., Response of citric acid levels of normal adults and children to intramuscular injection of epinephrine, *J. Clin. Invest.* **28**:741–745 (1949).

60. Natelson, S., Pincus, J. B., and Lugovoy, J. K., Response of citric acid levels to oral administration of glucose. I. Normal adults and children, II. Abnormalities observed in the diabetic and convulsive state, *J. Clin. Invest.* **27**:446–453 (1948).

61. Gottfried, S. P., Natelson, S., and Pincus, J. B., Response of serum citric acid levels in schizophrenics to the intramuscular administration of insulin, *J. Nerv. Mental Dis.* **117**:59–64 (1953).

62. Natelson, S., Pincus, J. B., and Rannazzisi, G., A rabbit serum calcium lowering factor from the pituitary, *Clin. Chem.* **9**:631–636 (1963).

63. Natelson, S., Rannazzisi, G., and Pincus, J. B., Effect of ACTH and Vasopressin on serum calcium and citrate levels in the rabbit, *Endocrinology* **77**:108–113 (1965).

64. Natelson, S., Walker, A. A., and Pincus, J. B., Chlordiazepoxide and diphenyl hydantoin as antagonists to ACTH effect on serum calcium and citrate levels, *Proc. Soc. Exp. Biol. Med.* **22**:689–692 (1966).

65. Vaughan, O. W. and Filer, L. J., Jr., The enhancing action of certain carbohydrates on the intestinal absorption of calcium in the rat, *J. Nutr.* **71**:10–14 (1960).

66. Charley, P. and Saltman, P., Chelation of calcium by lactose. Its role in transport mechanisms, *Science* **139**:1205–1206 (1963).

67. Natelson, S., Klein, M., and Kramer, B., The effect of oral administration of

calcium fructose diphosphate on serum organic phosphate, inorganic phosphate, calcium, protein and citric acid levels, *J. Clin. Invest.* **30**:50–54 (1951).

68. Wasserman, R. H., Comar, C. L., Schooley, J. C., and Lengemann, F. W., Interrelated effects of L-lysine and other dietary factors on the gastrointestinal absorption of calcium 45 in the rat and chick, *J. Nutr.* **62**:367–375 (1957).

68a. Parra-Covarrubias, A., Hypocalcemia during the infusion of arginine to a group of obese adolescents, *Arch. Inv. Med. (Mex.)* **2**:107–114 (1971).

69. Albright, F. A. and Reifenstein, E. C., *The Parathyroid Gland and Metabolic Bone Disease*, Williams & Wilkins, Baltimore, Maryland (1948).

70. Demetrious, J. A. and Beattie, J. M., Electrophoretic separation on agarose thin film of isoenzymes of alkaline phosphatase from human serum and tissue, *Clin. Chem.* **17**:290–295 (1971).

71. Albright, F., Note on the management of hypoparathyroidism with dihydrotachysterol, *J. Am. Med. Assoc.* **112**:2592–2593 (1939).

72. Barney, J. D. and Sulkowitch, H. W., Progress in the management of urinary calculi, *J. Urol.* **37**:746–762 (1937).

73. Nordin, B. E. C., Primary and secondary hyperparathyroidism, *Adv. Intern. Med.* **9**:81–105 (1958).

74. Gordan, G. S., Cantino, T. J., Erhardt, L., Hansen, J., and Lubich, W., Osteolytic sterol in human breast cancer, *Science* **51**:1226–1229 (1966).

75. McGreal, D. A., Idiopathic hypercalcemia syndromes of infancy, *Lancet* **2**:101–110 (1954).

76. Eisenberg, E. and Pimstone, B., Hypophosphatasia in an adult. A case report, *Clin. Orthop.* **52**:199–212 (1967).

77. Whedon, G. D., Osteoporosis, atrophy of disuse, in *Bone as a Tissue*, Rodahl, K., Nicholson, J. T., and Brown, E. M., Eds., McGraw-Hill, New York (1960), pp. 67–82.

78. Bronsky, D., Kushner, D. S., Dubin, A., and Snapper, I., Idiopathic hypoparathyroidism and pseudohypoparathyroidism: Case reports and review of the literature, *Medicine* **37**:317–352 (1958).

79. Lee, J. B., Tashjian, A. H., Jr, Streeto, J. M., and Frantz, A. G., Familial pseudohypoparathyroidism. Role of parathyroid hormone and thyrocalcitonin, *New Eng. J. Med.* **279**:1179–1184 (1968).

79a. Balsan, S. and Garabedian, H., 25-Hydroxycalciferol, a comparative study in deficiency rickets and different types of resistant rickets, *J. Clin. Invest.* **51**:749–759 (1972).

79b. Harrison, H. E. and Harrison, H. C., Dihydrotachysterol, a calcium active steroid not dependent upon kidney metabolism, *J. Clin. Invest.* 1919–1922 (1972).

80. Campbell, P. G. and Moosa, G. M., Hypomagnesemia and magnesium therapy in protein caloric malnutrition, *J. Pediat.* **77**:709–714 (1970).

81. Wilhelm, G., Vitamin D deficiency rickets (a review), *Z. Kinderheilk.* **117**:653–659 (1969).

82. Pincus, J. B., Gittleman, I. F., Sobel, A. E., and Schmerzler, E., Effects of Vitamin D on the serum calcium and phosphorus levels in infants during the first week of life, *Pediatrics* **13**:178–184 (1954).

82a. Tsang, R. C., Light, I. J., Sutherland, J. M., and Kleinman, L. I., Possible pathogenic factors in neonatal hypocalcemia of prematurity, *J. Pediat.* **82**:423–429 (1973).

83. Rogers, M. C. and Bergstrom, W. H., Diet induced hypoparathyroidism. A model for neonatal tetany, *Pediatrics* (Suppl.) **47**:207–210 (1971).

84. Coenegracht, J. M. and Houben, H. G. J., Idiopathic hypomagnesemia with hypocalcemia in an adult, *Clin. Chim. Acta* **50**:349–357 (1973).

85. Levy, H. M. and Ryan, E. M., Evidence that the contraction of actomyosin requires the reaction of adenosine triphosphate and magnesium at two different sites, *Biochem. Z.* **345**:132–147 (1966).

86. Michlrad, A., Kovacs, M., and Hegyi, G., The role of Mg^{++} in the contraction and adenosine triphosphatase activity of myofibrils, *Biochem. Biophys. Acta* **107**:567–578 (1965).

87. Herzberg, L. and Bold, A. W., A sex difference in mean magnesium levels in depression, *Clin. Chem. Acta* **39**:229–231 (1972).

88. Raut, S. J. and Viswanathan, R., Distribution of magnesium in body fluids, *Indian J. Med. Res.* **60**:1272–1277 (1972).

89. Vallee, B. L., Wacker, W. E. C., and Ulmer, D. D., The magnesium deficiency tetany syndrome in man, *New Eng. J. Med.* **262**:155–161 (1960).

90. Smith, W. O., Hammarsten, J. F., and Eliel, L. P., The clinical expression of magnesium deficiency, *J. Am. Med. Assoc.* **174**:77–78 (1960).

91. Battifora, H., Eisenstein, R., Laing, G. H., and McCreary, P., The kidney in experimental magnesium deprivation. A morphologic and biochemical study, *Am. J. Pathol.* **48**:421–428 (1966).

92. Robinson, R. R., Murdaugh, H. V., Jr., and Peschel, E., Renal factors responsible for the hypermagnesemia of renal disease, *J. Lab. Clin. Med.* **53**:572–576 (1959).

93. Prasad, A. S., Flink, E. B., and Zinneman, H. H., The base binding property of the serum proteins with respect to magnesium, *J. Lab. Clin. Med.* **54**:357–364 (1959).

94. Frizel, D. E., Malleson, A. G., and Marks, V., Measurement of plasma ionized calcium and magnesium by ion exchange strip, *Clin. Chim. Acta* **16**:45–66 (1967).

95. Schilli, W., Ochs, G., and Eschler, J., The influence of Mg ions on the viscosity of the human saliva and activation of mucolytic enzymes, *Z. Laryng. Rhinol.* **45**:110–112 (1966).

96. Stachura, J., Morphological investigations of magnesium deprivation and magnesium load in the rat, *Patol. Pol.* **22**:41–53 (1971).

97. Rojo Ortega, J. M., Brecht, H. M., and Genest, J., Effects of magnesium deficient diet on the thyroid C cells and parathyroid gland of the dog, *Virchow's Arch. Ab T. B. Zell. Pathol.* **7**:81–89 (1971).

98. Stokstad, E. R. and Britton, W. M., Aorta and other soft tissue calcification in the magnesium-deficient rat, *J. Nutr.* **100**:1501–1505 (1970).

99. Bunce, G. E., Price, N. O., and Hall, B. L., Reduction in kidney calcification of magnesium deficiency by administration of chlorpromazine, chlorquine, or acetyl salicylic acid, *Nutr. Rep. Int.* **2**:145–152 (1970).

100. Heeschen, W. H., Reichmuth, J., and Tolle, A., Magnesium content of cow's milk, *Wien Tierärtzl. Z.* **60**:55–59 (1973).

101. Salet, J. and Fournet, J. P., Magnesium deficiency in children past the neonatal period, *Sem. Hop. Paris* **47**:39–45 (1971).

102. Kyung, S. C., Studies of serum magnesium levels in healthy children and in various disease states, *Korea Univ. Med. J.* **9**:325–333 (1972).

103. Gittleman, I. F., Pincus, J. B., Kramer, B., Sobel, A. E., and Schmerzler, E., Citric acid metabolism in infants during the neonatal period, *Pediatrics* **15**:124–134 (1955).

104. Lim, P. and Jacob, E., Magnesium status of alcoholic patients, *Metabolism* **21**:1045–1051 (1972); Magnesium deficiency in liver cirrhosis, *Q. J. Med.* **41**:291–300 (1972).

105. Cronberg, S. and Caen, J. P., Mg^{++} and Ca^{++} induced platelet aggregation and ADP, *Thromb. Diath. Haemorrh.* **24**:432–437 (1970).

106. Helve, O., Blood phosphorus distribution in certain internal diseases, *Acta Med. Scand.* **125**:505–522 (1946).

107. Kerr, S. E. and Daoud, L., A study of the organic acid-soluble phosphorus of the erythrocytes of various vertebrates, *J. Biol. Chem.* **109**:301–315 (1935).

108. Wooton, I. D. P. and King, E. J., Normal values for blood constituents. Inter-hospital differences, *Lancet* **1**:470–471 (1953).

109. Spivek, M. L., Microchemical blood standards for normal five-day old newborn infants, *J. Pediat.* **48**:581–587 (1956).

110. Till, U., Brox, D., and Frunder, H., Orthophosphate turnover in the extra-cellular and intracellular space of mouse liver, *Eur. J. Biochem.* **11**:541–548 (1969).

111. Green, D. E. and Brucker, R. F., The molecular principles of biological membrane construction and function, *Biol. Sci.* **22**:13–19 (1972).

112. Brierley, G., Murer, E., Bachmann, E., and Green, D. E., Studies on ion transport. II. Accumulation of inorganic phosphate and magnesium by heart mitochondria, *J. Biol. Chem.* **238**:3482–3489 (1963).

113. Hatase, O., Wakabayashi, T., and Green, D. E., On the correlation of configurational changes in the inner mitochondrial membrane with energization, *J. Bioenergetics* **2**:183–195 (1971).

114. Cori, C. F., Cori, G. T., and Green, A. A., Crystalline muscle phosphorylase III. Kinetics, *J. Biol. Chem.* **141**:39–55 (1943).

115. Murad, F., Strauch, B. S., and Vaughan, M., The effect of gonadotropins on testicular adenyl cyclase, *Biochim. Biophys. Acta* **177**:591–598 (1969).

116. Reiss, E., Canterbury, J. M., Sercovitz, M. A., and Kaplan, E. L., The role of phosphate secretion of parathyroid hormone in man, *J. Clin. Invest.* **49**:2146–2149 (1970).

117. Webster, G. D., Mann, J. B., and Hills, A. G., The effect of phosphate infusions upon renal phosphate clearance in man: Evidence for tubular phosphate secretion, *Metabolism* **16**:797–814 (1967).

118. Yamahiro, H. S. and Reynolds, T. B., Phosphate excretion in normal and hyperparathyroid subjects with controlled phosphate intake, *Metabolism* **11**:213–225 (1962).

119. Smith, D. A. and Nordin, B. S. C., The effect of high phosphorus intake on total and ultrafiltrated plasma calcium and on phosphate clearance, *Clin. Sci.* **26**:479–481 (1964).

120. Agus, Z. S., Puschett, J. B., Senesky, D., and Goldberg, M., Mode of action of parathyroid hormone and cyclic adenosine 3′,5′-monophosphate on renal tubular phosphate reabsorption in the dog, *J. Clin. Invest.* **50**:617–626 (1970).

121. Kramer, B., Shear, M. J., and Siegel, J., Mechanism of healing in low phosphorus rickets, *J. Biol. Chem.* **91**:271–290 (1930).

122. Baretrop, D. and Oppe, T. E., Dietary factors in neonatal calcium homeostasis, *Lancet* **2**:1333–1335 (1970).

123. Vincke, H., Hereditary hypophosphatasia, *Acta Paediat. Belg.* **24**:131–138 (1970).

124. Guibaud, P. and Larbre, F., A case of benign form of hypophosphatasia treated with phosphate, *Pediatrics* **25**:319–330 (1970).

125. Lievre, J. A., Chigot, P. L., and Camus, J. P., Primary hyperparathyroidism symptomatology in a series of 100 cases, *Ann. Med. Interne. (Paris)* **124**:1–8 (1973).

125a. Morris, R. C., Jr., Renal tubular acidosis mechanisms: Classification and implications, *New Eng. J. Med.* **281**:1405–1413 (1969).

125b. Nash, M. A., Torrado, A. D., Greifer, I., Spitzer, A., and Edelmann, C. M., Jr., Renal tubular acidosis in infants and children. Clinical course, response to treatment and prognosis. *J. Pediat.* **80**:738–748 (1972).

126. Irvin, G. L., 3rd, Cohen, M. S., Moehris, R., and Mintz, D. H., Primary hyperparathyroidism. Current diagnosis, treatment and results, *Arch. Surg.* **105**:738–740 (1972).

127. Stamp, T. C. B. and Stacey, T. E., Evaluation of theoretical renal phosphorus threshold as an index of renal phosphorus handling, *Clin. Sci.* **39**:505–516 (1970).

128. Brown, R. G., Changes in aortic extensibility found in sulfate-deprived rats, *J. Nutr.* **92**:399–402 (1967).

129. Herbai, G., Effect of age, sex, starvation, hypophysectomy and growth hormone from several species on the inorganic sulphate pool and on the incorporation *in vivo* of sulphate into mouse costal cartilage. An attempt to study sulphation factor activity *in vivo*, *Acta Endocrinol.* **66**:333–351 (1971).

130. Pulkkinen, M. O., Sulphate conjugation during development, in human, rat and guinea pig, *Acta Physiol. Scand.* **66**:115–119 (1966).

131. Skipski, V. P., Smolowe, A. F., and Barclay, M., Separation of neutral glycosphingolipids and sulfatides by thin-layer chromatography, *J. Lipid Res.* **8**:295–299 (1967).

132. Miraglia, R. J. and Martin, W. G., The synthesis of taurine from sulfate. II. Chick liver phosphoadenosinephosphosulfate transferase, *Proc. Soc. Exp. Biol. Med.* **132**:640–644 (1969).

133. Brown, F. C. and Gordon, P. H., Cystathionine synthase from rat liver: partial purification and properties, *Can. J. Biochem.* **49**:484–491 (1971).

134. Patel, V., Tappel, A. L., and O'Brien, J. S., Hyaluronidase and sulfatase deficiency in Hurler's syndrome, *Biochem. Med.* **3**:447–457 (1970).

135. Miller, E., Hlad, C. J., Jr., Levine, S., Holmes, J. H., and Elrick, H., Use of radioisotopes to measure body fluid constituents. I. Plasma sulfates, *J. Lab. Clin. Med.* **58**:656–661 (1961).

136. Kleeman, C. R., Taborsky, E., and Epstein, F. N., Improved method for determination of inorganic sulfate in biologic fluids, *Proc. Soc. Exp. Biol. Med.* **91**:480–488 (1956).

137. Cuthbertson, D. F. and Tompsett, S. L., Inorganic sulfate content of the blood and its determination, *Biochem. J.* **25**:1237–1243 (1931).

138. Reed, L. and Denis, W., The distribution of the nonprotein sulfur of the blood between serum and corpuscles, *J. Biol. Chem.* **73**:623–626 (1927).

139. Schmidt, E. G., McElvain, N. F., and Bowen, J. J., Plasma amino acids and ether soluble phenols in uremia, *Am. J. Clin. Pathol.* **20**:153–261 (1960).

140. Natelson, S. and Sheid, B., X-ray spectroscopy in the clinical laboratory. II. Chlorine and sulfur. Automatic analysis of microsamples, *Clin. Chem.* **6**:292–313 (1960).

141. Natelson, S. and Sheid, B., X-ray spectroscopy in the clinical laboratory. III. Sulfur distribution in the electrophoretic protein fractions of human serum: Abnormalities observed in certain disease states, *Clin. Chem.* **6**:314–326 (1960).

142 Natelson, S. and Stellate, R., Sulfur content of paper electrophoretic fractions in human serum, *Clin. Chem.* **13**:626–649 (1967).

143. Papadopoulou, D. B., Urinary sulfur partition in normal men and in cancer patients, *Clin. Chem.* **3**:257–262 (1957).

144. Ittyerah, T. R., Urinary excretion of sulfate in kwashiorkor, *Clin. Chim. Acta* **25**:365–369 (1969).

145. Haux, P. and Natelson, S., A method for the determination of cystine in urine, *Clin. Chem.* **16**:366–369 (1970).

146. Frimpter, G. W., Horwith, M., Furth, E., Fellows, R. E., and Thompson, D. D., Insulin and endogenous amino acid renal clearances in cystinuria. Evidence for tubular secretion, *J. Clin. Invest.* **41**:281–288 (1962).

147. Haux, P. and Natelson, S., Identification of renal calculi on micro samples by infra red analysis, *Microchem. J.* **15**:126–137 (1970).

148. Berzofsky, J. A., Peisach, J., and Horecker, B. I., Sulfheme proteins. IV. The stoichiometry of sulfur incorporation and the isolation of sulfhemin, the prosthetic group of sulfmyoglobulin, *J. Biol. Chem.* **247**:3783–3791 (1972).

149. Cohen, H. J. and Fridovich, I., Hepatic sulfite oxidase. The nature and function of the heme prosthetic groups, *J. Biol. Chem.* **246**:367–373 (1971).

150. Hall, M. O. and Straatsma, B. R., The synthesis of 3'-phosphoadenosine 5'-phosphosulfate by retinae and livers of normal and vitamin A-deficient rats, *Biochim. Biophys. Acta* **124**:246–253 (1966).

151. Koizumi, T., Suematsu, T., Kawasaki, A., Hirdmatsu, K., and Iwabori, N., Synthesis and degradation of active sulfate in liver, *Biochim. Biophys. Acta* **184**:106–113 (1969).

152. Bostrom, H., Friberg, U., Larsson, K. S., and Nilsonne, U., *In vitro* incorporation of S^5-sulfate in chondrosarcomatous tissue, *Acta Orthop. Scand.* **39**:58–72 (1968).

153. Sass, N. L. and Martin, W. G., The synthesis of taurine from sulfate. III. Further evidence for the enzymatic pathway in chick liver, *Proc. Soc. Exp. Biol. Med.* **139**:755–761 (1972).

154. Pleasure, D. E. and Prockop, D. J., Myelin synthesis in peripheral nerve *in vitro*: Sulphatide incorporation requires a transport lipoprotein, *J. Neurochem.* **19**:283–295 (1972).

155. Herschkowitz, N., McKhann, G. M., Saxena, S., Shooter, E. M., and Herndon, R., Synthesis of sulphatide containing lipoproteins in rat brain, *J. Neurochem.* **16**:1049–1057 (1969).

156. Dreyfus, J., Characterization of a sulfate and thiosulfate transporting system in *salmonella typhinurium*, *J. Biol. Chem.* **239**:2292–2294 (1964).

157. Singh, S. P. and McKenzie, J. M., ^{35}S sulfate uptake by mouse harderian gland: Effect of serum from patients with Graves' disease, *Metabolism* **20**:422–427 (1971).

158. Robinson, J. W. L., Species differences in the intestinal response to sulphate ions, *FEBS Letters (Amst.)* **5**:157–160 (1969).

159. Berry, R. K., Hansard, S. K., Ismail, R. J., and Wysocki, A. A., Absorption deposition and placental transfer of sulfate sulfur by gilts, *J. Nutr.* **96**:399–408 (1969).

160. Hall, K. and Uthne, K., Some biological properties of purified sulfation factor (SF) from human plasma, *Acta Med. Scand.* **190**:137–143 (1971).

161. Van Den Brande, J. L., Van Wyk, J. J., Weaver, R. P., and Mayberry, H. E., Partial characterization of sulphation and thymidine factors in acromegalic plasma, *Acta Endocrinol. (Kbh.)* **66**:65–81 (1971).

162. Jerfy, A. and Roy, A. B., The sulphatase of ox liver. XII. The effect of tyrosine and histidine reagents on the activity of sulphatase A, *Biochim. Biophys. Acta* **175**:355–364 (1969).

163. Bowen, D. M. and Radin, N. S., Hydrolase activities in brain of neurological mutants: cerebroside galactosidase, nitrophenyl galactoside hydrolase, nitrophenyl glucoside hydrolase and sulphatase, *J. Neurochem.* **16**:457–460 (1969).

164. Notation, A. D. and Ungar, F., Rat testis steroid sulfatase. II. Kinetic study, *Steroids* **14**:151–159 (1969).

165. Marcilese, N. A., Ammerman, C. B., Valsecchi, R. M., *et al.*, Effect of dietary molybdenum and sulfate upon urinary excretion of copper in sheep, *J. Nutr.* **100**:1399–1406 (1970).

166. Cohen, H. J., Fridovich, I., and Rajagopalan, K. V., Hepatic sulfite oxidase. A functional role for molybdenum, *J. Biol. Chem.* **246**:374–382 (1971).

167. Hallick, R. B. and DeLuca, H. F., Metabolites of dihydrotachysterol in target tissues, *J. Biol. Chem.* **247**:91–97 (1972).

168. Schimpff, R. M., Quantitative determination of somatomedin in human serum (^{35}S uptake by embryonic chick cartilage), *Biomedicine* **19**:142–147 (1973).

169. Uthne, K., Human somatomedins. Purification and studies of their biological actions, *Acta Endocrinol. (Suppl.)* **175**:1–35 (1973).

Mechanisms for Maintenance of a Steady State Between Plasma and Interstitial Fluid

The interstitial fluid bathing the cells must be constantly replenished and readjusted in its composition, since it functions as a source of nutrition for the cells. The circulating blood supplies the means for this adjustment by supporting continuous exchange of fluid and solutes between blood and interstitial fluid in the capillary bed.[1-3] The mechanisms for this action will now be examined.

8.1 THE CAPILLARY BED

Large blood vessels are rather impermeable, even to water, but the capillary behaves as a semipermeable membrane. The distinctive patterns of the capillary bed vary in different regions of the body, but usually begin with an arteriole lined by smooth muscle cells.[4] From these vessels, fine capillaries branch out haphazardly, which collect and feed into small venules. These in turn communicate with a larger vein, lined on the inside with valves to prevent backflow of blood.

The fine capillaries are permeable to water and salts, and approximately 0.2% of the plasma proteins diffuse here into the interstitial fluid.[3a] This is represented schematically in Figure 8.1.

In certain organs, especially liver and spleen, there are well-

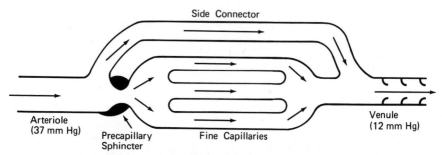

Figure 8.1   Schematic representation of capillary bed.

defined, capillary-like central channels serving as thoroughfares from arteriole to venule. In addition, side connectors in which flow is regulated by the precapillary sphincter tone also link the arteriole to the venule.[4] Flow though the capillary bed is regulated by vasomotion.[5,6] The precapillary sphincters open and close, depending upon pressure in the feeding arteriole, blood volume, and accumulation of metabolites or variation in O_2 tension in the tissues. The myriad of fine capillaries provide a high ratio of surface area to blood flow, thus supporting an active exchange of fluid and solute.

Larger arterioles are highly elastic and can vary their diameter by as much as 20%. The fine capillary, however, behaves as a rigid cylinder, varying its diameter in response to changes in blood flow and pressure by less than 0.2%.[2]

The major forces involved in the movement of fluid and solute out of and back into the capillary lumen are *blood pressure, the difference in osmotic pressure on either side of the capillary wall, and tissue tension* applied by the interstitial tissue and fluid which encases the capillary. This concept, presented by Starling,[7] is illustrated in Figure 8.2. Here the pressure decreases progressively along the course of the capillary until it becomes lower at the venous end. Thus the amount of ultrafiltrate entering the interstitial space decreases progressively as one moves from the arteriole to venule.[3] Tissue tension may be assumed to remain constant, and opposes the outflow of fluid with the same force all along the capillary. The osmotic pressure of the ultrafiltrate and plasma is essentially the same as far as salts are concerned. However, we have a concentration of 7% protein within the capillary blood and only approximately 0.02% protein in the ultrafiltrate. For this reason a differential of colloid osmotic pressure of about 22 mmHg exists, tending to effect a movement of water into the capillary.

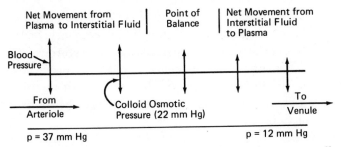

Figure 8.2 Balance between blood pressure and osmotic pressure allowing an ultrafiltrate of plasma to enter the interstitial space and return them to the blood capillary.

The net force F of the flow to the interstitial fluid is equal to the blood pressure BP plus the colloid osmotic pressure of the interstitial fluid π_i, minus the colloid osmotic pressure of the blood π_B, minus the tissue tension T:

$$F = BP + \pi_i - \pi_B - T$$

From Figure 8.2 it is apparent that a point is reached where the flow out of the capillary is equal to the flow back into the capillary. The mean systemic arterial blood pressure is normally 90–100 mmHg in the larger arteries, but is reduced to 37 mmHg by the time blood enters the fine capillary. Nevertheless, it still exceeds the colloid osmotic pressure. When a balance is reached between opposing forces, the net flow is zero. Beyond this point the fluid flows in the opposite direction, back into the capillary lumen.

The return of the plasma ultrafiltrate from the interstitial fluid is not entirely complete, and if no additional system existed, most of the water in blood vessels would end up in the interstitial fluid and blood would become a sludge. *Some mechanism must exist to return this excess water back to the blood stream.* This is done by a *lymphatic system* of drainage.

8.2 THE LYMPHATIC SYSTEM

The interstitial fluid represents the medium in which the body cells are suspended. It donates nutrients and maintains a constant environment around these cells.[18] This fluid also removes protein and foreign substances which find their way into the interstitial spaces.[8,9]

To perform these functions, a fine network of thin-walled tubules, called lymph capillaries, is dispersed in the blood capillary beds.[10,10a,11] Their porosity to relatively large particles, even erythrocytes, has been demonstrated with the electron microscope.[12] A pressure differential exists between the fluid and the lymph capillary. In the mouse's ear, pressure has been measured to be 1.9 cm H_2O in the interstitial fluid and 1.2 cm H_2O in the lymph capillary.[13]

These fine lymph capillaries contain valves which prevent backflow of fluid, so that alternate pressure and release of pressure from the pulsating action of the arterioles and muscular movement causes a peristaltic movement of fluid in the lymph capillaries.[10,11] In this way, they act in a manner similar to the sampler of the Autoanalyzer instrument.

The majority of the lymph channels in the body converge to form the thoracic duct, which empties via a valvular arrangement into the subclavian veins in the upper thorax. Additional small lymphatico-venous communications occur throughout the body, especially in the area of the kidney.[10] The major functions of the lymphatic system are to provide avenues of drainage, absorption, and transport throughout the body, as well as to allow a continuous circulation of the interstitial fluid. A schematic representation of this process is shown in Figure 8.3.

At various points in its course, the contents of these lymph capillaries percolate through bean-shaped structures called lymph nodes (Figures 8.4 and 8.5).

8.2.1. *Drainage*

As pointed out above, under normal conditions, a certain amount of protein and plasma ultrafiltrate enters the interstitial tissue spaces by diffusion from the blood capillaries. If this material were allowed to accumulate, mechanical back pressure would develop in the interstitial spaces. This circumstance can be readily produced experimentally by blocking the thoracic duct, resulting in pileup of interstitial fluid and stoppage of normal filtration of the plasma, producing a condition called *edema* (see page 205).

In the normal individual, drainage of the interstitial spaces by the

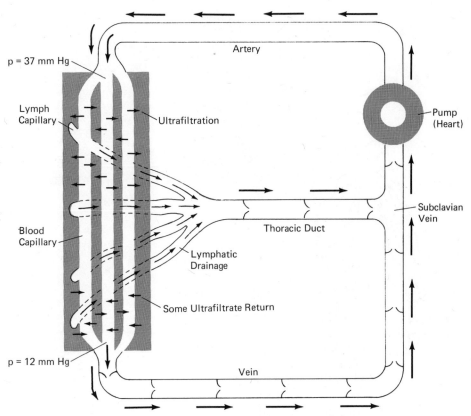

FIGURE 8.3 Schematic representation of the circulation of the interstitial fluid. Ultrafiltrate from the proximal capillary wall and cell secretions and excretions make up the interstitial fluid. Some interstitial fluid ultrafiltrate returns to blood circulation at the distal end of the capillary. Excess interstitial fluid is taken up by the porous lymph capillary and returned to the blood at the subclavian vein.

lymphatic system is sufficient to prevent the accumulation of interstitial fluid, or edema. If one considers that approximately 10% of the fluid passing through the arteries does not enter the veins directly, but travels via the interstitial spaces and the lymphatic system, the significance of this route of drainage becomes apparent.[11] When the lymphatic drainage of the kidney is interrupted, for example, the interstitial protein concentration of the renal medulla rises sharply, and renal function is rapidly impaired. Similarly, the function of other organs is compromised with impaired lymph drainage.

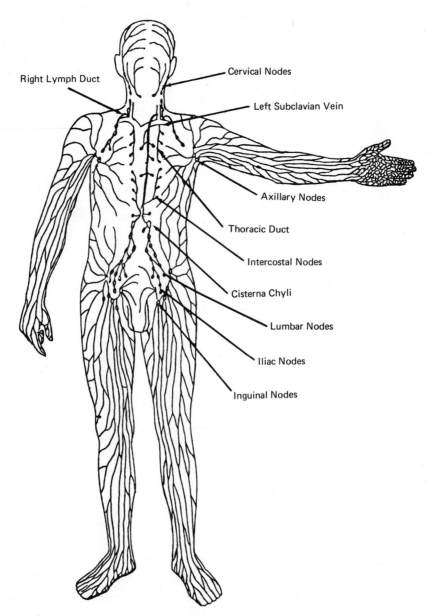

FIGURE 8.4   The lymphatic drainage system.

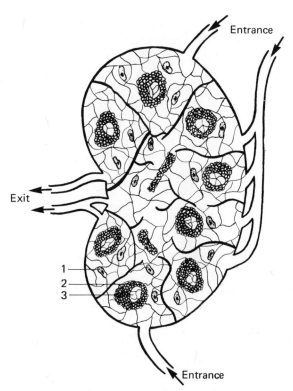

FIGURE 8.5   Schematic representation of a lymph node, the functional unit of the lymphatic system. Cells of the reticuloendothelial system are seen to line the coarse meshwork within the node. They serve to cleanse the lymph as it flows through the node on its way to the thoracic duct and to generate the lymphocytes. The germinal centers surrounded by the many lymphocytes they have produced are also shown. 1. Reticuloendothelial cell. 2. Germinal center. 3. Lymphocytes.

8.2.2   *Absorption*

The lymph ducts that enter the intestinal villi become enlarged and are called lacteals (Figure 8.6). They provide the principal route of absorption of long-chain fatty acids and cholesterol from the intestinal lumen.[14-16] As the digest enters the lacteal, stimulation by an intestinal hormone, *villikinin*, causes the tubular walls to contract, forcing the mixture toward the thoracic duct. Normally, the fluid in the thoracic duct is clear, but after a meal, it becomes milky, due to the absorption

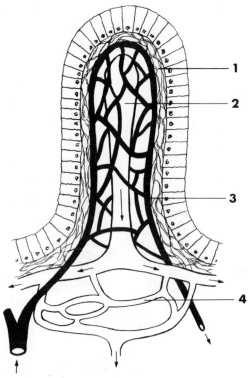

FIGURE 8.6 An intestinal villus, showing its single lacteal draining to a lymphatic plexus on its way to the lymph nodes and finally the thoracic duct. The rich network of capillaries which permeates the villus is also shown. 1. Connective tissue. 2. Lacteal. 3. Capillary network. 4. Lymphatic plexus.

of triglyceride-containing chylomicrons. The short-chain fatty acids can be directly absorbed into the blood stream.

Tests for the integrity of this absorptive system are the *Fat* and *Vitamin A* absorption tests. Lymphatic absorption also is important after intramuscular and subcutaneous injections of substances, such as high molecular weight proteins, that are too large to be absorbed directly through the capillary walls.[3a,14,15]

8.2.3 *Transport*

The lymphatic system also functions in the transport of endocrine secretions, particularly from the thyroid gland, and perhaps may also carry renin from the juxtaglomerular apparatus of the kidney into the blood stream.[17,18]

Transport of antigens to the antibody-forming sites located in the lymph nodes is also performed by the lymphatic system. The lymph nodes (Figure 8.5) are specialized filters that are saturated with phagocytic cells, and contain the *germinal centers*, which are the major sites of lymphocyte formation in the body. If a bacterial cell enters the lymph, it is carried to the lymph node where, after lysis of the cell by the phagocytic cells, antibodies against its constituent proteins are made.

The average rate of flow of lymph in the thoracic duct is 100 ml/hr. Lymphatic flow can be stimulated by muscular contractions or elevated tissue pressures, and directly by epinephrine and norepinephrine acting on the lymphatic channels.[19]

The molecular concentrations in lymph are similar to those found in an ultrafiltrate of plasma, but the average protein content is greater, being 3–4 g/100 ml.[22,23]

The composition of the lymph varies in different parts of the body. Generally, it resembles the interstitial fluid of the area it drains. From the liver, where the proteins are synthesized, and where the central capillary channels of the blood stream are very porous, the protein composition is from 60 to 80% of the protein content of the plasma. From the intestines, it reflects the composition of the meal eaten. From the muscles and skin, its protein content is approximately 40% of that found in the plasma. Fasting thoracic duct lymph containing a mixture from various sources resembles dilute plasma. The electrophoretic and immunoelectrophoretic patterns of lymph proteins are almost identical with those obtained with serum.[19]

Turbid intestinal lymph collected during the absorption of a fatty meal is called *chyle*. It is rich in protein-bound lipids and colloidal particles called *micelles*, which contain bile salts, lipid, and protein. Damage to the thoracic duct can cause this material to empty into the potential spaces of the pleura, peritoneum, pericardium, or to the skin, instead of into the veins. This can result in large protein losses. These *chylous effusions*, as they are called, usually contain 3–4% protein and are rich in lipids. These are often referred to as *exudates*, as distinct from fluids that have passed a membrane, called *transudates*. Transudates would necessarily have a low protein content; total protein is often called for in order to help to explain the origin of the fluid.

TABLE 8.1

Composition of Thoracic Duct Lymph in a Fasting Human[20–23]

Component	Concentration (Range)	Component	Concentration (Range)
Na	127 (118–132) mEq/liter	Cholesterol (T)	72 (34–106) mg/100 ml
K	4.7 (3.9–5.6) mEq/liter	Cholesterol (free)	34 (0.5–51) mg/100 ml
Cl	97 (87–193) mEq/liter	Total lipid	1100 mg/100 ml ± 10%
Ca	4.2 (3.4–5.6) mEq/liter	Glucose	136 mg/100 ml ± 10%
P (inorganic)	2.5 (2.0–3.6) mEq/liter	Alkaline phosphatase	1.9 U/ml (Bodansky) ± 15%
Acid phosphatase	0.7 U/ml (Bodansky)	Transaminase (GOT)	<12 U/ml (Reitman)
Aldolase	10 U/ml (Friedman) ± 15%	Nitrogen (nonprotein)	46 (14–139) mg/100 ml
Bilirubin	<0.8 mg/100 ml	Creatinine	3.0 (0.8–8.9) mg/100 ml
Uric acid	4.5 (1.7–10.8) mg/100 ml	Total protein	4.9 (2.9–7.3) g/100 ml

Typical Electrophoretic Protein Pattern (Paper)[23]

Albumin, %	Globulins, %			
	α_1	α_2	β	γ
58.3	6.4	6.2	9.9	19.2

Table 8.1 lists typical values found in the thoracic duct lymph in the fasting human.

8.3 *ABNORMAL DISTRIBUTION OF EXTRACELLULAR FLUID (EDEMA)*

Normally the interstitial fluid exists in three phases. One is the *liquid phase*, which includes fluids such as synovial fluid, cerebrospinal fluid, the aqueous humor of the eye, and the lymph. The second phase is the ground substance found in cartilage, bony matrix, and connective tissue, and made up of polymers of carbohydrates complexed with protein.[24] This material can absorb large quantities of fluid and vary greatly in the degree of hydration.[25] This is the *solid phase* of the interstitial fluid. A third phase could be considered as the *gel phase*. Fluid in the vitreous humor of the eye and in the umbilical cord (Wharton's jelly) are examples of this type. The major gelling agent is hyaluronic acid and related polysaccharides combined with protein. All of these three phases of interstitial fluid are normally isotonic to each other and remain fairly constant in the normal individual, both in volume and distribution.

Any major increase in volume or interference with the normal circulation of the interstitial fluid will cause it to pool noticeably, either in a particular area or more diffusely throughout the body tissues.[25,26] This accumulation of interstitial fluid is called *edema*.

There are numerous causes for localized edema. One example is *lymphatic obstruction*. If this should occur in one lower extremity, the leg will become swollen. A blow in a particular area, resulting in rupture of blood and lymph capillaries, may result in edema localized in the area of the blow. Tissue tension will also be reduced, in this case, encouraging excessive capillary effusion.

Any form of generalized edema usually requires salt and, hence, water retention by the kidney, but does not necessarily imply intrinsic renal disease.

Cirrhosis of the liver results in a decreased albumin synthesis and a lowering of the plasma proteins, particularly albumin. As discussed above (Figure 8.2), plasma colloid osmotic pressure is a major factor promoting the normal return of interstitial fluid into the blood capillary

TABLE 8.2

Major Causes for Edema Formation in the Human

Type	Nature of the cause	Clinical condititon
A. Hydrostatic	*Interference with normal venous return* results in an increased venous pressure; although a lesser percentage of the capillary filtrate is returned in the lymph, the absolute return may be greater	1. Varicosities of the lower extremities 2. Occlusion of the hepatic veins (Budd–Chiari syndrome) 3. Portal venous hypertension and post-sinusoidal block in hepatic cirrhosis
B. Osmotic	*Hypoproteinemia*, causing increased diffusion from arterial capillary and decreased reabsorption in the venous capillary	1. Nephrotic syndrome 2. Malnutrition 3. Protein-losing enteropathy 4. Hepatic cirrhosis
C. Increased resorption or decreased sodium excretion by the renal tubules	*Increased total body sodium*, resulting in an increased sodium content of the extracellular fluid and increased water retention in order to maintain osmotic balance	1. Reduced glomerular filtration rate with or without parenchymal renal disease 2. Nephrotic syndrome 3. Congestive heart failure 4. Hepatic cirrhosis

D.	Lymphatic obstruction	*Decreased lymphatic drainage* from the affected area	1. Acquired and congenital lymphangiectasis
E.	Change in ground substance hydrophilia	*Increased hydration of ground substance* due to impaired synthesis of polysaccharide–protein complex resulting in local increase in osmostic pressure and decreased return to lymph; can be caused by local injury	1. Myxedema and thyrotoxicosis 2. Edema during menstrual cycle
F.	Hormonal	*Secretion of estrogens during menstrual cycle,* excessive secretion of aldosterone causing an increase in the distal tubular resorption of sodium	1. Observed during menstrual cycle 2. Hyperaldosteronism
G.	Neurogenic	*Damage to the osmoreceptors and stretch sensors* of the brain and centers controlling hormone secretion	1. Generalized edema due to damage to hypothalamus
H.	Idiopathic	Unexplained origin	1. Commonly occurs in women, probably hormonal in origin

at its distal portion. Albumin, at 40 g/liter, accounts for a major proportion of the colloid osmotic pressure. Thus, with decreased plasma albumin, the increased capillary effusion places a greater load on the lymphatic drainage system.[25,27]

Generalized edema frequently results from *hypoalbuminemia*.[28] This occurs because the increase in interstitial fluid results at the expense of a reduction in circulating blood volume. The subsequent hypovolemia reduces renal blood flow and glomerular filtration rate, decreasing the filtered load of sodium. The reduction in blood volume then stimulates secretion of *aldosterone* from the adrenal cortex. This mineralocorticoid acts on distal renal tubule to promote sodium and hence water reabsorption. Extreme edema may result before blood volume is able to approach normal.[29]

A similar circumstance occurs with the *nephrotic syndrome*, where marked protein loss takes place in the urine. Protein loss in stool may also be of sufficient magnitude to reduce the plasma protein concentration, resulting in generalized edema.[30–32] Aside from primary intestinal disease, such a condition also can occur in certain patients with congestive heart failure and constrictive pericarditis. Here increased venous pressure can interfere with the normal discharge of the lymph into the venous blood, simulating partial lymphatic obstruction. Moreover, the increased pressure in the lymph ducts will cause dilatation of the lacteals in the intestinal villi, interfering with the absorption of protein and causing excessive loss of albumin in the stool.

The mechanism of edema in heart failure is complex. Reduction of glomerular filtration rate, arterial shunts in the kidney, and elevated circulating aldosterone levels all play a role, the net result being sodium and water retention.

Certain normal processes affect the accumulation and distribution of edema. During the menstrual cycle, secretion of the female sex hormones will initially result in edema. Estrogens will stimulate release of histamine in the uterus, causing vasodilation and local edema. This is normally useful in storing salts and water for the impending fetus. Progesterone is an antagonist of aldosterone and its administration will increase sodium excretion and thus water loss.

It is interesting to note that certain hormones will cause changes in the concentration and nature of polysaccharide–protein complexes supporting the tissues, causing increased retention of water in the ground substance. The *thyroid stimulating hormone* (TSH) has this effect, and this type of edema occurs in *myxedema and thyrotoxicosis*.[33] Ascites may also result from severe myxedema.[34]

In certain instances no cause for periodic edema can be identified. These are referred to as the *"idiopathic type"* and are probably related to certain, presently undefined hormonal changes.[35]

In Table 8.2 the basic causes of edema are classified. One should keep in mind that separation of the various types of edema is difficult, because extensive overlap exists in their pathophysiology.

8.4   RECAPITULATION

At the capillary bed, an ultrafiltrate of plasma is forced against the colloid osmotic pressure into the interstitial spaces. A portion of this returns at the distal end of the capillary bed, where the colloid osmotic pressure is greater than the blood pressure. The net flow is positive to the interstitial spaces. Porous lymph capillaries disposed in the interstitial spaces serve to return the excess filtrate to the blood stream via the thoracic duct. The net result is a balance in the volume of interstitial fluid. Proteins, synthesized especially in the liver, reach the blood stream via the lymph system. Colloidal particles absorbed by the villi in the intestinal tract also reach the blood stream via the lymphatics. *A closed system thus exists for maintenance of balance in the composition and volume of the interstitial fluid.*

8.5   SELECTED READING—STEADY STATE BETWEEN PLASMA AND INTERSTITIAL FLUID; LYMPH, EDEMA

Rusznyak, I., Földi, M., and Szabó, G., *Lymphatics and Lymph Circulation. Physiology and Pathology*, 2nd Ed., Pergamon, New York (1968).

Mayerson, H. S., Ed., *Lymph and the Lymphatic System* (Proc. Conf. on Lymph and the Lymphatic System), C. C. Thomas, Springfield, Illinois (1968).

McMaster, P. D., Ed., Lymph, *Ann. N.Y. Acad. Sci.* **46** (8):679–882 (1946).

Yoffey, J. M. and Courtice, F. C., *Lymphatics, Lymph and Lymphoid Tissue*, Harvard Univ. Press, Cambridge, Massachusetts (1956).

Moyer, J. H. and Fuchs, M., Ed., *Edema, Mechanisms and Management*, Saunders, Philadelphia, Pennsylvania (1960).

Battezzati, M. and Donini, I., *Lymphatic System* (translated from Italian), Halsted Press, Wiley, New York (1973).

8.6 *REFERENCES*

1. Zweifach, B. W. and Intaglietta, M., Fluid exchange across the blood capillary interface, *Fed. Proc.* **25**:1784–1788 (1966).
2. Burton, A. C., Role of geometry, of size and shape, in the microcirculation, *Fed. Proc.* **25**:1753–1760 (1966).
3. Renkin, E. M., Capillary blood flow and transcapillary exchange, *Physiologist* **9**:361–366 (1966).
3a. Arturson, G., Areskog, N. H., and Arfors, K., The transport of macromolecules across the blood lymph barrier. Influence of capillary pressure on macromolecular composition of lymph. *Bibl. Anat.* **10**:228–233 (1969).
4. Rhodin, J. A. G., The ultrastructure of mammalian arterioles and precapillary sphincters, *J. Ultrastruct. Res.* **18**:181–223 (1967).
5. Mellander, S., The regulation of blood flow, *Proc. Roy. Soc. Med.* **61**:55–61 (1968).
6. Johnson, P. C. and Wayland, H., Regulation of blood flow in single capillaries, *Am. J. Physiol.* **212**:1405–1415 (1967).
7. Starling, E. H., On absorption of fluids from the connective tissue spaces, *J. Physiol.* **19**:312–326 (1896).
8. Hyman, C., Physiologic function of the lymphatic system, *Cancer Chemotherap. Rep.* **58**:25–30 (1969).
9. Langgard, H., The subcutaneous absorption of albumin in edematous states, *Acta Med. Scand.* **174**:645–650 (1963).
10. Threefoot, S. A., Gross and microscopic anatomy of the lymphatic vessels and lymphaticovenous communications, *Cancer Chemotherap. Rep.* **58**:1–20 (1968).
10a. Lauweryns, J. M., The blood and lymphatic microcirculation of the lung. *Pathology Annual* **6**:365–415 (1971).
11. Mayerson, H. S., Lymph and Lymphatics, *Circulation* **28**:839–842 (1963).
12. Courtice, F. C., Harding, J., and Steinbeck, A. W., The removal of free red blood cells from the peritoneal cavity of animals, *Aust. J. Exp. Biol. Med. Sci.* **31**:215–225 (1953).
13. McMaster, P. D., Conditions in the skin influencing interstitial fluid movement, lymph formation and lymph flow, *Ann. N.Y. Acad. Sci.* **46**:743–787 (1946).
14. Ludwig, J., Linhart, P., and Baggenstoss, A. H., Hepatic lymph drainage in cirrhosis and congestive heart failure, *Arch. Pathol.* **86**:551–562 (1968).
15. Bloom, B., Chaikoff, I. L., and Reinhardt, W. O., Intestinal lymph as pathway for transport of absorbed fatty acids of different chain lengths, *Am. J. Physiol.* **166**:451–455 (1951).
16. Hellman, L., Frazell, E. L., and Rosenfeld, R. S., Direct measurement of cholesterol absorption via the thoracic duct in man, *J. Clin. Invest.* **39**:1288–1294 (1960).
17. Daniel, P. M., Gale, M., and Pratt, O. E., Hormones and related substances in the lymph leaving four endocrine glands—The testis, ovary, adrenal and thyroid, *Lancet* **1**:1223–1234 (1963).
18. Lever, A. F. and Peart, W. S., Renin and angiotensin-like activity in renal lymph, *J. Physiol.* **160**:548–563 (1962).

19. Doemling, D. B. and Steggerda, F. R., Stimulation of thoracic duct lymph flow by epinephrine and norepinephrine, *Proc. Soc. Exp. Biol. Med.* **110**:811–813 (1962).
20. Werner, B., The biochemical composition of the human thoracic duct lymph, *Acta Clin. Scand.* **132**:63–76 (1966).
20a. Ianlez, L. E., Guimaraes, F. R., and Sabbaga, E., Biochemical composition of thoracic duct lymph fluid in chronic renal patients, *Rev. Hosp. Clin. Fac. Med. S. Paulo* **28**:141–146 (1973).
21. Szabó, G., Gergely, J., and Magyar, Z., Immunoelectrophoretic analysis of the lymph, *Experientia* **19**:98–99 (1963).
22. Blomstrand, R., Nilsson, I. M., and Dahlback, O., Coagulation studies on human thoracic duct lymph, *Scand. J. Clin. Lab. Invest.* **15**:248–254 (1963).
23. Bartlett, R. H., Falor, W. H., and Zarafonetis, C. J. D., Lipid studies on human lymph and chyle, *J. Am. Med. Assoc.* **187**:126–127 (1964).
24. Gersh, I. and Catchpole, H. R., The nature of ground substance of connective tissue, *Perspect. Biol. Med.* **3**:282–319 (1960).
25. Seki, K., Yamane, Y., Shinoura, A., Koide, K., Uechi, M., Mori, K., Nagasaki, M., and Yoshitoshi, Y., Experimental and clinical studies on the lymph circulation, *Am. Heart. J.* **75**:620–629 (1968).
26. Losowsky, M. S., Alltree, E. M., and Atkinson, M., Plasma colloid osmotic pressure and its relation to protein fractions, *Clin. Sci.* **22**:249–257 (1962).
27. Hollander, W., Reilly, P., and Burrows, B. A., Lymphatic flow in human subjects as indicated by the disappearance of I[131] labeled albumin from the subcutaneous tissues, *J. Clin. Invest.* **35**:713 (1956).
28. Castell, D. O., Ascites in cirrhosis. Relative importance of portal hypertension and hypoalbuminemia, *Am. J. Dig. Dis.* **12**:916–922 (1967).
29. Wolff, H. P., Blaise, B. H., and Düsterdieck, G., Role of aldosterone in edema formation, *Ann. N.Y. Acad. Sci.* **139**:285–294 (1962).
30. Van Tongeren, J. H. and Reichert, W. J., Demonstration of protein-losing gastroenteropathy. The quantitative estimation of gastrointestinal protein loss using ^{51}Cr-labeled plasma proteins, *Clin. Chim. Acta* **14**:42–48 (1966).
31. Kerr, R. M., Du Bois, J. J., and Holt, P. R., Use of ^{125}I and ^{51}Cr labeled albumin for the measurement of gastrointestinal and total albumin catabolism, *J. Clin. Invest.* **46**:2064–2082 (1967).
32. Dobbins, W. O., Diseases associated with protein losing enteropathy: Electromicroscopic studies of the intestinal mucosa, *South. Med. J.* **60**:1077–1081 (1967).
33. Kocen, R. S. and Atkinson, M., Ascites in hypothyroidism, *Lancet* **1**:527–531 (1963).
34. Thorn, G. W., Approach to the patient with "idiopathic edema" of "periodic swelling," *J. Am. Med. Assoc.* **206**:333–338 (1968).
35. Vorburger, C., Pathogenesis of edema, *Angiology* **19**:362–371 (1968).

Abnormal Blood pH
(Acidosis and Alkalosis)

This chapter logically follows Chapter 4 on the maintenance of constant pH in the human. However, it has been put here because it also leads naturally into the next chapter, which concerns the practical application of the foregoing material. For this reason, it is recommended that Chapter 4 be re-read before proceeding.

The blood pH normally ranges from 7.38 to 7.42.[1] As refinements are made in blood pH measurements, this range becomes narrower. For example, a glass electrode maintained in the carotid artery of a dog showed a pH constant to ± 0.001 pH units during and in between meals.[2] If respiration is impaired, CO_2 will accumulate in the blood and the pH will fall. The pH will be of the order of 6.4 within 5 min after respiration ceases. *This emphasizes the role of the lungs in maintaining constant pH.*

If an individual deliberately hyperventilates, the pH of the blood will rise to 7.6–7.7, and if this is prolonged, he will lose consciousness. With high blood pH, we call the condition *alkalosis* or *alkalemia*. With low blood pH, we use the terms *acidosis or acidemia*. In cases where the cause is due to abnormal respiration, we designate the condition *respiratory acidosis or alkalosis*.

When acidosis or alkalosis is due primarily to disproportion between sodium and chloride ions in the plasma and/or an accumulation of anions other than bicarbonate, the condition is said to be metabolic acidosis or alkalosis, depending upon the pH. This may be due to diarrhea, decreased intake of fluids, vomiting, renal disease, excessive use of diuretics, diabetes mellitus, or losses caused by continuous intestinal drainage.[3,4]

These conditions are classified as *metabolic acidosis or alkalosis* in order to distinguish them from acidosis or alkalosis due to impaired respiration. It must be stressed that *in metabolic acidosis the problem is a deficit in Na and K ion concentration required to neutralize the anions, mainly chloride, phosphate, and organic acids. In metabolic alkalosis the deficit is mainly in chloride ion.* In respiratory acidosis the problem lies in CO_2 being retained due to some defect in respiratory exchange. Respiratory alkalosis is due to hyperventilation, with reduced p_{CO_2} levels in the blood, as occurs with brain damage or early salicylate intoxication.

In diabetic coma, organic acids may be present in such high concentration that blood pH is lowered markedly.[5] This is an example of *metabolic acidosis*. This condition is usually accompanied by a disproportion between sodium and chloride ions due to excretion by the kidney of extra amounts of sodium with the organic acids. The elevated plasma chloride and depressed plasma sodium concentrations in these patients are the results of the Na^+ being excreted along with the organic acids. The Cl^- ion competes with organic acids for excretion sites and is therefore retained, resulting in elevated plasma Cl^- concentration.

The problem in metabolic acidosis often simplifies to restoring a balance between the chloride and sodium ions. If the pH is brought within range of the normal level (7.40), then with the return of normal metabolism, accumulated organic acids will be metabolized or excreted. Where a metabolic defect exists, such as diabetes mellitus or kidney disease, other measures have to be taken. For example, with diabetes, insulin would be administered in addition to the administration of bicarbonate solutions.

With moderate disproportion between sodium and chloride levels, change in respiration can adjust the bicarbonate level so that the pH is brought to approximately normal levels. This condition is called *compensated metabolic acidosis or compensated metabolic alkalosis*. When this occurs in diabetics with elevated acetoacetic acid, hydroxy butyric acid, and acetone levels, the patient is said to be in *ketosis* but not in acidosis.[6] Some authors call a disproportion between Na and Cl ions compensated for by a change in bicarbonate levels *acidosis*. They use the term *acidemia* to indicate reduced blood pH and *alkalemia* to indicate elevated blood pH. Generally, the terms acidosis and acidemia are now used interchangeably.

From the above discussion, it is apparent that correction in blood pH is mainly a correction of the sodium-to-chloride ratio, except in those cases where organic acids and phosphates are present in high concentration.[7] *In acidosis, correction of sodium deficits and concomitant lowering of the chloride ion concentration by dilution will often adjust the blood pH to normal.*

Historically, in the absence of a rapid method for sodium determination, estimates of "base deficit" were made indirectly. Van Slyke introduced the method of *CO_2 combining power*.[8] In this method, gas at p_{CO_2} of 40 mmHg (residual air from the human lungs) was blown through plasma. A CO_2 determination was then performed. If CO_2 combining power was high, then an excess of base was indicated, and the patient was presumed to be in alkalosis. If it was low, then the patient was in acidosis.

The CO_2 combining power determination had one practical advantage. It permitted the drawing of blood in an ordinary tube, centrifuging, and drawing off the serum or plasma under ordinary conditions. This was acceptable, since the plasma was subsequently being exposed to expired air or equilibrated with a gas containing a partial pressure of CO_2 of 40 mmHg.

The CO_2 combining power determination was misleading in respiratory conditions, unless the *CO_2 content* was measured at the same time. For example, a patient arrives in the hospital hyperventilating because of a severe anxiety problem, unrelated to abnormal metabolism. The Na level is 140 mEq/liter, the K level is 4.6 mEq/liter, and chloride is at 103 mEq/liter. However, CO_2 content is 8 mmol/liter and blood pH is 7.65. If plasma from this patient is removed and expired air from the technician is blown through, the CO_2 content will return to normal (27 mmol/liter) and the pH will be 7.4. If the original CO_2 content determination had not been performed, it would be difficult to interpret the results. *The CO_2 content* is the total CO_2 present in the plasma.

As another example, consider the case of a child who swallows a penny, which lodges in the trachea. This child, if rushed to the hospital, might show normal electrolyte levels and a CO_2 content of 45 mmol/liter and a pH of 7.1. If a CO_2 combining power test is done, *pH and CO_2 content will be normal*, because the technician will blow off the excess CO_2. The CO_2 combining power was designed to eliminate effects on the CO_2 content of plasma due to respiratory difficulty. Both of the examples above illustrate this point.

The main inorganic components of blood are sodium, potassium, and chloride. If one adds the potassium and sodium concentrations, one obtains $140 + 4.6 = 144.6$ mEq/liter. If one then subtracts the chloride concentration, one obtains $144.6 - 103 = 41.6$ mEq/liter. This value is due mainly to the sum of the protein and bicarbonate ion concentrations ($16 + 25.6 = 41.6$ mEq/liter) and represents the variable buffering capacity of the blood. Van Slyke and Hastings call this the buffer base (BB) of the blood.[8,9] They calculate the value as 40.8 using all the fixed anions and cations at a pH of 7.40, a temperature of 37°C, and p_{CO_2} of 40 mmHg. For each gram of hemoglobin oxygenated, one loses 0.36 mEq/liter of bicarbonate ion, since oxyhemoglobin is the stronger acid. This has been explained before. In order to maintain the same pH, one needs to add more sodium ion. Thus, using their value for buffer base, we obtain, for 14 g of hemoglobin, $40.8 + (14 \times 0.36) = 40.8 + 5.04 = 46.2$ mEq/liter. Astrup calls this the *normal buffer base* and uses the symbol NBB.[10]

Note that with fully oxygenated blood, a difference of 5.04 mEq/liter is obtained as compared to fully reduced blood. Venous blood is approximately 40% saturated with oxygen. This difference therefore reduces to less than 3 mEq/liter when comparing arterial to venous blood.

Generally, in doing CO_2 combining power, the effect of hemoglobin was ignored, introducing a moderately significant error in calculating base requirements in acidosis. For this reason, Astrup introduced the concept of *standard bicarbonate*. This is the bicarbonate concentration of plasma from fully oxygenated blood that has been equilibrated with a gas where the p_{CO_2} is 40 mmHg at 37°C and p_{O_2} is 100 mmHg. *This differs from a CO_2 combining power in that the gas is blown into the blood before the erythrocytes are separated, rather than after.* The pH of the blood is measured, the cells are separated, and the total CO_2 is estimated. The various methods of CO_2 analysis are compared in Table 9.1.

If CO_2 content, CO_2 combining power, and standard bicarbonate determinations are carried out on the same blood exposed to gases of varying p_{CO_2} levels, the results obtained would be as shown in Table 9.1.

The amount of gas that dissolves in a solvent is a function of the partial pressure of that gas in equilibrium with the solvent (Henry's law). For this reason, the variation with different p_{CO_2} levels of CO_2

TABLE 9.1

Comparison of CO_2 Combining Power and Content and Standard Bicarbonate After Equilibration of the Original Blood with Gases of Different p_{CO_2} Values[14]

	p_{CO_2}, mmHg		
	20	40	80
Oxygenated blood			
CO_2 content, mmol/liter	16.8	22.2	30
CO_2 combing power, mmol/liter	18.0	22.2	27.5
Standard bicarbonate, mmol/liter	21.0	21.0	21.0
Reduced blood			
CO_2 content, mmol/liter	19.6	25.7	34.8
CO_2 combining power, mmol/liter	21.0	25.7	31.9
Standard bicarbonate, mmol/liter	21.0	21.0	21.0

content of the plasma in Table 9.1 is to be expected. Variation between the values observed for CO_2 content with oxygenated and reduced hemoglobin result, not only from the effect of oxygenation of hemoglobin already discussed, but also from the shift of chloride from the erythrocyte to the plasma when the blood is oxygenated (see the discussion of chloride shift in Section 4.6).

When blowing air at p_{CO_2} of 40 mmHg through whole blood or plasma, excess dissolved CO_2 due to previous exposure to gas of p_{CO_2} of 80 mmHg is blown off. On the other hand, amounts of dissolved CO_2 lost due to exposure to p_{CO_2} at 20 mmHg are replaced. In order to bring the reaction to equilibrium rapidly, carbonic anhydrase needs to be present. This is supplied by the erythrocytes in the standard bicarbonate procedure, but not with the CO_2 combining power determination, unless some hemolysis has taken place.

The standard bicarbonate remains constant, since the blood attains equilibrium with a gas of constant composition (100 mmHg p_{O_2} and 40 mmHg p_{CO_2} in each case) and constant degree of Hb oxygenation.

From the Table 9.1 the effect of the separation of plasma from blood of different degrees of oxygenation is apparent in the *different CO_2 combining power values* obtained. This should have been approximately 3 mmol/liter as calculated above. The values suggest the effect of the chloride shift and that equilibrium has not yet been established.

Certain terms which will be used in subsequent discussions are defined in Table 9.2.

TABLE 9.2

Definition of Expressions Used in Blood pH Studies

Term	Units	Definition
Total plasma CO_2 (CO_2 content)	mmol/liter	All the CO_2 liberated on adding acid to serum or plasma; determined experimentally on a gasometer
p_{CO_2} (CO_2 partial pressure or CO_2 tension)	mmHg	The partial pressure of CO_2 in a gas, or the vapor tension of CO_2 from blood or plasma in equilibrium with that gas
Blood pH	pH units	The negative logarithm to the base 10 of the hydrogen ion concentration (e.g., when $[H^+] = 10^{-7}$, the pH = 7)
Carbonic acid concentration	mmol/liter	$[H_2CO_3] = 0.03 p_{CO_2}$; the undissociated carbonic acid in the blood is calculated from the p_{CO2} by this equation
CO_2 combining power	mmol/liter	The total CO_2 in plasma after separating the erythrocytes and equilibrating the plasma with normal expired air of a p_{CO_2} value of 40 mmHg
Plasma bicarbonate	mmol/liter (mEq/liter)	The bicarbonate concentration $[HCO_3^-]$ derived from the equation pH = 6.1 + $[HCO_3^-]/0.03 p_{CO_2}$, where pH and p_{CO_2} are measured experimentally. Since total $CO_2 = [HCO_3^-] + 0.03 p_{CO_2}$, the $[HCO_3^-]$ can be calculated by either measuring total CO_2 and p_{CO2} or total CO_2 and pH.
Standard bicarbonate	mEq/liter	The blood drawn from the patient is equilibrated with a gas containing p_{O_2} at 100 mmHg and p_{CO_2} at 40 mmHg. The pH and p_{CO2} are measured on this blood and the bicarbonate calculated as for the plasma bicarbonate above. The CO_2 content of the plasma from the fully oxygenated blood is the standard bicarbonate.
Buffer base (BB)	mEq/liter of plasma	The concentration of the cations (Na + K + Ca + Mg) minus the chloride + phosphate ions; in effect it is the sum of the titration equivalent concentration of the protein plus the bicarbonate concentration; it varies with the hemoglobin concentration and the extent of its oxygenation

TABLE 9.2.—*continued*

Term	Units	Definition
Base excess or base deficit	mEq/liter	Base excess is zero with blood at pH 7.40, $p_{CO_2} = 40$ mmHg, and 37°C; the base excess is a positive number and the base deficit is a negative number obtained when one subtracts the normal buffer base from the actual buffer base; BE = BB(actual) − BB(normal), where the normal buffer base is corrected for Hb concentration.
Respiratory acidosis	—	CO_2 retention with lowering of blood pH
Respiratory alkalosis	—	Excessive ventilation with loss of CO_2 resulting in increased blood pH
Metabolic acidosis	—	Reduced pH due to relative increase in anions other than HCO_3^- (Cl^-, phosphate, organic acids) or relative decrease in cations (mainly Na and K ions)
Metabolic alkalosis	—	Increased pH due to increase in cations (Na + K) relative to anions or relative decrease of anions (mainly chloride ions)

9.1 *METABOLIC ACIDOSIS*

9.1.1 *Calculation of Base Excess and Base Deficit*

9.1.1.1 *From Na^+, K^+, and Chloride Concentrations*

It was pointed out that at pH 7.4, p_{CO_2} of 40 mmHg, and 37°C, a buffer base value of 40.8, according to Hastings, or 41.7, according to Astrup and Siggaard-Andersen, is considered as the normal value, and represents zero base excess or deficit. This value is also obtained by subtracting the normal values for the fixed anions (mainly chloride) from the fixed cations (sodium plus potassium). This yields a practical method for estimating base excess or deficit, namely by comparing this difference to the value of 41.6 found in the normal by subtracting the normal value for chloride from the sum of the sodium and potassium values.

Example 1. The values found in the patient are, in mEq/liter, Na 120, K 5.0, chloride 110; buffer base = 120 + 5 − 110 = 15 mEq/liter. But the normal value should be 41.6. The base deficit is therefore 15 − 41.7 = −26.6, the minus sign indicating a deficit and suggesting a metabolic acidosis.

Example 2. The values found in the patient are, in mEq/liter, Na 155, Cl 89, K 2.9. The base deficit is given by buffer base = 155 + 2:9 − 89 = 68.9 mEq/liter. Thus we have 68.9 − 41.6 = +27.3 mEq/liter. Since the value obtained is positive, this is a base excess, and the patient is in metabolic alkalosis.

The general formula used in this procedure is $[Na^+] + [K^+] − [Cl^+] − 41.6$ = base excess or deficit, depending upon the sign.

The advantages of this procedure are obvious. The Na, K, and chloride values may be obtained rapidly in the modern laboratory. This was not the case during almost all of the first half of this century, and this gave rise to the elaborate procedures used to measure the sodium concentration indirectly.

The main disadvantages of this approach are that it does not take into account the accumulation of other anions, such as phosphate in uremia, nor of organic acids in diabetes. Phosphate levels that are normally approximately 2 mEq/liter (3 mg %) may readily rise to six times that value (11 mEq/liter or 18 mg/100 ml). In diabetic coma, the accumulation of acetoacetic and hydroxybutyric acids often reaches levels in excess of 10 mEq/liter. Attempts to correct for this by taking into account lowering of the CO_2 content of serum does not solve this problem, since, as the condition progresses and the patient's condition deteriorates, CO_2 levels may actually rise, confusing the situation, a condition which will be discussed further below.

At present it is the accepted procedure to follow electrolyte changes during fluid therapy and make adjustments to bring them to the normal range if practicable. In many cases this also solves the problem of acidosis or alkalosis as the case might be.

9.1.1.2   *From Bicarbonate Base Levels*[9–13]

a.   *Method of Singer and Hastings.* In the absence of rapid direct methods for sodium estimations, Bohr, Van Slyke, and others attempted to evaluate the patient's condition by measuring carbon dioxide levels, since gasometers were available even before the turn of the century. The method also required small quantities of blood. As a result, the "CO_2 combining power" test was devised, along with colorimetric procedures for estimating pH and titrimetric procedures for chloride estimation. The Henderson–Hasselbalch equation was then used to evaluate the acidity of the blood and base deficits or excesses. The term

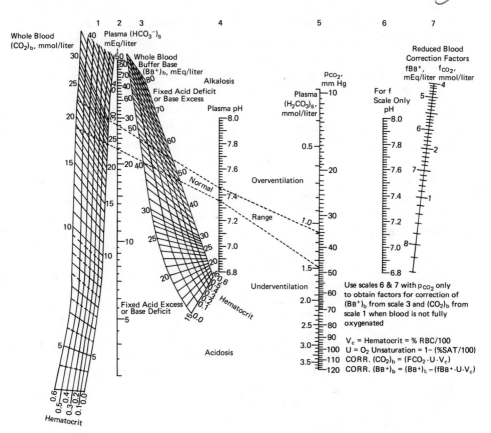

FIGURE 9.1   Nomogram for the acid–base balance of human blood at 37°C. For oxygenated blood (scales 1–5) a straight line through points given on two of the scales intersects the remaining scales at simultaneously occurring values of the other variables. The position of this line indicates the kind and magnitude of any disturbance of the acid–base balance. The seven scales in the figure are (1) whole blood CO_2 content, (2) plasma bicarbonate, (3) *whole blood buffer base*, (4) plasma pH, and (5) p_{CO_2}. Scales 6 and 7 are the same as Figure 3.6 to correct for the extent that the blood is not completely oxygenated.[9]

"*alkaline reserve*" was used to indicate the bicarbonate level, or, in effect, how much sodium ion was available for further buffering action.

These experiments were summarized in a paper by Singer and Hastings[11] in which a nomogram was proposed for estimating all of the various parameters by making only measurements readily available at that time, the CO_2 content of whole blood, hematocrit, and plasma pH (Figure 9.1).

The assumption was made that one was not dealing with a micro or macrocytic anemia, and therefore the hematocrit served as a measure of hemoglobin estimation. The objective of the nomogram was to obtain the buffer base level. A line drawn from the whole blood CO_2 value (choosing a vertical line with the hematocrit value of the blood under test), and the pH value gives the buffer base value if the blood was fully oxygenated. This is directly useful with arterial blood (90% saturated), but for venous blood, a correction factor needs to be applied from scales 6 and 7. The application of this correction factor is described in Figure 3.6.

For example, for the normal, assume a hematocrit value of 40%, whole arterial blood CO_2 content of 21 mmol/liter, and a pH of 7.4. Draw a line between these two points. One obtains a p_{CO_2} value of 40 mmHg (which could not be directly measured readily at that time) and a buffer base value of 43 mmol/liter, which these authors apparently considered the mean of the normal range. Plasma bicarbonate is then 25 mmol/liter.

Assuming the same hematocrit value for a patient, one obtains a whole arterial blood CO_2 content of 10 mmol/liter and a pH of 7.2. One then obtains a p_{CO_2} of 28 mmHg and a buffer base of 31. Now, $31 - 43 = -12$ mEq/liter, and this would be the base deficit. The p_{CO_2} of 28 also shows hyperventilation from the figure. The buffer base deficit shows a *metabolic acidosis*.

For another patient, let us assume the same hematocrit value, an arterial blood CO_2 content of 30 mmol/liter, and pH of 7.2. One sees immediately that the buffer base is in the normal range, but p_{CO_2} is 90 mmHg, and the patient is underventilating; since there is no base excess or base deficit, this is a *respiratory acidosis*.

In a similar fashion the diagram can be used to illustrate a *metabolic alkalosis* (whole blood $CO_2 = 30$, pH 7.6) and a *respiratory alkalosis* (whole blood $CO_2 = 11$, blood pH 7.55).

With the development of direct and rapid methods for p_{O_2} and p_{CO_2} measurement, one could directly measure the parameters of blood that are affected by gas exchange, and thus decide directly the extent of involvement of respiration in acidosis or alkalosis. However, there remained the need for measurement of base deficit or excess, since direct measurement of sodium, potassium, and chloride did not yield information on the level of organic acids, phosphates, and other anions, other than chloride.

b. *Method of Hastings, Astrup, and Siggaard-Andersen.*[10,14,15] The Henderson–Hasselbalch equation can be set up in the form of a linear equation, $y = mx + b$, where y is the pH, x is $\log p_{CO_2}$, and b is the y intercept.

$$pH = 6.1 + \log \frac{[HCO_3{}^-]}{0.03 p_{CO_2}} \tag{9.1}$$

and

$$pH = 6.1 + \log[HCO_3{}^-] - \log 0.03 - \log p_{CO_2} \tag{9.2}$$

Then

$$pH = -\log p_{CO_2} + (6.1 - \log 0.03 + \log[HCO_3{}^-]) \tag{9.3}$$

The y intercept is equal to $6.1 - \log 0.03 + \log[HCO_3{}^-]$ at a particular p_{CO_2} level chosen, and the slope is always -1, or $45°$ descending. This can be seen in Figure 9.2. All one needs to do is measure the p_{CO_2} of the blood equilibrated with a gas to fully oxygenate the blood, and with a p_{CO_2} of 40 mmHg. One then measures the pH. A point will be found and a line drawn at a $45°$ angle with the y axis, which will generate the line. As an alternative, the blood may be equilibrated first with a p_{CO_2} of 20 mmHg and then a p_{CO_2} of 40 mmHg, measuring the pH in each case. One then draws the line from these two points. This is the method chosen by Astrup for generating the line. Fully saturating the blood with oxygen eliminates any variation due to variation in the percentage of oxygenation, and the bicarbonate values as indicated in Figure 9.2 would be derived from standard bicarbonate values.

However, another variable exists, that is, the hemoglobin level. Siggaard-Andersen calculates the mean normal buffer base at a p_{CO_2} of 40 mmHg and a pH of 7.38 to be 44.4 mEq with 10 g of hemoglobin and 48 mEq with 20 g of hemoglobin. If one then plots the lines of Figure 9.2, the slope will vary somewhat with hemoglobin content.[16] To avoid this effect, Siggaard-Andersen took advantage of the fact that since the slope was steeper at high Hb values and less steep at lower Hb values, a point of intersection existed where the *hemoglobin concentration had no effect*. This can be seen in Figure 9.3 (compare Figure 9.3 with Figure 3.4).

Connecting these points, one could set up a base excess and a base deficit curve. This can be seen in Figures 9.3 and 9.4.[14] Note that in these figures p_{CO_2} is plotted as the ordinate and pH as the abscissa. Also note that a logarithmic scale is used for p_{CO_2}, rather than plotting the

$\log p_{\text{CO}_2}$ as in Figure 9.2. This is done to eliminate the need for taking the log of the p_{CO_2} value.

The measurement of the pH at two values of p_{CO_2} requires the saturation with gases at these two values, and the instrumentation was cumbersome for this technique. Siggaard-Andersen[15] therefore simplified the procedure by producing a nomogram somewhat similar to that of Singer and Hastings, but avoiding the need for subtracting the buffer base value obtained from a normal buffer base value (Figure 9.5). Only two measurements are required, pH and p_{CO_2}. A line drawn from the pH to the p_{CO_2} scale intersects a line at a particular hemoglobin content of blood, to yield the base excess. The bicarbonate and total CO_2 are also displayed, calculated from the Henderson–Hasselbalch equation. For example, pH = 7.3, p_{CO_2} = 30 mmHg, and Hb is 15 g. The base deficit (minus value) is 11 mEq, bicarbonate is 14.5 mEq, and

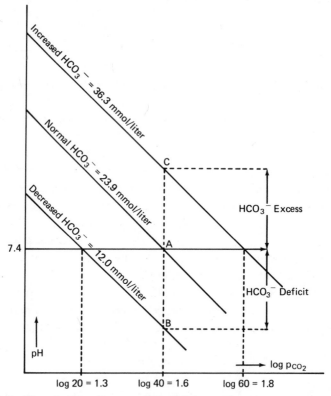

FIGURE 9.2   Plot of pH vs. $\log p_{\text{CO}_2}$. Base deficit or excess is measured by displacement of the line.

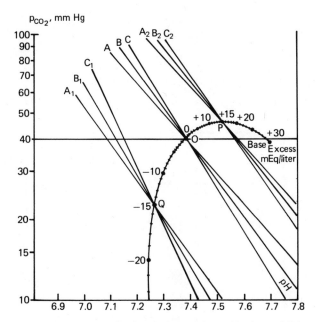

Figure 9.3 Displacement of pH/log p_{CO_2} lines (*A, B, C*) after addition of 15 mEq acetic acid (A_1, B_1, C_1) and 15 mmol/liter sodium carbonate (A_2, B_2, C_2) per liter of blood. The points of intersection for these lines are all on the base excess curve. At point *O* the base excess is zero, at point *P* the base excess is $+15$ mEq/liter blood, and at point *Q* the base excess is -15 mEq/liter blood, *independent of the hemoglobin concentration.*[14]

total CO_2 is 15 mmol, in round figures; the patient is hyperventilating, since p_{CO_2} is substantially lower than 40 mmHg.

Note that the extent of oxygenation is being ignored, since it amounts to less than 3 mmol/liter.

Instrumentation is now available to read all of these parameters simultaneously. The p_{CO_2}, p_{O_2}, and pH are measured and the bicarbonate, base excess, and total CO_2 are displayed with digital figures (Radiometer, Corning, I.L. Instr. Co.).

9.1.2 Direct Calculation of Total Sodium Bicarbonate Required to Bring to Normal pH

The amount of sodium bicarbonate or sodium lactate required to bring a patient in metabolic acidosis to normal pH can be calculated directly from the Henderson–Hasselbalch equation.[17]

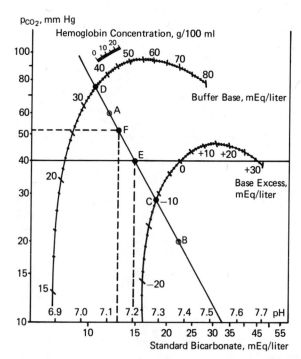

FIGURE 9.4   Nomogram for calculating base excess or deficit by the method of Astrup. The pH is determined at two p_{CO_2} values to obtain the line such as *FE*. Intercept at point *C* on the calibrated curve yields the base deficit as approximately 10 mEq/liter.[14]

The procedures described heretofore are designed to determine the bicarbonate concentration after the blood has been equilibrated with gas at p_{CO_2} of 40 mmHg and subtracting from that value a normal value, the sign determining whether it is base excess or deficit. The problem can be approached directly by asking the question, "How much sodium bicarbonate needs to be added to bring the blood to pH 7.4?"

Hastings raised the questions, "What do we mean by compensated acidosis. Do we mean compensated to normal pH or normal p_{CO_2}?" It can readily be shown that the system is under pH control, and not p_{CO_2} control, and therefore our objective is to bring the blood to normal pH.

In the equation $pH = 6.1 + (\log[HCO_3{}^-])/0.03p_{CO_2}$, it has been shown earlier that if the total CO_2 and pH are measured, the p_{CO_2} and $[HCO_3{}^-]$ levels can be calculated. If the p_{CO_2} and pH are measured, the other factors can also be calculated. The problem requires that we now add sodium bicarbonate until the blood pH comes to normal. If we do

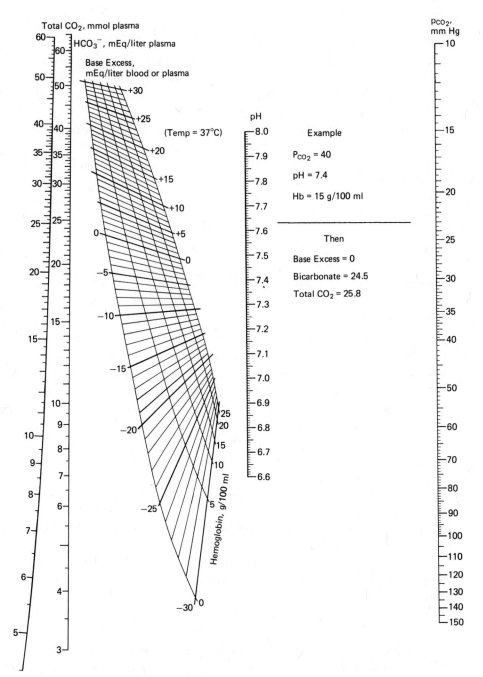

FIGURE 9.5   Nomogram for relationship among p_{CO_2}, pH, total CO_2, bicarbonate and hemoglobin concentration.[23] (Radiometer Co., Copenhagen.)

this, we are increasing the numerator of the fraction in the equation without changing the denominator, provided no change in p_{CO_2} takes place during the procedure. By administering NH_4Cl, we are neutralizing the Na ion and in effect subtracting the bicarbonate ion. The ammonia is then excreted. When adding sodium lactate, we expect the lactate to be metabolized and, in effect, we are adding $NaHCO_3$.

Let us compare the conditions before and after adding the bicarbonate. Let us assume that we have measured the pH and the total CO_2 of the patient, and find a pH value of 7.1 and a total CO_2 of 10 mmol/liter. Then we have three equations,

$$\text{before correction:} \quad 7.1 = 6.1 + \log \frac{[HCO_3^-]}{0.03\, p_{CO_2}} \tag{9.4}$$

$$\text{total } CO_2 = 10 \text{ mmol/liter} = [HCO_3^-] + 0.03 p_{CO_2} \tag{9.5}$$

$$\text{after correction:} \quad 7.4 = 6.1 + \log \frac{[HCO_3^-] + [HCO_3^-]_{\text{added}}}{0.03 p_{CO_2}} \tag{9.6}$$

The values for the concentrations are all in mmol/liter.

Note that we have specified for the present that the p_{CO_2} will remain constant. We also assume that during the process the patient will not disturb the system by metabolism of organic acids or by urination. This means that there will be no change in the rate of ventilation or blood flow to the lungs. We have three equations and three unknowns (original bicarbonate level, added bicarbonate, and p_{CO_2}). The problem is therefore readily solvable. The solution (see Appendix, p. 380 for derivation) is shown in equation (9.7).

$$Q = \frac{(\text{total } CO_2)\ 19.95 - \text{antilog (pH} - 6.1)}{1 + \text{antilog (pH} - 6.1)} \tag{9.7}$$

Values for Q (added bicarbonate) obtained by applying equation (9.7) at different pH values and CO_2 contents are given in Table 9.3.

Let us now apply this approach to a specific case. The patient weighs 70 kg and assuming 20% extracellular fluid, we are dealing with 14 liters. From Table 9.3, we see that with a CO_2 content of 10 mmol/liter and pH 7.1 we require 9.05 mmol of $NaHCO_3$/liter. Our total requirement is then 9.05 × 14 = 126.7 mmol. *If this amount is administered to the patient in practice and the patient ceases to hyperventilate, the pH will fall short of reaching the normal level.* This amount will be found to be inadequate to correct the acidosis.

TABLE 9.3

Relationship Between pH, CO₂ Content, and Base Deficit or Excess in Acidosis or Alkalosis from the Equation;

$$\text{Base deficit or excess} = T[19.95 - \text{antilog}\,(pH - 6)][1 + \text{antilog}(pH - 6.1)]^a$$

pH	CO₂ content mmol/liter											
	5	10	15	20	**25**	30	35	40	45	50	55	60
6.90	9.35	18.7	28.11	37.4	46.8	56.1	65.5	74.8	84.2	93.5	102.9	112.2
6.95	7.86	15.7	23.6	31.4	39.3	47.2	55.0	62.8	70.7	78.6	86.5	94.3
7.00	6.70	13.4	20.1	26.8	33.5	40.2	46.9	53.6	60.3	67.0	73.7	80.4
7.05	5.57	11.1	16.7	22.3	27.9	33.4	39.0	44.5	50.1	55.7	61.3	66.8
7.10	4.53	9.05	13.6	18.1	22.7	27.1	31.7	36.1	40.7	45.3	49.8	54.3
7.15	3.57	7.14	10.7	14.3	17.9	21.4	25.0	28.5	33.3	35.7	39.3	42.8
7.20	2.71	5.42	8.13	10.8	13.6	16.3	19.0	21.6	24.3	27.1	29.8	32.5
7.25	2.06	4.12	6.18	8.24	10.3	12.4	14.4	16.4	18.5	20.6	22.7	24.7
7.30	1.22	2.43	3.66	4.86	6.08	7.29	8.51	9.72	10.9	12.2	13.4	14.6
7.35	0.580	1.16	1.74	2.32	2.90	3.48	4.06	4.64	5.22	5.80	6.38	6.96
7.40	0	0	0	0	0	0	0	0	0	0	0	0
7.45	−0.52	−1.04	−1.56	−2.08	−2.60	−3.12	−3.64	−4.16	−4.68	−5.20	−5.72	−6.24
7.50	−0.99	−1.98	−2.97	−3.96	−4.95	−5.94	−6.93	−7.92	−8.91	−9.90	−10.9	−11.9
7.55	−1.41	−2.82	−4.23	−5.64	−7.05	−8.46	−9.87	−11.3	−12.69	−14.1	−15.5	−16.9
7.60	−1.79	−3.58	−5.37	−7.16	−8.95	−10.7	−12.5	−14.3	−16.11	−17.9	−19.7	−21.5
7.65	−2.13	−4.26	−6.39	−8.52	−10.7	−12.8	−14.9	−17.0	−19.2	−21.3	−23.4	−25.6
7.70	−2.44	−4.87	−7.31	−9.74	−12.2	−14.6	−17.1	−19.5	−21.9	−24.3	−26.8	−29.2
7.75	−2.71	−5.42	−8.13	−10.8	−13.5	−16.2	−18.9	−21.7	−24.4	−27.0	−29.8	−32.4

ᵃ Values represent mEq of sodium bicarbonate ion to be added to bring to pH 7.4 in acidosis, per liter of extracellular fluid (5 kg of body weight). Negative values represent the mEq of acid (NH₄Cl) needed to be added to bring to normal pH in alkalosis. Interpolate for CO₂ content from the adjacent value on the pH line. Example: For CO₂ content of 18 mmol/liter and pH 7.20, base requirement is 8.13/15 × 18 = 9.76.

The pH to which the patient's blood will move on addition of 9.05 mmol of $NaHCO_3$ per kg of extracellular fluid can be calculated as follows. From Table 2.3, with a pH of 7.1 and total CO_2 of 10 mmol/liter, p_{CO_2} is 30.3 mmHg. When respiration is restored to normal, p_{CO_2} will be 40 mmHg. In the equation pH = 6.1 + $(\log[HCO_3^-])/0.03 p_{CO_2}$, the denominator will change from 30.3 × 0.03 = 0.909 to 40 × 0.03 = 1.2. At pH 7.1 the value of the log of the fraction is equal to 1.0, so that the fraction has the value of 10.0. At the outset, then, we have 9.09/0.909 as the fraction. From Table 9.3, we need to add 9.05 to the numerator to correct to pH 7.4 at a CO_2 content of 10 mmol/liter, making the numerator equal to 9.09 + 9.05 = 18.14. If the denominator remained at 0.909, the value of the ratio would be approximately 19.95 and the pH would be 7.4. But the denominator changes to 1.2. Therefore the pH = 6.1 + log 18.14/1.2 = 7.27. The patient's pH would then be only 7.27, and more $NaHCO_3$ needs to be administered.[1]

The reason for this is apparent if one looks across the 7.10 pH line in Table 9.3. *The lower the CO_2 content, the less sodium bicarbonate is required. This is a demonstration of respiratory compensation.* By acidotic (Kussmaul) breathing, the patient is blowing off CO_2 and lowering p_{CO_2} levels, thus partly correcting the pH. This results in a decreased sodium bicarbonate requirement. For example, at CO_2 content of 10 mmol/liter, as compared to the normal level (25 mmol/liter), the patient has decreased his deficit from 22.7 to 9.05 mmol/liter. This is a substantial difference.

If the patient's respiratory exchange is impaired so that p_{CO_2} is

---

[1] Using the data from the 25 mmol/liter column in Table 9.3 in calculating the amount of $NaHCO_3$ to be added assumes a p_{CO_2} of 40 mmHg will be reached when pH is restored. A ratio of 23.94/1.2 is assumed in the fraction and the sum of the numerator and denominator is equal to 25 in round numbers. The CO_2 content column under 25 is emphasized to indicate that this is the deficit corrected for in metabolic acidosis uncomplicated by respiratory abnormalities.

  If CO_2 combining power is performed by equilibrating the plasma with gas at p_{CO_2} = 40 mmHg and the total CO_2 and pH are measured on this plasma, then respiratory effects are eliminated and Table 9.4 can be used directly. If this blood, formerly in equilibrium with a p_{CO_2} of 30 mmHg, is now equilibrated to p_{CO_2} = 40 mmHg, then more CO_2 will dissolve, CO_2 content will rise, and pH will fall. One would then find that CO_2 combining power is 11.7 mmol/liter and pH is 7.05. From Table 9.3, 22.7 mEq of $NaHCO_3$/liter of extracellular fluid is then required, which is closer to the correct value.

greater than 40 mmHg, and at pH 7.1, one finds 30 mmol/liter for total CO_2, the deficit indicated would be 27.1 mmol/liter, of which approximately 5 mEq/liter is the effect due to respiratory acidosis. In general, if p_{CO_2} is measured and differs substantially from 40 mmHg, one is dealing with increased or decreased ventilation rate in the lungs, depending upon whether the p_{CO_2} is less or greater than 40 mmHg.

The patient would go to normal pH with 9.05 mmol of $NaHCO_3$ per liter of extracellular fluid, provided he remained at a CO_2 content of 10 mmol/liter. Since the rate of ventilation in the lungs decreases as the acidosis is corrected and breathing returns to normal, we require *the amount of bicarbonate to correct the deficit at the CO_2 content when a pH of 7.4 is reached.*

At first, this would appear to be unpredictable. Severe acidosis for a substantial period of time may result in some impairment of the central nervous system, and when fluids are administered, the ventilation and thus p_{CO_2} may not return to normal. Second, *since the fluids are administered over a substantial period of time, the patient may metabolize and excrete the organic acids and excrete some phosphate to significantly diminish the requirements.*

As one corrects the acidosis, depth and rate of breathing are gradually adjusted and the p_{CO_2} and total CO_2 continue to change. Experimentally, a mean value of CO_2 content of 25 mmol/liter when substituted in equation (9.12) was found to yield the required amount of bicarbonate for blood pH corrections for individuals in acidosis, uncomplicated by respiratory conditions. This column is therefore singled out in Table 9.4.

The process has been followed by titration *in vivo*, adding sodium bicarbonate and taking repetitive blood pH measurements during the process. Two typical cases are shown in Table 9.4. The theoretical value is calculated on the basis of determining the amount of $NaHCO_3$ necessary to bring to the particular pH indicated from the initial value as calculated above.

Comparison of the data shown in Table 9.4 with the theoretical titration curve shown in Figure 10.3 indicates that these patients followed the theoretical curve fairly well.

On the assumption that the total CO_2 in the equation of Table 9.3 is 25 mmol/liter, the theoretical amount of alkali can be read off directly. How efficient this system operates in estimating the alkali requirements in acidosis is shown in Table 9.5. It is apparent that

TABLE 9.4

In Vivo Titration of Blood of Patients with Severe Acidosis and Anuria[17]

Case 1 (age 2 yr, weight 13.7 kg)						
pH (obs)	7.04	7.08	7.16	7.20	7.30	7.42
$NaHCO_3$ (added) g	—	1.0	2.6	3.4	5.1	6.9
$NaHCO_3$(calc),[a] g	—	1.1	2.9	3.7	5.5	6.8[b]
Case 2 (age 10 months, weight 5 kg)						
pH	6.94	7.12	7.25	7.32	7.37	
$NaHCO_3$, g	—	2.0	2.8	3.1	3.6	
$NaHCO_3$(calc),[a] g	—	1.9	2.8	3.3	3.4[b]	

[a] Calculated to bring to the pH observed, assuming 20% extracellular fluid volume or one liter of extracellular fluid per 5 kg of body weight.
[b] Calculated to bring to pH 7.40.

TABLE 9.5

Effect of Sodium Bicarbonate Administered to Patients in Accordance with the Equation of Table 9.4 on Blood pH and CO_2 Content[17]

Case No.	Diagnosis	Weight, kg	CO_2 content, mmol/liter		pH		$NaHCO_3$, g	
			Initial	Final	Initial	Final	Administered	Calculated[a]
1	Diarrhea	7	20.3	31	7.03	7.43	3.6	3.7
2	Partial respiratory obstruction	68	33.5	41	7.23	7.36	15.0	13.5
3	"Status asthmaticus"	60	29.3	34	7.25	7.41	10.0	10.0
4	Nephritis	65	14.4	23	7.20	7.35	18.0	15.3
5	Salicylism	5	8.1	25	7.28	7.45	0.65	0.65
6	Diarrhea	15	18.9	26	7.15	7.39	4.5	4.7
7	Cardiac decompensation (uremia)	72	20.3	25	7.23	7.36	16.3	14.5
8	Salicylism	53	18.0	20	7.10	7.23[b]	11.3	10.7
9	Diarrhea	11	7.7	23	7.23	7.40	2.54	2.24
10	Diarrhea	13	12.6	27	7.24	7.45	2.34	2.34
11	Diarrhea	1.75	1.0	22	6.92	7.42	1.15	1.30

[a] Calculated to bring to pH 7.40.
[b] Further treatment with $NaHCO_3$ discontinued. Calculated to bring to pH 7.23.

Table 9.3 serves a useful purpose in estimating roughly the patient's alkali requirements in metabolic acidosis.

Table 9.3 serves several purposes in the management of the routine laboratory. The blood in the sealed tube will not change its CO_2 content or p_{CO_2} value. For this reason, a rough check on the accuracy of results obtained can be made by analyzing the data obtained from the laboratory. For example, the technician reports Na 140 mEq/liter, Cl 110 mEq/liter, pH 7.20, and a CO_2 content of 15 mmol/liter. Examination of Table 9.3 shows that at a pH of 7.20 and a CO_2 content of 15 mmol/liter, the bicarbonate deficit is 8.13 mEq/liter. The chloride excess in the patient is $110 - 103 = 7$ mEq/liter, and is close enough to 8.13 to be acceptable. The organic acid and phosphate excess would then be $8.13 - 7 = 1.13$ mEq/liter, assuming all values are accurate.

As pointed out before, the organic acids normally found in blood are relatively strong acids ($pK \approx 3$) as compared to carbonic acid ($pK = 6.1$). For this reason, the organic acids can be considered similar in action to HCl with regard to their ability to displace bicarbonate ion. In addition to the organic acids, a contribution is made by phosphate, which is usually elevated in metabolic acidosis. With a serum phosphorus level of 10 mg % (7 mg % excess), this contributes approximately 4 mEq/liter to the base deficit.

In diabetic acidosis, a rough estimate of organic acid levels can be obtained in a similar fashion. A diabetic in coma shows total CO_2 = 20 mM/liter, pH = 7.15, Na = 136, and Cl = 109 mEq/liter. From Table 9.3 the base required to bring to normal pH is 14.3 mEq/liter. The chloride excess is $6 + 4 = 10$ mEq/liter, the 4 mEq/liter being added because of the lowered sodium value. The sum of the organic acids and phosphate excess is $14.3 - 10 = 4.3$ mEq/liter above the normal level of these components in the blood. *Table 9.3 includes the deficit caused by chloride excess and elevated organic acid and phosphate levels.*

In the case listed above, the CO_2 level was given as 20 mmol/liter. The diabetic patient excretes large volumes of urine in order to clear high glucose levels. Excessive amounts of sodium ion over chloride ion are then excreted with accumulated organic acids. The extracellular fluid volume shrinks, and the patient becomes dehydrated. Cellular metabolism is partially interrupted, and the organic acid intermediates pour into the blood, aggravating the condition. The patient hyperventilates to expel CO_2 to compensate for the lowering of blood pH. Total CO_2 levels drop progressively, results as low as 5 mmol/liter being observed. If untreated, the condition of the patient, now in coma, continues to deteriorate. At this point, along with altered metabolism, respiration is interfered with and gas exchange slows down. Soon the patient is gasping intermittently. The CO_2 *level, which had been lowered, now proceeds to rise and levels as high as 30 mmol/liter along with a severe metabolic acidosis are not uncommon.* The pH level drops further, due to increase in blood p_{CO_2} (the denominator in the Henderson–Hasselbalch equation). Assuming a pH level of 7.0, at this stage, the base required from Table 9.3 would be 40.2 mmol/liter. *When CO_2 levels are not markedly lowered in severe diabetic acidosis, the patient has passed the point of compensation his metabolic acidosis being aggravated by CO_2 retention.*

As pointed out above, the figures in Table 9.3 correct not only for excess of chloride over sodium but also for excess organic acids and phosphate which are usually present. It also assumes that the patient is anuric. If the patient is then brought to normal pH, one will find that the organic acids will be metabolized and excreted with excess phosphate in the urine. The patient will be left with an excess of sodium ion in the blood and go into alkalosis. This phenomenon can be seen from the data in Table 9.6.

The two cases in Table 9.6 are each chosen to illustrate a point. Cases 1 and 2 are the same as cases 1 and 2 in Table 9.4. In case 1 the patient has passed the point of compensation by respiration (p_{CO_2} was 70 mmHg), and CO_2 content had risen to 23.5 mmol/liter, aggravating the acidosis. The patient was severely dehydrated (elevated protein and hematocrit) and in coma and was breathing irregularly.

Disparity between sodium and chloride was only 5 mEq/liter. Base deficit from Figure 9.1 was 10.2 mEq/liter. From Table 9.3 sodium bicarbonate requirement to bring to pH of 7.40 was 27.9 mmol/liter or per 5 kg of body weight.

In view of the severity of the condition, the patient was titrated with sodium bicarbonate as indicated in Table 9.3.

As the patient's pH approached the normal level, the patient began breathing regularly and normally and began to urinate. At this point CO_2 content was 38.2 mmol/liter, p_{CO_2} was 60 mmHg, and the pH was 7.42. Base excess according to Figure 9.1 was 10 mEq/liter.

When normal pH was reached, the disparity between Na and Cl levels suggested an impending alkalosis when organic acids and phosphate would be removed by metabolism and excretion. Continuous

TABLE 9.6

Demonstration that Correction of Bicarbonate Deficit Due to Both Chloride and Organic Acid Excess Will Eventually Result in Alkalosis Due to Metabolism and Excretion of Organic Acids: Electrolyte Levels Before and After Treatment with $NaHCO_3$

	pH	CO_2 Content, mmol/liter	Cl, mEq/liter	Na, mEq/liter	K, mEq/liter	Protein, g/100 ml	Hematocrit, %	Urea, mg/100 ml	P, mg/100 ml	p_{CO_2}
Case 1 (13.7 kg)										
Initial	7.04	23.5	92	134	4.4	8.1	47	69	8.0	70
After 6.9 g NaHCO₃[a]	7.42	38.3	83	137	4.2	5.8	36	63	6.8	61
After 8 hrs[b]	7.52	31.1	94	139	2.5	4.8	33	19.2	5.0	40
Case 2 (5 kg)										
Inital	6.94	8.3	85	118	5.0	5.2	37.2	88	—	33
After 3.5 g NaHCO₃[c]	7.37	21.1	88	138	4.6	4.9	33.0	60	—	37
After 12 hr[d]	7.48	18.9	89	131	2.2	4.6	43.0[d]	24	—	26

[a] Administered in 200 ml of 5% glucose in water mixed with 200 ml of "normal saline."

[b] Received 50 mEq of sodium chloride, made up to 600 ml with 5% glucose in water. Excreted 400 ml of urine.

[c] Administered in 100 ml of "normal saline" mixed with 100 ml of 5% glucose in water.

[d] Received 150 ml of "normal saline" mixed with 300 ml of 5% glucose in water followed by 150 ml of whole blood. Excreted 380 ml of urine.

infusion of NaCl did not completely correct this tendency after 8 hr. Subsequently, the patient was brought back to pH 7.4 with NaCl and recovered uneventfully. Note the p_{CO_2} of 61 at pH 7.4 showing a lag in adjustment of respiration resulting in a mismatch between ventilation rate and blood flow to the lungs because of the rapid correction of blood pH.

In case 2, sodium deficit is only 4 mEq/liter as compared to chloride. From Figure 9.1, bicarbonate deficit is 26 mEq. From Table 9.3, 39.3 mEq of sodium bicarbonate was suggested and administered. The p_{CO_2} in this case was 34 mmHg. This, then, is a case where hyperventilation was able to compensate for a total of $39.3 - 15.7 = 23.6$ mEq of acid, of which only 4 mEq was due to chloride ion. Organic acids and phosphate accounted for 23.6 mEq. From Table 9.6, case 2, the sodium value of 138 mEq/liter and chloride value of 88 mEq/liter indicate that alkalosis will probably follow. Anticipating this, saline was administered and, as indicated in the table, the patient was in alkalosis at 12 hr. However, administration of KCl and NH_4Cl corrected the condition. Note that the blood administered contained sodium citrate, which also contributed to the alkalosis.

Cases 1 and 2 illustrate the basic principle, generally accepted, that treatment of a metabolic acidosis should be conservative. In both of these cases, it would have been advantageous to only partially correct the condition with sodium bicarbonate, and when the patient was out of danger (e.g., pH 7.30), to permit normal metabolism to complete the removal of the organic acids. Present technology permits the ready measurement of all the necessary parameters rapidly, and periodic measurements during the administration of fluids serve to avoid overtreatment.

Table 9.3 can also be used to calculate the number of millimoles of sodium lactate required to correct an acidosis in a patient, provided the acidosis is not severe. In mild acidosis, sodium lactate is usually administered on the basis of the fact that the lactate will be metabolized or excreted by the kidney, permitting the sodium to combine with bicarbonate to correct the acidosis. The pH of $\frac{1}{6}$ M lactate is 6.8, since excess lactic acid is added as a preservative.[2] Administration of this

---

[2] Historically, all lactic acid was prepared by fermentation and was therefore L-lactate. Recently, much commercially available lactate has been prepared synthetically and is therefore D,L-lactate. D-lactate is not as readily metabolized as L-lactate. Fluids need to be prepared from L-lactate.

solution to patients in severe acidosis will only lower the pH and aggravate the condition. Generally, lactate is recommended in acidosis when blood pH is higher than 7.25, although some use sodium bicarbonate whenever acidosis is being treated.

Equations (9.4) and (9.6) can be used to solve for the bicarbonate required to be added in order to obtain the relationship of equation (9.7) in terms of p_{CO_2} and pH in place of total CO_2 and pH. Since p_{CO_2} is readily measured, this equation has advantages. This simplifies to

$$\text{mmol } HCO_3 \text{ to be added/liter}$$
$$= Q = p_{CO_2} [0.5985 - 0.03 \text{ (antilog pH} - 6.1)] \quad (9.8)$$

Since the antilog (pH − 6.1) for various pH values are given in the Appendix (Table A.5, p. 381), one can readily calculate the base excess from the p_{CO_2} and pH values. For the example given above (pH = 7.2, CO_2 content = 15 mmol/liter), the deficit from Table 9.3 is 8.13 mEq/liter. From Table 2.3, the p_{CO_2} of this patient = 36.8 mmHg. Substituting in equation (9.8), we obtain

$$Q = 36.8 [0.5985 - 0.03(12.59)] = 8.13$$

As has been discovered before, this value assumes no adjustment in ventilation rate. From Table 9.3, when hyperventilation to compensate for the acidosis ceases with pH adjustment, the requirement increases to 13.6 mEq/liter (CO_2 = 25 mmol/liter). From Figure 9.1 which also assumes correction to normal ventilation (normal p_{CO_2}) this is estimated as 13.0 mEq/liter for this patient. Thus, hyperventilation produced an effect equivalent to adding approximately 13 − 8 = 5 mEq/liter of base per liter of extracellular fluid.

Equation (9.8) gives the alkali requirements, assuming no adjustment of rate of ventilation. In applying this equation, one uses the amount indicated, again obtains the blood pH and p_{CO_2}, and recalculates for additional sodium bicarbonate to be administered. In this way, gradual pH adjustment takes place, guided by measurement of the blood changes.

In answer to the question as to whether one should bring the patient to normal pH, normal bicarbonate level, or normal p_{CO_2}, it is apparent from Table 9.6 *that the objective is to bring to normal pH.* Only when the pH is at normal levels will normal function be restored to the lungs and kidneys, and metabolism proceed to eliminate the excess organic acids. Note that in case 1, Table 9.6, p_{CO_2} was 61 mmHg at normal pH when the patient was on the way to recovery.

9.2 RESPIRATORY ACIDOSIS

Values for p_{CO_2} significantly above 40 mmHg and associated with an acid pH and an elevated total CO_2 content indicate respiratory acidosis. While base deficits can be read from Figure 9.4 or Table 9.3, these patients' abnormality *can only be corrected with improved gas exchange.* This can be illustrated by a case with a gunshot wound to the head, which produced irreversible damage to the cerebral cortex. The patient was breathing, however, spontaneously, and was maintained by oxygen insufflation and an endotracheal tube, but not by artificial respiration. Concentrated NH_4OH (50 ml) was dissolved to 1 liter in 1:1, 5% glucose and normal saline solution. The objective was to see whether CO_2 could be removed by conversion to urea and excreted by the kidneys in order to ameliorate the condition and to correct any metabolic acidosis. Dilute NaOH solutions could not be added because the veins collapse and thrombose. The 500 ml of ammonia solution was administered over a 1½ hr period. Vigorous intermittent positive pressure breathing was applied for a 7-min period. Additional NH_4OH was administered and the process repeated. Table 9.7 shows the data obtained.

TABLE 9.7

Demonstration That Alkali Administration Serves No Purpose Toward Adjusting Blood pH in Respiratory Acidosis in Vivo, Only Increasing CO₂ Retention[17]

	Initial	After 500 ml 0.8 M NH₄OH[a]	After 7 min ventilation	After 500 ml additional 0.8 M NH₄OH	After 7 min ventilation	After 30 min
pH	6.80	6.81	7.30	6.82	7.35	6.80
CO₂ content, mmol/liter	60	77	31	95	36	95

[a] The patient received a total of 50 ml of conc. NH₄OH, or 800 mEq of base.

Table 9.7 illustrates clearly that *where acidosis is induced primarily by* CO_2 *retention, addition of alkali has no effect on the pH and serves only to aggravate the* CO_2 *retention.* Only correction of gas exchange can help in these cases. If the respiratory acidosis is accompanied by a metabolic acidosis, the patient may benefit by treatment of the metabolic condition only.

9.3   *METABOLIC ALKALOSIS*

Metabolic alkalosis is less common than metabolic acidosis, since disease will most commonly result in dehydration and acidosis. The most common cause of metabolic alkalosis is intestinal obstruction with excessive vomiting or prolonged aspiration of gastric secretions. Excessive loss of chloride ion occurs resulting in an excess of sodium over chloride ion in the plasma. Respiration now changes in the direction of *decreased gas exchange to retain* CO_2 *and compensate for this loss.* Increase in the p_{CO_2} of the plasma serves to raise the value of the denominator in the $[HCO_3{}^-]/0.03p_{CO_2}$ fraction, resulting in a decreased pH. In this way, we have *respiratory compensation for the alkalosis.* Some of the retained carbon dioxide forms $HCO_3{}^-$ which also raises the numerator of the fraction. The value of the fraction, however, is reduced.

Table 9.3 calculates the amount of bicarbonate ion that should be subtracted from the numerator to give the fraction a value of 19.95 and thus normal pH of 7.4. This subtraction can be performed by adding chloride ion (NH_4Cl) to neutralize the excess Na ion. This liberates $HCO_3{}^-$ to combine with the hydrogen ion formed in the reaction

$$2NH_4Cl + CO_2 \rightarrow urea + H_2O + 2H^+ + 2Cl^- \qquad (9.9)$$

$$H^+ + HCO_3{}^- \rightarrow H_2O + \uparrow CO_2 \qquad (9.10)$$

This subtracts from the numerator of the $[HCO_3{}^-]/0.03p_{CO_2}$ fraction, thus lowering the pH. NaCl or KCl is also used, either with or without NH_4Cl, depending upon the severity of the alkalosis and whether a K deficiency exists.

The rationale of giving NaCl or KCl is based on the fact that the kidney will preferentially secrete the cation and retain the chloride ion in alkalosis. The ratio of Na to Cl in the body is normally 140:103. Adding NaCl where the sodium to chloride ratio is 1:1 also serves to lower this ratio and thus correct the high value of the ratio, which is the underlying cause of the metabolic alkalosis.

In severe alkalosis, NH_4Cl is preferred because its solution is at a pH of 4.5. Thus it is a direct acidifying solution and acts more quickly. A substantial part of the ammonium ion is rapidly converted to urea in the liver. NH_4Cl, in being converted to urea, generates hydrogen ion [by equations (9.9) and (9.10)]. Ammonium ion is also readily excreted in the urine as such. All of these phenomena result in an effect as though HCl had been administered. *HCl itself cannot be administered because the veins will collapse and thrombose even with concentration as low as 0.02 N.*

Referring to Figure 4.2, the central nervous system controls the metabolic rate of the cells and synchronizes this with the action of the respiratory system. *Alkalosis serves to stimulate cell metabolism in order to generate more CO_2* so as to attempt to adjust to normal pH, while gas exchange at the lungs decreases in order to retain the CO_2. In alkalosis, the secretory glands, such as those of the stomach and the saliva, are therefore stimulated to a higher level of activity, thus manufacturing, in addition to CO_2, more fluid. "Kidney activity" increases, and if the patient is not severely dehydrated, urinary output increases. *This results in a pouring out of fluids from the body.* With a tube in the patient's stomach, gastric and intestinal secretion rises markedly. If NaCl solutions are administered to compensate for the fluid loss, the situation is aggravated, since a greater quantity of chloride is removed from the stomach than sodium. *Thus sodium chloride administration results in increase in base and aggravates the alkalosis.* Correction of blood pH will reduce the volume of fluids being secreted and excreted.[18] This can be seen from Figure 9.6.

It is not uncommon to see cases similar to that of Figure 9.6 where large volumes of drainage fluids are collected in increasing quantities daily. The volume of fluids administered was also increased daily to compensate for the fluid loss.

Increased amounts of fluids administered only resulted in greater increase of fluids lost. In Figure 9.6, in a child with an extracellular fluid volume of 2.4 liters, daily losses reached 4.5 liters, or almost twice the extracellular fluid volume. On administration of NH_4Cl and some KCl to adjust to normal blood pH and electrolyte levels, fluid excretion and secretion dropped markedly and when brought to normal pH, fluid losses ceased to be excessive. Since all fluids lost contained potassium, large losses of potassium were noted and plasma potassium levels decreased. This was corrected for by administration of KCl solutions.

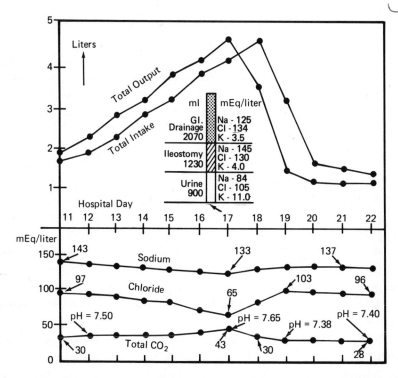

FIGURE 9.6 Effect of blood pH adjustment on fluid output in a child weighing 12 kg. Block diagram shows composition of fluid output for the 17th day before NH$_4$Cl administration. Lower chart shows changing blood electrolyte levels. *Clinical abstract*: Following surgery for an inflamed, perforated appendix this 4-yr-old female became distended and was unable to retain oral feedings. Wangensteen suction did not reduce the distention and an ileostomy was performed on the 11th postoperative day. Feeding by intravenous route was regulated in accordance with the analysis of the total output as to total volume, and electrolyte (Na, K, Cl) and nitrogen content. Blood was administered from time to time to maintain the erythrocyte count. Although the child was disoriented from the 11th to the 19th day, no obvious clinical signs could be elicited to indicate brain damage to a particular area. In spite of increased fluid intake, a rising voluminous output could not be matched. On the 17th day, when the patient's condition was grave, Ca gluconate and 10 g of NH$_4$Cl were added to the fluids, and 2 g of NH$_4$Cl given on the 18th day. A prompt decrease in fluid output was observed. Ileostomy output was 250 ml on the 18th day, less than 25 ml on the 19th day, and by the 21st day the ileostomy had sealed itself. The patient's condition was sufficiently improved so that she could be returned to surgery where adhesions, which had caused the ileum to be potentially obstructed in several places, could be divided. Recovery was uneventful. (Case of T. F. Krauss, M.D., Rockford Memorial Hospital.)

A similar phenomenon occurs in those cases where the alkalosis is a result of brain damage resulting in sodium retention without concomitant retention of chloride. *In these cases administration of NaCl aggravates the alkalosis, increasing water retention, which results in edema of the brain, aggravating the condition.*[18]

In summary, correction of an alkalosis promptly results in decrease of fluid output and permits more practicable management of the patient.[18]

It was advocated by some that the low plasma K levels in alkalosis were due to the fact that in potassium deficiency, movement of K out of the cells takes place in exchange for hydrogen ions[19]; thus the plasma would become more alkaline. It was advocated that K deficiency was the primary cause of the alkalosis. That this is not the case could easily be demonstrated by the fact that potassium acetate or carbonate would aggravate the alkalosis. Only KCl would correct it.[20–22] Some potassium was utilized to correct the potassium deficiency, but the bulk of it was found in the urine, the chloride being retained to correct the alkalosis.

KCl is more efficient than NaCl in correcting an alkalosis in that, as pointed out in Chapter 11 on the kidney, potassium is more readily excreted by the kidney than sodium, the chloride ion being retained. In severe alkalosis, with low potassium levels, where large quantities of chloride are required, many prefer to give both NH_4Cl and KCl, the KCl in amounts designed to correct only the potassium deficiency. As has been pointed out, NH_4Cl has a pH of 4.5 and serves to correct the alkalosis directly.

Mild alkalosis does occur with elevated potassium levels. Due to vomiting, such as with intussusception of the bowel or pyloric stenosis, infants develop an alkalosis and become dehydrated, with high potassium levels. Hypotonic saline is often used in these cases. With the conditions mentioned, *these infants may eventually become acidotic if untreated, even though they continue to vomit.* Laboratory data must be obtained before treatment is instituted.

In summary, metabolic alkalosis will cause the brain to signal all cells to produce more CO_2. As a result, more products of these cells are also produced other than CO_2, such as secretory and excretory fluids. Large losses of electrolytes will necessarily follow. Replacement therapy is facilitated if the blood pH is brought promptly to the normal range.

Conversely, *cell metabolism is slowed down in acidosis* in order to produce less CO_2 and ameliorate the acidosis. For this reason, patients in acidosis have reduced saliva production and their mouths are dry. They become oliguric and even anuric. Bringing to normal pH and hydration will cause the saliva and urine flow to return to normal, if underlying kidney disease is not a factor.

9.4 *RESPIRATORY ALKALOSIS*

Alkaline blood pH and values for total CO_2 of less than 25 mmol/liter, with low p_{CO_2}, mean a respiratory alkalosis, usually due to hyperventilation. This occurs when the central nervous system control of respiration has been impaired. Impairment of the central nervous system control can occur in cerebrovascular disease, head injury, or due to certain drugs (e.g., salicylate intoxication).

Excessive dosage of salicylates will first result in hyperventilation and respiratory alkalosis.[23] If the condition is prolonged, excess sodium over chloride will be excreted in the urine and a metabolic acidosis will ensue. For this reason, treatment of salicylism usually includes administration of mildly alkaline fluids so as to correct the impending acidosis, the hydration with fluids and alkaline pH stimulating urine excretion carrying along the salicylate. If kidney function is normal, the urine will contain salicylate in much higher concentration than in the plasma.

In deep anxiety, uncontrolled hyperventilation can proceed to the point where the patient lapses into coma, continuing to hyperventilate. In these cases the patient is caused to rebreathe, by placing a bag over his head, to raise the p_{CO_2} level and lower the pH.

Many patients with acute cerebrovascular pathology hyperventilate, causing a respiratory alkalosis. If saline is administered to these patients, they will often tend to retain the sodium and excrete the chloride ion, developing, in addition, a metabolic alkalosis. For this reason blood pH, p_{CO_2}, and p_{O_2} measurements need to be followed in these patients.

Just as it has been shown above that fluid therapy will not benefit a condition where respiratory acidosis is the underlying cause, so it is with respiratory alkalosis.

9.6 *RECAPITULATION*

An alkalosis or acidosis due to disparity between Na and Cl levels in the plasma is called a metabolic acidosis or alkalosis. Where the acidosis is complicated by the presence of organic acids, as in diabetes, or excess phosphate, as in uremia, we call this also a metabolic acidosis.

Where the acidosis or alkalosis is due primarily to an abnormality in pulmonary ventilation, this is called respiratory acidosis or alkalosis. In these latter cases, repair fluids can only serve to correct any secondary metabolic alkalosis or acidosis. The respiratory acidosis or alkalosis can only be corrected by proper pulmonary ventilation. The p_{CO_2} levels substantially greater or lower than 40 mmHg suggest respiratory involvement in the acidosis or alkalosis.

For repair by fluid administration, for metabolic acidosis or alkalosis, one analyzes the blood for Na, K, Cl, p_{CO_2}, pH, p_{O_2}, and total CO_2 content. From Figure 9.5, measurement of any two of the parameters p_{CO_2}, pH, and total CO_2 will permit the estimation of the third, and also bicarbonate levels and base excess or deficit. A value for base excess or deficit due to disparity between the sodium and chloride ions can be calculated directly from the sodium and chloride values. The amount of sodium bicarbonate necessary to bring to normal pH, in a metabolic acidosis, can be obtained also from Table 9.3 when the CO_2 content and pH are measured. To calculate the sodium bicarbonate required to bring to normal pH, one assumes a total CO_2 of 25 mmol/liter and obtains the mEq of sodium bicarbonate or sodium lactate required per liter of extracellular fluid from the pH. This assumes that respiration will be restored to normal when the blood pH is corrected (see discussion of calculation of fluid deficits in Chapter 10).

For metabolic alkalosis, Table 9.3 or Figure 9.5 is used to estimate the mEq of NH_4Cl required to bring to normal pH. KCl is also used if a potassium deficiency exists (see calculation of fluid deficits in Chapter 10). In either case, chloride ion is used to correct the alkalosis. Mild alkalosis is corrected with NaCl solutions.

Alkalosis stimulates fluid production by secretory and excretory organs, resulting in large fluid losses. Acidosis tends to decrease secretion and excretion of fluids in the body. Correction to normal pH will correct these tendencies.

9.6   SELECTED READING—ABNORMAL BLOOD pH: ACIDOSIS AND ALKALOSIS

Goldberger, E., *A Primer of Water, Electrolyte and Acid–Base Syndromes*, 4th ed., Lea & Febiger, Philadelphia, Pennsylvania (1970).

Peters, J. P. and Van Slyke, D. D., *Quantitative Clinical Chemistry* Vol. I, *Interpretations*, Williams & Wilkins, Baltimore, Maryland (1931).

Dickens, M., *Fluid and Electrolyte Balance, A Programmed Text*, Davis Co., Philadelphia, Pennsylvania (1967).

Siggaard-Andersen, O., *Acid–Base Status of the Blood*, 3rd ed., Williams & Wilkins, Baltimore, Maryland (1974).

Olszowka, A. J., Rahn, H., and Farhi, L. E., *Blood Gases: Hemoglobin, Base Excess and Maldistribution. Nomograms for Normal and Abnormal Bloods; Effect of Maldistribution*, Lea & Febiger, Philadelphia (1973).

9.7 REFERENCES

1. Natelson, S. and Tietz, N., Blood pH measurement with the glass electrode, *Clin. Chem.* **2**:320–327 (1956).
2. D'Elseaux, F. C., Blackwood, F. C., Palmer, L. E., and Sloman, K. G., Acid-base equilibrium in the normal, *J. Biol. Chem.* **144**:529–535 (1942).
3. Gessler, U., Clinical disturbances of electrolyte and water disturbances in neuropathies. II. Experimental and clinical data, *Klin. Chem.* **1**:144–148 (1963).
4. Hartmann, A. F. and Smyth, F. S., Chemical changes in the body occurring as the result of vomiting, *Am. J. Dis. Child.* **32**:1–28 (1926).
5. Butler, A. M., Diabetic Coma, *New Eng. J. Med.* **243**:648–659 (1950).
6. Williams, R. N., Ketosis, *Arch. Intern. Med.* **107**:69–74 (1961).
7. Seligson, D., Bluemle, L. W., Jr., Webster, G. D., Jr., and Senesky, D., Organic acids in body fluids in the uremic patient, *J. Clin. Invest.* **38**:1042–1043 (1959).
8. Van Slyke, D. D. and Cullen, G. E., Studies of acidosis. I. The bicarbonate concentration of the blood plasma, its significance and its determination as a measure of acidosis, *J. Biol. Chem.* **30**:289–346 (1917).
9. Hastings, A. B. and Shock, N. W., Studies of the acid–base balance of the blood. Nomogram for calculation of acid–base data for blood, *J. Biol. Chem.* **104**:575–584 (1934).
10. Astrup, P., Siggaard-Andersen, O., Jørgensen, K., and Engel, K., The acid base metabolism. A new approach, *Lancet* **1**:1035–1039 (1960).
11. Singer, R. B. and Hastings, A. B., An improved clinical method for the estimation of disturbances of the acid–base balance of human blood, *Medicine* **27**:223–242 (1948).
12. Sinclair, M. J., Hart, R. A., Pope, M., and Campbell, E. J. M., The use of the Henderson–Hasselbalch equation in routine medical practice, *Clin. Chim. Acta* **19**:63–69 (1968).
13. Suero, J. T. and Woolf, C. R., An equation for calculating "derived" acid base parameters, *Can. J. Physiol. Pharmacol.* **45**:891–895 (1967).
14. Astrup, P., A new approach to acid–base metabolism, *Clin. Chem.* **7**:1–15 (1961).
15. Siggaard-Andersen, O., The acid-base status of the blood, *Scand. J. Clin. Lab. Invest.* **15**:(Suppl.) 70–134 (1963).
16. Christiansen, J., Douglas, C. C., and Haldane, J. S., The absorption and

dissociation of carbon dioxide by human blood, *J. Physiol.* (*London*) **48**:244–277 (1914).

17. Natelson, S. and Barbour, J. H., Equation and nomogram for approximation of alkali requirements in acidosis, *Am. J. Clin. Pathol.* **22**:426–439 (1952).

18. Natelson, S., Chronic alkalosis with damage to the central nervous system, *Clin. Chem.* **4**:32–42 (1958).

19. Sanslone, W. R. and Muntwyler, E., Muscle cell pH in relation to chronicity of potassium depletion, *Proc. Soc. Exp. Biol. Med.* **122**:900–902 (1966).

20. Schwartz, W. B., Van Ypersele de Strihon, C., and Kassirer, J. P., Role of anions in metabolic alkalosis and potassium deficiency, *New Eng. J. Med.* **279**:630–639 (1968).

21. Kasserer, J. P., Berkman, P. M., Lawrenz, D. R., and Schwartz, W. B., The critical role of chloride in correction of hypokalemic, hypochloremic alkalosis in man, *Am. J. Med.* **38**:172–189 (1965).

22. Struyvenberg, A., de Graeff, J., and Lamijer, I. D. F., Role of chloride in hypokalemic alkalosis in rat, *J. Clin. Invest.* **44**:326–338 (1965).

23. Eichenholz, A., Mulhausen, R., and Redleaf, P., The nature of the acid–base disturbance in salicylate intoxication, *J. Lab. Clin. Med.* **58**:816 (1961).

Fluid and Electrolyte Applications

The ultimate objective in studying fluid and electrolyte balance in the human is to place treatment in the diseased state on a rational basis. The many variables and areas of uncertainty make any method of calculation necessarily a rough estimate. However, if major deficits are corrected, normal function may be restored, in which case the homeostatic mechanisms of the body can take over and complete the adjustment. For example, at a pH of 7.1, in the adult, cell metabolism is severely impaired, and the central nervous system ceases to perform its normal function. Restoration to a pH of 7.4 and correction of other major deficits will often restore this function and reestablish a drive toward normal metabolism. *The primary objective of fluid therapy is to restore the normal function of the homeostatic mechanisms.*

Of major importance are the patient's history and physical examination. The decision as to what fluids to administer, if any, will be influenced by whether the patient is edematous or dehydrated as judged by clinical observation. The state of respiration is of prime importance. Of the many examples which could be chosen, a simple example will suffice: Hyperventilation, without apparent physiological cause, may result from salicylate intoxication. Numerous other factors, including body temperature, prior fluid therapy, or history of diabetes mellitus, can serve to guide the physician in interpreting laboratory data in terms of therapy.

The immediate purpose of administering intravenous fluids to a patient can be summarized as follows:

1. To correct any electrolyte and pH abnormalities.
2. To correct osmotic pressure and volume of extracellular and intracellular fluids.

3. To supply water and electrolyte requirements on a 24-hr basis.
4. To approach caloric needs.
5. To supply vitamins and trace elements in order to maintain normal metabolism.

In order to carry out these objectives, certain information needs to be at hand. Laboratory data for pertinent components of blood must be available initially and obtained at regular intervals determined by the condition of the patient. These include, as a minimum, Na, K, Cl, hematocrit, protein, pH, CO_2 content, glucose, and urea. In addition, Ca, Mg, p_{CO_2}, p_{O_2}, and other components may need to be determined from time to time.

Urine and other fluids, such as drainage fluids from nasogastric suction or an ileostomy, should be collected on a 24-hr basis and the volume recorded. Knowledge of Na, Cl, K, and N losses in these fluids permit proper calculation of the necessary replacements.

10.1 WATER REQUIREMENTS

There are three procedures commonly employed for expressing water losses. These are *by body weight*, *by surface area*, and *by caloric expenditure*. These will now be presented.

10.1.1. By Body Weight

Requirements are expressed as milliliters per pound or kilogram of body weight. This method is used commonly in adults, although arguments have been raised indicating that it is also best to express water requirements in children based on body weight.[1]

From Table 10.1 one can readily see that prior to the onset of puberty, fluid requirements are calculated in the same manner for both males and females. For both, the volume is reduced progressively from 60 to 35 ml/lb at age 12. Subsequently, the requirement for females drops more precipitously (between the ages of 12–16) than for males, probably because they mature more rapidly. After the age of 18 the requirements are calculated in the same manner, commonly quoted at approximately 43 ml/kg.

TABLE 10.1

Approximate Fluid Requirements Based on Weight and Age[1]

	Fluid requirements									
	Males and females					Males			Females	
Age, yr	< 1	1–5	6–8	9–11	12	13–15	16–17	> 18	13–14	15–17	> 18
ml/kg	132	110	99	88	77	60	50	43	60	50	43
ml/lb	60	50	45	40	35	27	23	20	27	23	20

The figures in Table 10.1 must be taken as a rough approximation since many factors will determine the exact water needs. Wide variation will be observed even in the normal, depending upon the temperature, salt intake, and other factors.

The kidney has a high capacity to conserve or dispose of salt and water. Thus the volume and salt concentration of the urine are direct functions of the water and salt intake. In severe dehydration and renal failure, urine volume can drop to zero, or, in diabetes insipidus, can rise to levels as high as 28 liters/24 hr. Salt intake normally ranges from 5 to 15 g in the adult, with the mode being at 7–8 g daily. On any particular day, intake may be a small fraction of, or many times, these figures.

Perspiration on a warm day can bring the loss through sweat glands to as high as 5 liters per day. Several liters can be lost through the stool in diarrhea. Thus many factors affect intake, which under normal circumstances is regulated by the thirst mechanism sensitive to changes in osmotic pressure and other stimuli in the body.[2]

10.1.2 *Water Replacement*

Water replacement is calculated from losses that normally take place from the lungs, skin, via the feces, and kidneys and also those measured in drainage fluids. Water loss is normally balanced by water intake. This can be seen from Table 10.2, where the results are given of a balance study performed on a 70-kg normal male over a 24-hr period.

The 70-kg man referred to in Table 10.2 was 5 ft, 10 inches tall, with a surface area of 1.87 m². The surface area for the "normal" adult is 1.73 m².

TABLE 10.2

Water Balance in a 70-kg Healthy Male Measured and Calculated for a 24-hr Period

	Water loss		Water intake	
	ml/kg/24 hr	ml/70 kg		ml/70 kg/24 hr
Via the skin (insensible perspiration)	11	770	As liquid	1700[a]
Via the lungs	7	490	With food	900[a]
Via the kidney	22	1540[a]	Due to oxidation of food	305
Via the feces	1.5	105[a]		
Total	—	2905	Total	2905

[a] Measured, other values were calculated.

His total water loss based on weight was $2905/70 = 41.5$ ml/kg/24 hr. Compare this value with Table 10.1, where a figure of 43 ml/kg is recommended for the adult.

The volume of water lost through the lungs per day can be calculated as follows. At rest, a 70-kg man breathes 12–14 times per minute and exchanges approximately 500–600 ml at each respiration. This is roughly 8 liters of gas expelled per minute. The gas is saturated with water vapor, $p_{H_2O} = 47$ mmHg at 37°C. Assuming atmospheric pressure at 760 mmHg, we have $47/760 \times 8 = 0.495$ liter of pure water vapor per minute. Correcting to 0°C (273°K), we have $273/310 \times 0.495 = 0.436$ liter/min of pure water vapor at 0°C and 760 mmHg. Multiplying by 1440 min, we have $1440 \times 0.436 = 627.8$ liter/24 hr. Since 22.4 liters of H_2O vapor is equivalent to 18 ml (gram molecular volume), we have

$$(627.8/22.4) \times 18 = 504 \text{ ml of } H_2O \text{ for 24 hr}$$

This value needs to be corrected for water vapor taken in from the air. This will give a variable correction factor depending upon the humidity. Dividing 504 by 70, we find ~ 7.2 ml/kg/day. We usually use the round number of 7 ml/kg/day. This is markedly increased with exercise or fever. With fever and acidosis, water loss of from 1000 to 2000 ml/24 hr may be attained via the lungs.

10.1.3 *The Infant*

Of special interest is the fluid requirement of newborn infants for the first 30 days of life. An experiment to determine these values is illustrated in Figure 10.1. In this experiment infants on breast milk were weighed before and after each feeding at first in the hospital and then in the home.[3] The results shown in the figure are plotted after adjusting the values obtained to the weights of the infants. In the figure the results are being compared to two feeding formulas, one with a high albumin-to-casein ratio and the other with a high casein-to-albumin ratio. The protein content of the formulas was 1.2% in order to simulate the concentration in breast milk. The volumes of intake are not very different on the three methods of feeding.

Note from Figure 10.1 that breast-fed infants obtain very little fluid for the first three days of life before mature milk begins to flow from the mother. However, intake is higher in these children than in formula-fed infants for the next 30 days, partly because of the lower caloric content of mature breast milk as compared to artificial milk formulas employed.

After the 20th day of life the intake drops gradually and stabilizes

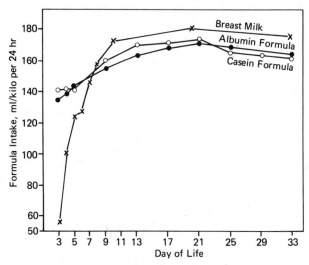

FIGURE 10.1 Volume of breast milk taken daily by normal infants during the first month of life compared to a high-albumin and high-casein feeding formula of approximately the same caloric content as the breast milk.[3] The composition of the breast milk is shown in Table 10.3. There were 15 women in each group.

TABLE 10.3

Comparison of Breast-Milk Composition on the 7th and 30th Days After Parturition with that of a Synthetic Formula and Cow's Milk as Purchased in the Market

Feeding formula	Protein, %	Carbohydrate, %	Fat, %	Na, mEq/liter	K, mEq/liter	Cl, mEq/liter	Ca, mg/100 ml	P, mg/100 ml	Mg, mg/100 ml
Breast milk, 7 days post partum									
Mean	1.42	6.98	2.92	14.7	18.1	18.9	26.3	18.0	4[a]
SD±	0.31	0.78	0.54	3.10	1.93	2.91	2.12	2.44	
Ext. range	1.1–1.9	5.4–8.2	2.4–3.1	9.2–19.5	14.8–20.8	14.6–24.5	24.0–32.0	14.5–23.3	
Breast milk, 30 days post partum									
Mean	1.03	6.37	2.98	9.94	14.5	13.6	31.0	16.7	4[a]
SD±	0.11	0.43	0.41	3.84	0.83	4.5	3.54	1.28	
Ext. range	0.78–1.31	5.2–7.3	1.4–4.5	5.9–20.0	13.2–16.5	8.5–26.0	25.3–37.6	14.7–18.3	
Typical synthetic (Similac)	1.55	7.1	3.6	12	26	17	60	44	4
Cow's milk (commercial)	3.25	5.0	3.5	21.7	38.4	30	110	100	10

[a] Estimated from analysis of other samples of breast milk.

at from 130 to 140 ml/kg for the first nine months, unless weaned prior to that time. The data in Figure 10.1 were obtained from a white population in a small city in Illinois surrounded by a farm community during the summer months. Table 10.3 compares the composition of the breast milk with that of cow's milk and an artificial feeding formula.

10.1.4 *Surface Area*

This system proposes that the body fluid requirements be calculated on the basis of body surface area rather than on body weight. Proponents of this system point out that infants become dehydrated more readily than adults, because of higher surface area to body weight ratio in infants. The body surface area of an infant is 15% of the adult, whereas his body weight is only 5% of that of an adult. Calculated by body weight, different figures for water needs are obtained for infants and adults (Table 10.1). However, calculated by surface area, this difference is markedly decreased.[4,5]

It is this concept, namely that requirements are the same for children and adults when calculations are based on surface area, that has influenced many pediatricians to estimate fluid requirements on the basis of surface area. This has been questioned by some who point out that the figures obtained by DuBois[6] for surface area (see Tables A.1 and A.2 in the appendix) were obtained on abnormal individuals, and are in error by as much as 18% in infants and 8% in adults.[7]

The rationale behind the use of surface area for calculation of fluid requirements is based on the following two principles:

1. Insensible loss of fluid and electrolyte is theoretically more closely related to body surface area than to body weight. For example, excretion from the lungs, kidneys, and sweat glands is related to surface area and only indirectly to body weight.
2. The basal metabolic rate in infants is the same as (or slightly less than) in adults, when compared on the basis of body surface area. This ratio is much greater in children when compared on the basis of body weight.

The validity of these two principles may be explored with a specific example. Referring to Table 10.2, the total loss via the skin, lungs, and feces in the adult is 1365 ml. Dividing by the surface area, we have a loss of $1365/1.87 = 730$ ml/m². The insensible loss in the infant is substantially higher, being estimated at approximately 1000 ml/m².

Newborn infants are less responsive to the antidiuretic hormone for approximately the first 30 days of life and are less able to excrete a highly concentrated urine when required to conserve water. While the adult under water deprivation requires only 0.7 ml of water/mosmol of salts, the infant requires twice that amount, or 1.4 ml of water/mosmol. Normal salt excretion required on a normal diet for both infants and adults is approximately 600 mosmol/m²/24 hr.[8] From these data, comparison may be made of the minimum water requirement per square meter for the adult as compared to the infant. This is given in Table 10.4.

Table 10.4 can be compared with Table 10.2 for the adult. In Table 10.2 for 1.87 m² of surface area, 2905 ml was excreted. This represents $2905/1.87 = 1555$ ml/m² as compared to the value of 900 ml/m² in Table 10.4 indicated as an obligatory minimum. On the other hand, the figure of 1570 is also a minimum for the infant. Most pediatricians recommend from 1800 to 2200 ml/m² for normal maintenance of newborn infants. For adults the figure of 1500 ml/m² is recommended. It is therefore apparent that in spite of the claims that fluid requirements based on surface area are the same for adults and infants, it is simply not so. *Generally, the figure of 1500 ml/m² is used in adults and 1800 ml/m² is used in infants* as an approximation.

TABLE 10.4

Comparison of Minimum Daily Water Losses for Healthy Adults and Infants on a Normal Diet[5]

	Adult, ml/m²/day	Newborn, ml/m²/day
1. Insensible loss (breath, sweat, stool)	750	1000
2. Renal (minimum)	420[a]	840[a]
3. Water gained by oxidation of foods	(−270)	(−270)
Net *minimum* total requirement per square meter per day	900	1570

[a] These figures are calculated from the observation that both newborns and adults, on a normal diet, present approximately 600 mosmol m²/day for renal excretion. The most concentrated urine in adults is 0.7 ml/mosmol, as compared to 1.4 ml/mosmol in newborns. Thus for adults one has $600 \times 0.7 = 420$, and for newborns $600 \times 1.4 = 840$, as the minimum daily water requirement for kidney function per square meter of surface area.

10.1.5   *By Caloric Expenditure*

This method is used mainly for infants and was originally estimated as 150 ml/100 cal expended.[9] However, this figure was suitable for newborns in the first 30 days of life but is too high for older children, and 125 ml of fluid/100 cal is closer to the actual value required before puberty. After puberty the value drops to approximately 100 ml/100 cal expended.[10]

Table 10.5 indicates how the figure of 125 ml/100 cal in infants is distributed among the losses in the lungs, skin, urine, and stool.

Of the procedures described, by weight, by surface area, and by caloric expenditure, the latter seems to be the most rational. It has the advantage that it automatically includes increased fluid needs due to exercise and increase in body temperature. Unfortunately it is often difficult to calculate caloric expenditure.

Table 10.6 compares the values obtained for the fluid requirements when the three procedures, weight, surface area, and caloric expenditure are compared. The maintenance fluids calculated from the weight of the patient are based on fluid requirements per kilogram of body weight listed in Table 10.5. The maintenance fluids from caloric expenditure are based on the recommended caloric intake of the Canadian Council of Nutrition.[11] The value of 125 ml/100 cal is used for children and 100 ml/100 cal for those beyond puberty. The value of 1800 ml/m^2 is used in calculating the maintenance fluids for individuals of all ages.

The large caloric intake for boys from 16 to 19 yr of age reflects

Table 10.5

Water and Electrolyte Losses per 100 cal Metabolized in Infants[9]

Route	H_2O, ml	Na, mEq	K, mEq
Lungs	15	0	0
	(10–60)		
Skin	40	0.15	0.4
	(20–100)	(01–3.0)	(0.2–1.5)
Urine	65	3	2
	(0–400)	(0.2–3.0)	(0.4–3.0)
Stool	5	0.2	0.25
	(0–50)	(0.1–0.4)	(0.2–0.3)
Total (mean)	125	3.35	2.65

Table 10.6

Comparison of Fluid Requirements Based on Body Weight, Surface Area, and Caloric Expenditures[a]

Sex	Age	Weight, kg	Height, cm	Recommended total calories per 24 hr	Surface area, m²	From weight	Maintenance fluids, ml/24 hr		
							From caloric expenditure, 125 ml/100 cal	From surface area, 1800 ml/m²	
M,F	0–1	3.2–9.2	50–75	360–900	0.20–0.42	423–1214	450–1125	360–765
M,F	1–2	9.1–11.8	75–86	900–1200	0.42–0.52	1001–1298	1125–1500	756–936
M,F	2–3	11.8–14.0	86–95	1200–1400	0.52–0.60	1298–1540	1500–1750	936–1080
M,F	4–6	15.6–19.0	100–114	1400–1700	0.65–0.78	1716–1881	1750–2125	1080–1404
M,F	7–9	21.4–26.0	118–130	1700–2100	0.83–0.96	2118–2288	2125–2625	1494–1728
M,F	10–12	28–37	135–146	2100–2500	1.07–1.24	2464–2849	2625–3125	1926–2232
M	13–15	39–52	149–160	2500–3100	1.30–1.54	2340–3120	2500–3100[b]	2340–2772
M	16–17	55–65	165–170	3100–3700	1.60–1.75	2750–3250	3100–3700[b]	2880–3150
M	18–19	67–72	174–179	3700–3800	1.79–1.91	2881–3096	3700–3800[b]	3222–3438
M	25–29	73–78	179–180	2800–2850	1.94–2.00	3139–3354	2800–2850[b]	3492–3600
F	13–15	41–53	152–157	2500–2600	1.31–1.54	2460–2650	2500–2600[b]	2358–2772
F	16–17	54–56	162–165	2350–2400	1.56–1.60	2700–2800	2350–2400[b]	2808–2880
F	18–19	57–58	165–167	2400–2450	1.63–1.66	2451–2494	2400–2450[b]	2934–2988
F	25–29	59–61	167–168	2350–2400	1.67–1.70	2537–2623	2350–2400[b]	3006–3060

[a] The recommended caloric requirements listed are based on nutrition standards set by the Canadian Council on Nutrition.[11]
[b] 100 ml/100 cal.

their occupation with vigorous exercise in the sports in which they engage at that age. Note that this is reflected in increased water intake based on caloric intake but not on weight or surface area, which do not take the increased caloric consumption into account.

The figures for fluid intake for younger children seem to be too low when based on surface area. For this reason some use 2000 ml/m² for children under 10 yr of age, 1800 ml/m² for older children, and 1500 ml/m² for adults.

In view of the fact that one is only attempting to obtain an order of magnitude of a volume in treating a patient who is ill or on intravenous fluids, any of the three methods will yield a suitable value for practical purposes if properly calculated.

10.2 RECIRCULATING FLUIDS

Large volumes of water are normally recirculated in the body. For example, saliva is swallowed and the water reabsorbed. Most of the water is reabsorbed from gastric juice, intestinal juice, and pancreatic juice. In the kidneys, 125 ml/min of fluid is filtered or $1440 \times 125 = 180,000$ ml/24 hr. Of this 180 liters, only 1.5 liters appears in the urine after reabsorption by the renal tubules.

From the above, it is apparent that any break in the reabsorption cycle can result in massive water and salt losses. This can occur, for example, with vomiting, diarrhea, intestinal drainage from nasogastric suction or fistulas, and with diabetes insipidus and diabetes mellitus. Table 10.7 summarizes the possible loss of water from fluids being recirculated.[12]

TABLE 10.7

Possible Losses from Recirculation Fluids

	ml/24 hr
Saliva	500–1500
Gastric juice	1000–5000
Bile	100–1000
Intestinal juice	700–3000
Pancreatic juice	700–1000
Kidney (failure of reabsorption)	to 28 liters/24 hr

From the above, it is apparent that these losses can only be estimated by measurement. It is obvious that *rational management of fluid therapy cannot be attained unless the drainage fluids and urine are carefully collected, their volume measured, and losses of electrolytes calculated after analysis of aliquots of these fluids by the laboratory.* These data are essential in order to calculate replacement needs.

10.3   *MAINTENANCE FLUIDS*

The fluids administered to the patient comprise the *repair fluids*, to correct any significant variation from the normal, and the *maintenance fluids*, or those designed to support the patient until he can resort to the normal oral route. This may be for a 24-hr period or for many weeks, depending on the condition of the patient.

10.3.1   *The Adult*

The *volume of fluid* for the adult for 24 hr has been discussed and summarized in Table 10.2. While the figure of 43 ml/kg is used for adults in preparing Table 10.6, this is only a rough approximation and a figure of 40 ml/kg is generally used. For the adult one can also use 1500 ml/m^2 when calculating on the basis of surface area or 1 ml/cal expended (100 ml/100 cal). These are essentially minimum figures and other factors, such as fever, will increase this requirement.

The basic ingredients to be added for this volume include salt (NaCl), potassium, calcium, magnesium, protein, nitrogen, and phosphate. The amount of these components to be added to the maintenance fluids can be deduced by examination of the amounts excreted by the three major routes, normally the urine, feces, and sweat. This is summarized in Table 10.8. The amounts vary widely since the diet of the adult varies widely. On salty foods much more salt will be excreted than on a bland diet. The protein content of the food will vary widely, which will determine the amount of nitrogen and phosphate excreted.

The percentage of nitrogen in protein varies from 17% in albumin to approximately 15% in the globulins. Using a figure of 16%, the

TABLE 10.8

Excretion of Electrolytes, Nitrogen, and Phosphorus in the Normal Adult[a]

Pathway	Na	K	Cl	Ca	Mg	Nitrogen	Phosphorus
Urine	1.10–4.10	0.40–1.45	1.10–5.10	0.03–0.42	0.035–0.20	120–250	9–15
Feces	0.05–0.10	0.10–0.20	0.05–0.15	0.25–0.50	0.13–0.30	11–36	0.01–0.02
Sweat[b]	0.10–0.60	0.05–0.30	0.10–0.60	0.005–0.20	0.001–0.003	2–6	0.001–0.005
Total	1.25–4.8	0.55–1.95	1.25–5.75	0.285–1.12	0.166–0.503	133–292	9.01–15.025

[a] Electrolyte values mEq/kg/24 hr, nitrogen and phosphorus values mg/kg/24 hr.
[b] Calculated assuming excretion of 11 ml sweat/kg/24 hr.

nitrogen excreted represents from 0.83–1.83 g of protein/kg/24 hr. For a 70-kg man this represents 58–128 g of protein/day. For maintenance, with the patient hospitalized, the lower figure is taken and usually rounded to 50 g of protein as protein hydrolysate per 24 hr. In patients who are malnourished, some prefer to administer more protein hydrolysate daily, depending upon the weight and the condition of the patient.

Bone stores huge amounts of phosphate, and maintenance fluids designed for shorter periods do not contain phosphate except as it occurs in the protein hydrolysate. A 70-kg man would require at least $9 \times 70 = 630$ mg of phosphorus per 24 hr. One liter of protein hydrolysate will supply approximately half of this amount. Patients on I.V. fluids may become oliguric. In this case it is wise to avoid phosphate replacement at first and depend on the bone stores. If the patient is on I.V. fluids for extended periods of time, phosphate needs to be considered a necessary additive.

The sodium chloride excretion in Table 10.8 varies from approximately 1.25 to 5 mEq/kg for the 70-kg individual. This is equivalent to from 87.5 to 350 mEq of NaCl or from 5 to 20 g of salt per day. These are extreme figures and a round figure of 2 mEq/kg is often employed. This calculates to 140 mEq or 8.2 g of salt per 24 hr for the 70-kg man. In practice, this value varies widely and it is not unusual to administer 2 liters of saline per 24 hr (308 mEq or 18 g of NaCl) to a 70-kg adult during a 24-hr period.

From Table 10.8 it is apparent that excretion of potassium also varies widely with diet, from 0.55 to 1.95 mEq/kg, or from 38.5 to 136 mEq/24 hr, excreted mainly in the urine. Maintenance fluids are usually calculated at 1 mEq/kg, or 70 mEq for the 70-kg man, per 24 hr. Amounts ranging from 20 to 100 mEq are usually easily tolerated and excess over that required is readily excreted in the urine.

Like phosphate, substantial amounts of Ca and Mg are present in the body stores, mainly in the bones. For a 70-kg individual, Ca excretion is, from Table 10.8, 20–78 mEq/24 hr or from 0.4 to 1.5 g of Ca^{2+} per 24 hr. For maintenance one would have to administer at least 38.5 mEq of Ca per day. For this relative large amount of Ca^{2+}, at the neutral or alkaline pH of the maintenance fluids, the calcium would precipitate as the hydroxide. In Ringer's lactate some calcium is complexed and kept in solution (see Table 10.10) and also in the protein hydrolysates. By the use of the gluconate some calcium is also kept dissolved. In general, for short periods of time, calcium requirements

TABLE 10.9

Maintenance Fluid and Nutrients for the Normal Adult for 24 hr (Approximate Values)

Component	Per kg	Per m²	Per 70-kg man
Water, ml	40	1500	2800
Salt (NaCl), mEq	2	75	140
			(8.2 gs)
Potassium, mEq	1.0	37	70
Protein, g	1.0	27	70
Caloric requirement	40	1500	2800

are not met. For longer periods one often resorts to administration by the I.M. route from time to time.

Magnesium losses in the urine, feces, and sweat range from 2 to 6 mg/kg as calculated from Table 10.8, or 140–420 mg (11.6–35 mEq) per 24 hr for the 70-kg man. As is the case for calcium, the usual parenteral fluids (Table 10.10) do not contain adequate amounts of magnesium for maintenance. This, like calcium, is usually administered from the ampoules listed in Table 10.11 or added to the fluids.

Table 10.9 summarizes the basis for calculation of the maintenance fluids for an adult based on the above discussion. The Mg and Ca figures are not included since they are usually not added initially unless there is a demonstrated deficiency. This will be discussed below. (See 10.4.1. Total Intravenous Alimentation.)

10.3.2 *Repair and Maintenance Solutions*

Repair and maintenance fluids of varied composition are commercially available. The composition is stated on the bottle, and calculations can be made therefrom. However, it is preferable to design the fluids for a particular patient. When this is done, only a limited number of solutions are required, simplifying the problem. These basic solutions are listed in Table 10.10.

Some materials, such as calcium, magnesium salts, bicarbonate solutions, and vitamins, create a condition of instability in the parenteral fluids administered. For this reason they are dispensed from ampoules separately and added to the fluids just before their application. This also permits I.M. or I.V. administration of these materials directly in

TABLE 10.10

Parenteral Fluids Used in Repair and Maintenance

Solution	Na, mEq/ liter	Cl, mEq/ liter	K, mEq/ liter	Carbo-hydrates, g	Calories per liter	Other
Normal saline (0.9%)	154	154	—	—	—	—
Normal saline $\bar{c}$ glucose (5%)	154	154	—	50	200	—
Hypotonic saline (0.45% + 5% glucose)	77	77	—	50	200	—
Hypertonic saline (3%)	513	513	—	—	—	—
Hypertonic Saline (5%)	855	855	—	—	—	—
Isotonic glucose (5%)	—	—	—	50	200	—
Glucose (10%)	—	—	—	100	400	—
Sodium lactate ($\frac{1}{6}$ M)	167	—	—	—	45	Lactate: 167 mEq
Fructose (5%)	—	—	—	50	200	—
Fructose (10%)	—	—	—	100	400	—
5% Glucose, 5% ethanol	—	—	—	50	650	50 ml ethanol
Ammonium chloride (2.14%)	—	400	—	—	—	NH_4^+: 400 mEq
Ringer's	147.5	156	4	—	—	Ca: 4.5 mEq
Ringer's lactate	130	109	4	—	8	Ca: 3 mEq Lactate 28 mEq
Protein hydrolysate[a] (5%) (Amigen) (Fortified casein hydrolysate)	35	22	19	—	200	Pr. Eq: 50 g Ca 5 meq/liter Mg 2 mEq/liter PO₄ 30 mEq/liter
Protein hydrolysate 5% + glucose	35	22	19	50	400	Pr. Eq: 50 g Ca 5 mEq/liter Mg 2 mEq/liter PO₄ 30 mEq/liter
Protein hydrolysate 5% + glucose 5% + ethanol 5%	35	22	19	50	850	Pr. Eq: 50 g 50 ml ethanol Ca 5 mEq/liter Mg 2 mEq/liter PO₄ 30 mEq/liter

[a] An alternative is Aminosol, which is a fibrinogen hydrolysate.

TABLE 10.11
Additives to Maintenance Fluids Available in Ampoules

Solution	Ampoule volume, ml	Content	Total content
Sodium bicarbonate (7.5%)	50	$NaHCO_3$	44.6 mEq
Potassium chloride	10	KCl	10, 20, or 40 mEq
Calcium gluconate (10%)	10	Ca^{2+}	4.47 mEq
Calcium chloride (10%)	10	$CaCl_2$	13.4 mEq (1g)
Magnesium sulfate (10%)	20	$MgSO_4$	16.2
Ascorbic acid	2	Ascorbic acid	500 mg and 1 g ampoules
Vitamin B complex (e.g., Solu-B)	10	B complex vitamins to be added to one liter of parenteral fluids	

special cases as required. The most commonly used ampoules are listed in Table 10.11.

In addition to the solutions listed in these tables, *one must consider blood and plasma as repair solutions*. Because of the sodium citrate added to blood, donor plasma sodium levels are of the order of 175 mEq/liter and chloride levels are approximately 90 mEq/liter. Thus, in donor's plasma, chloride is reduced by 13 mEq/liter and sodium is elevated by 35 mEq/liter from the normal. The total disparity is about $35 + 13 = 48$ mEq/liter. Since a unit of plasma is 300 ml, *administering a unit of plasma is equivalent to giving 15 mEq of NaHCO₃*. A unit of blood is 500 ml, of which 300 ml is plasma. Therefore, administering *a unit of whole blood is also equivalent to giving 15 mEq of NaHCO₃*.

To relieve tetany induced by moving from a severe acidosis to pH 7.4, some prefer calcium chloride solutions to calcium gluconate since the calcium content is three times as great for the same concentration. An ampoule (10 ml) of 10% calcium chloride solution contains 13.4 mEq of Ca while an ampoule of Ca gluconate contains 4.47 mEq (see Table 10.11). Calcium and magnesium solutions cannot be mixed with sodium bicarbonate solutions since they will precipitate due to the alkaline pH, as has been pointed out.

When the adrenal is perfused with cholesterol and ascorbic acid, the yield of corticoids becomes markedly elevated as compared to that obtained without ascorbic acid.[13] Since the patients being treated are under stress, it is advantageous to adminster ascorbic acid with the fluids. Ascorbic acid is readily oxidized in alkaline solution. It is preferred, therefore, to add the ascorbic acid to the neutral fluids. Since sodium lactate solutions are not alkaline (usually pH 6.8), ascorbic acid may be added to lactate solutions but preferably not to sodium bicarbonate solutions (pH 8.6).

Patients on maintenance fluids for extended periods of time require vitamin supplementation. The B-complex vitamins are all coenzymes for the anaerobic and aerobic systems of metabolism and their deficiency will prevent rapid establishment of electrolyte and fluid balance. The ampoules containing the vitamins listed in Table 10.11 are designed to be added to 1 liter of maintenance fluid.

A typical maintenance regimen for a 70-kg adult is listed in Table 10.12 to show how the principles discussed above might be applied.

It is apparent that caloric requirement will not be met, normally, unless one resorts to administration of hypertonic solutions or fat emulsions.[14] The use of fat emulsions is now not generally recommended and is disapproved by the Pure Food and Drug Administration, except for experimental purposes. Failure to meet caloric requirements does not become a serious problem unless the patient is maintained on I.V. fluids for periods of time exceeding 30 days, or unless the patient was admitted malnourished. For practical purposes, it is best to rely on body fat stores or to attempt oral tube feeding, if the patient is not suffering from bowel obstruction, for prolonged maintenance therapy. Some use 20% glucose in order to meet caloric requirements (see section 10.4.1).

From Table 10.9 one notes that recommended protein intake is 1 g/kg or 70 g for a 70-kg individual. Note also from the table that 1 g of protein is recommended for 40 cal expended $(2700/40 = 70)$.

Referring to Table 10.10, it appears that 1 liter of protein hydrolysate will approach protein requirements for the immobilized

TABLE 10.12

Typical Maintenance Regimen for a 70-kg Adult for a 24-hr Period

1000 ml 5% glucose in normal saline + 10 mEq KCl +	
500 mg ascorbic acid + one ampoule vitamin B complex	200 cal
1000 ml 5% protein hydrolysate + 5% glucose (19 mEq K$^+$)	400 cal
1000 ml 10% glucose + 10 mEq KCl	400 cal
Total	1000 cal

patient. This is not the case since protein hydrolysates are inefficiently utilized.[15] Each liter of protein hydrolysate contains 19 mEq of potassium. It is also advisable that glucose and saline be present, not only to meet the caloric requirement, but also because protein hydrolysates are more efficiently utilized in the presence of glucose, potassium, and sodium ion. Under normal circumstances, in the presence of normal serum potassium levels, additional potassium need not be added to the protein hydrolysate solution.

In blood the ratio of sodium to chloride is 140/103. It would be expected that excess sodium in the form of lactate would be recommended for the maintenance fluids. If this were done, the patient would eventually go into alkalosis. Experiments with dogs have shown that if they are given fluids with the Na/Cl ratio as it exists in the plasma, the animals develop an alkalosis. At a ratio of 116 mEq Na to 100 mEq Cl the animals will maintain their blood pH.

In Table 10.12 excess of Na^+ over Cl^- is not administered. With normal kidney function the kidney will excrete the sodium and chloride in approximately stoichiometric amounts and excess alkali is not essential. Some prefer to add 16 mEq of sodium lactate for every 100 mEq of sodium chloride administered. Note that in Ringer's lactate, the excess of sodium over chloride is 21 mEq and this was designed with the objective of preparing a fluid which would maintain blood pH at its normal level.

From the discussion above it is apparent that maintenance fluids in the adult can be readily calculated from Table 10.9 in combination with Table 10.6 and that the values used allow a fairly wide range of modification to adjust to the particular patient and his condition at the time the fluids are being administered.

In any case it is important *to monitor the effectiveness of the regimen by analysis for the electrolyte components of the blood, following also the serum protein and hematocrit levels* in order to discover any changes in the state of hydration of the patient and in electrolyte composition.

10.4 *MAINTENANCE FLUIDS IN INFANTS*

In the newborn, the only source of nutrition is the feeding formula and some added water. Balance studies are therefore less difficult to perform for the infant as compared to the adult.

It would be expected that measuring the volume and composition of breast milk consumed by infants who received all of their nourishment at the breast would define their maintenance requirements. The results of such a study are described in Figure 10.1 and Table 10.3. Table 10.3 also lists the composition of a typical synthetic infant feeding formula as compared to cow's milk. The cow's milk does not represent any particular breed of cow but is a composite as purchased in the market.

From the data in Table 10.3 the fluid and electrolyte intake of normal infants during the 21st day of life is computed and summarized in Table 10.13.

The values in Table 10.13 were arrived at as follows. From Figure 10.1, the infants were taking 179 ml/kg of milk. Since their mean weight was 3.25 kg, one obtains 581 ml as the total volume taken per day. From Table 10.3 one notes a mean sodium concentration for breast milk as 14.7 mEq/liter. The sodium content of the day's formula is then $0.581 \times 14.7 = 8.6$ mEq. Since the infant's surface area was 0.2 m², multiplication by five yields 43 mEq of Na/m². Dividing the total sodium intake by 3.25, which is the mean weight of the infants, yields $8.6/3.25 = 2.65$ mEq/kg/24 hr. Note that this is in the general range of that used for maintenance in the adult (2 mEq/kg/24 hr, see Table 10.9). The values for the other elements were calculated in a similar manner.

TABLE 10.13

Observed Intake of Fluid and Electrolytes on Newborns on the 21st Day of Life as Compared to Recommended Values Using a Synthetic Feeding Formula for Infants[a]

Component	Breast milk			Synthetic formula[b]		
	Total/24 hr	per m²	per kg	Total/24 hr	per m²	per kg
Water, ml	581	2900	179	439	2195	135
Na, mEq	8.60	43	2.65	5.27	26.4	1.62
Cl, mEq	11.1	55.8	3.42	7.46	37.3	2.30
K, mEq	10.6	52.9	3.26	11.4	5.70	3.51
P, mg	105.3	526.5	23.4	193	965	59.4
Ca, mg	153.9	769.2	47.4	263	1315	80.9
Mg, mg	23.4	117.0	7.20	17.6	88.0	5.42
Protein, g	8.25	41.25	2.53	6.80	34.0	2.09
Calories	347.9	1739	107	294	1470	90.5

[a] Mean weight was 3.25 kg with 0.2 m² surface area.
[b] Similac.

For the synthetic formula, the figure of 135 ml/kg was found to be taken by other infants on their 21st day of life. Using the figures for the composition of the synthetic formula from Table 10.3, calculation resulted in the figures of Table 10.13. Dividing the volume of water by the calories consumed, one obtains 149 ml/100 cal for the synthetic formula. A figure of 150 ml/100 cal is used by many in calculating maintenance fluids for infants. For breast milk this value calculates to 167 ml/100 cal consumed.

The intake on breast milk was unusually large in this study partly because it was carried out during the hot summer months in a rural area (no air conditioning). The caloric content of the breast milk was relatively low so that the calories consumed per kg (107) is in the range of, or even somewhat below, that usually recommended.

From Table 10.5 it is estimated that sodium excretion per 100 cal consumed in infants is 3.35 mEq/100 cal and for potassium is 2.65 mEq/100 cal. For 300 cal (approximately 24 hr) these figures become 10.05 mEq for Na and 7.95 mEq for K. These figures are in the general range of the requirements calculated from actual intake in Table 10.13.

Comparing Table 10.13 for infants with Table 10.9 for adults, we find that salt (NaCl) requirements are higher in the adult while potassium requirements are lower when based on surface area. For infants, a figure of 30–50 mEq/m²/24 hr for sodium, chloride, and potassium is widely quoted.[4,16] From the data in Table 10.13 a better figure would be 30–60 mEq/m²/24 hr for the infant.

It has been pointed out that a wide tolerance exists for Na^+, Cl^-, and K^+ excretion when adequate water is supplied, the figure being estimated as 10–250 mEq/m²/24 hr as extremes.[17-19]

Cow's milk has three to four times the calcium concentration of breast milk (Table 10.3). For this reason synthetic feeding formulas made from cow's milk are traditionally high in calcium content. This is apparent in Table 10.13. The same situation applies to phosphate and magnesium. Note that phosphate is expressed as mg of phosphorus to avoid ambiguity. Since phosphate exists as $H_2PO_4^-$, HPO_4^{2-}, and PO_4^{3-}, it is difficult to express its concentration in milliequivalents since this would change with change in pH and other factors. One can use millimoles but this is not done. Some express its concentration in fluids in milliosmoles, which is even more ambiguous since the milliosmoles change also with pH. What is usually meant is milliequivalents expressed as PO_4^{3-}, which is equivalent to $31/3 = 10.3$ mg of phosphorus. Note that for serum, a divisor of 1.8 is used (see p. 76), which is distinct from the calculation used in feeding formulas. As pointed out, there is no problem if the figure is expressed in mg phosphorus/100 ml.

In summary, maintenance I.V. fluids for the infant should contain from 30 to 60 mEq of NaCl and 30–60 mEq of K^+ per m², in a volume ranging from 1800 to 2200 ml/m², depending upon various factors, particularly the age of the infant. It is preferable to use the lower values initially so that one can increase the amounts if indicated. The figures in Table 10.6 and 10.13 are to be considered as rough approximations and the values need to be adjusted to the requirements of the particular infant treated.

If one were to set up fluids for the infants discussed (0.2 m²), then one would administer approximately 500 ml of fluid per 24 hr. This would contain (using a mean value) $45/5 = 9$ mEq of NaCl and 9 mEq of K^+. Note that normal saline contains 77 mEq of NaCl in 500 ml as compared to 9 mEq in this case. This solution would then be hypo-osmotic. For this reason it is important that the solution contain glucose to bring to approximately normal osmotic pressure. This would require that the 500 ml contain approximately 25 g of glucose. Approximately 100 mg of ascorbic acid and some B-complex vitamins would also be added.

Note that 25 g of glucose would yield only 100 cal, which is approximately one-fourth of the needs. With the solution being administered slowly, 10% glucose or fructose might be used to yield twice the number of calories.

The potassium administered needs to be balanced with a negative ion and one might use some K as the phosphate and some K as the chloride or lactate, depending upon the condition of the patient.

Note that all natural milks contain a higher chloride content than sodium (Table 10.3). Mechanisms for retention of excess sodium over chloride are normally in operation in both adults and children. For this reason maintenance formulas do not need to have a preponderance of sodium over chloride to simulate the plasma composition.

10.4.1. *Total Intravenous Alimentation*

The problem of total intravenous alimentation often occurs where intravenous feeding needs to be carried out for extended periods of time for any of numerous reasons, such as bowel obstruction, chronic diarrhea, malabsorption syndromes, and other reasons. Generally, in adults, stored body fat and minerals serve to supplement intravenous feeding, and with intravenous carbohydrate, protein hydrolysate, and vitamin and mineral supplements this does not present an insurmountable problem.[20] However, in premature infants and newborns, these stores are not available in substantial amounts. Further, the problem is compounded by the need to achieve growth and weight gain, which is not the problem in the adult.

The major problem was twofold. First, the need to maintain an indwelling catheter for extended periods of time without tissue necrosis. This has been achieved by the use of Silastic, a flexible and inert polysilicone material. This is inserted into the internal or external jugular vein and threaded into the superior vena cava, the tip of the catheter lying just above the right atrium.[21]

The second major problem was to achieve the caloric requirements of the infant, which must reach at least 110 cal/kg/24 hr. A solution of 20% glucose containing 3% protein hydrolysate contains $23 \times 4 = 92$ cal/100 ml. By infusing this solution at the rate of 130 ml/kg/day this yields a protein intake of 4 g/kg/day and 120 cal/kg/day. At this rate premature infants and newborns have been shown to be completely nurtured for periods in excess of 100 days, weight gain being as great as that observed in infants on oral feeding.[22] Most investigators in this field now use 2.5 g of protein hydrolysate per kg per day since it results in adequate growth with less stress on the kidneys in removing excess nitrogen.[22a]

A typical regimen would include in the fluids, in addition to the glucose and protein hydrolysate per kilogram per 24 hr, NaCl, 3–4 mmol; KH_2PO_4 2–3 mmol; calcium gluconate, 0.25 mmol; $MgSO_4$, 0.125 mmol. In addition, folic acid, 50–75 μg and B_{12}, 50 μg are given per day along with 1 ml of a multivitamin B complex and vitamin C preparation. The infusion also needs to be supplemented from time to time by blood infusions to maintain the hematocrit value and administer iron.

Problems of sterility need to be carefully maintained in preparing the fluids. As a further precaution the administered fluids are passed through a 0.22 μm millipore filter.

10.5   *REPAIR SOLUTIONS*

For rational fluid therapy one needs to bear in mind the following points:

1. Fluid therapy, if indicated clinically, should include both correction of deficits and supply of the normal requirements of the patient.
2. Laboratory data should be obtained initially and during treatment to evaluate the effectiveness of the treatment and avoid gross errors.
3. Treatment should be conservative to permit changes if they should be indicated.

Initially, plasma levels for sodium, potassium, chloride, pH, glucose, creatinine, urea N, total protein, total CO_2, p_{CO_2}, and p_{O_2} are the minimum amount of information necessary for evaluating plasma deficits with a view toward designing corrective fluids. From the physical examination and protein and hematocrit values, a rough estimate of the degree of dehydration is often possible. The creatinine and urea N levels are necessary to evaluate kidney function, especially when compared to the values obtained for those components after treatment. Glucose levels are useful in estimating whether hyper- or hypoglycemia is a contributing factor. Blood p_{CO_2} and p_{O_2} levels are important in evaluating respiratory exchange.

The objective is to obtain first an estimate of the electrolyte requirements in order to bring to normal pH and to correct deficits, and after correcting these deficits, to evaluate the need for further corrective and maintenance fluids. Calculation of deficits will be necessarily only approximate and require revision as one proceeds, evaluating the course with the aid of the laboratory data. One must be certain that maintenance fluid requirements are added to the repair solutions. In many cases this comprises a substantial or major part of the total fluids administered.[8] In presenting the method for calculation of deficits, a general formula will be presented for each component and then justified.

10.5.1   *To Correct Erythrocyte Deficit*

To raise the hematocrit value one unit, 2.2 ml of whole blood per kilogram of body weight is required. This is equivalent to using 1 ml of whole blood per lb of body weight. If packed cells are used, 0.4 ml/lb or 0.88 ml/kg of body weight is required to raise the hematocrit value by one unit.

Example. Patient weighs 154 lb (70 kg), hematocrit value is 30%. To find amount of blood to raise to 40%.

$$1 \times 154 \times 10 = 1540 \text{ ml of whole blood}$$
$$2.2 \times 70 \times 10 = 1540 \text{ ml of whole blood}$$
$$0.4 \times 154 \times 10 = 616 \text{ ml of packed cells}$$
$$0.88 \times 70 \times 10 = 616 \text{ ml of packed cells}$$

Assumptions. Donor's blood has a hematocrit of 40%. The blood volume of the patient will readjust itself to its original volume. The blood volume is 9% of the body weight. This approximate figure is valid for infants but is also applicable to adults for practical purposes (see Table 10.14).

Rationale. The amount of blood per kilogram of body weight is $0.09 \times 1000 = 90$ ml. If 1 ml of packed cells is added to 100 ml, the volume would be 101 ml. Stretch receptors would supply the signal to readjust the volume to 100 ml. The hematocrit would then be one unit higher. For 90 ml of blood, one would require $90/100 \times 1 = 0.9$ ml of packed cells. If whole blood containing 40% of red cells is used, we need $100/40 \times 0.9 = 2.25$ ml. Since there are 2.2 lb/kg, this calculates to *1 ml of blood per lb of body weight,* a very convenient number, if we use 2.2 in place of 2.25. Since the percentage of blood is somewhat less than 9% (see Table 10.14), and since we are dealing with an approximate calculation, this is justified. Calculating back from 1 ml/lb unit rise in hematocrit, we obtain 2.2 ml of blood/kg and 0.88 ml of packed cells per kg. Since we require 1 ml of blood/lb, this is equivalent to 0.4 ml of packed cells since we have assumed a 40% erythrocyte content. Table 10.14 details the whole blood, plasma, and erythrocyte percentages as related to body weight for adults of both sexes and for infants.

TABLE 10.14

Volume of Whole Blood, Erythrocytes, and Plasma in the Human[12]

		Mean, ml/kg	Range, ml/kg
Whole blood	Male (adult)	76	65–80
	Female (adult)	66	55–70
	Infant (newborn)	84	75–90
Erythrocytes	Male (adult)	31	28–34
	Female (adult)	23	21–25
	Infant (newborn)	43	35–51
Plasma	Male (adult)	45	34–58
	Female (adult)	43	32–59
	Infant (newborn)	41	32–50

10.5.2 *To Correct Na, K, and Cl Deficit*

To raise the ionic concentration of Na, K, or Cl in the plasma, we require 1 mEq of the ion for 5 kg of body weight.

Assumptions. Normal level for Na is 140 mEq/liter, for K is 4.6 mEq/liter, and for Cl is 103 mEq/liter. Extracellular fluid comprises 20% of the body weight, and one is correcting the overall level of the extracellular fluid. Extracellular fluid volume will readjust itself to its original volume. See Table 5.2 for actual extracellular fluid volume in adults and infants.

Example 1. Patient's plasma shows Na = 120, K = 3.6, Cl = 103 mEq/liter. Patient weighs 70 kg. The sodium level is 20 mEq/liter too low, the K level is 1 mEq/liter too low. The object is to find the total deficit for these ions.

Using the formula given above for ion deficits (1 mEq/5 kg body weight) we write

$$\text{total Na deficit} = 70/5 \times 20 = 280 \text{ mEq}$$

$$\text{total K deficit} = 70/5 \times 1 = 14 \text{ mEq}$$

Calculation can also be made from the assumptions. Since the extracellular fluid volume is 20% of the body weight, then the patient has a volume of $70 \times 0.2 = 14$ liters of extracellular fluid. Sodium deficit is 20 mEq/liter and for 14 liters the deficit is $14 \times 20 = 280$ mEq. For potassium the deficit is $14 \times 1 = 14$ mEq.

Example 2. Patient's plasma shows Na = 140, K = 5.0, Cl = 80 mEq/liter. Chloride deficit is $103 - 80 = 23$ mEq/liter. Patient weighs 70 kg. Total Cl deficit is $70/5 \times 23 = 322$ mEq. This is equivalent to $322/400 \times 1000 = 805$ ml of 2.14% NH_4Cl solution (see Table 10.10). The potassium level is in the upper limit of the normal range. Adjustment of this value would take place with subsequent maintenance fluids.

Example 3. Patient's plasma shows Na = 120, K = 5.0, Cl = 110 mEq/liter. The sodium deficit is 20 mEq/liter. The chloride excess is 7 mEq/liter. The object is to bring the ratio of sodium to chloride to 140/103. We therefore add the sodium deficit to the chloride excess and get 27 as the sodium deficit. If the extracellular fluid were so corrected, we would get a Na level of 147 and chloride would remain at 110. However, the elevated chloride usually shows dehydration and 5% glucose in water would also be administered, permitting adjustment to

140 and 103 by dilution. For the 70-kg individual, the total sodium deficit is then $70/5 \times 27 = 378$ mEq.

Rationale. If we assume that the extracellular fluid volume is 20% of the body weight, then we need 5 kg of body weight to obtain one liter of extracellular fluid. Adding 1 mEq of Na, K, or Cl per 5 kg of body weight would then raise the level by 1 mEq/liter of extracellular fluid. Stretch and osmoreceptors will adjust the blood volume to the original value. The figure of 20% is a rough approximation, as has been pointed out before (Table 5.2). The extracellular fluid volume varies with age and sex.

10.5.3 *To Correct Plasma Protein Deficit*

To raise the plasma protein level by 1 g/100 ml, 7 ml of plasma or 12 ml of whole blood is required per kilogram of body weight.

Assumptions. Blood volume will readjust itself to its original volume. Donor's plasma contains 7% protein and has a hematocrit value of 40%. Plasma comprises $\sim 5\%$ of the body weight (see Table 10.14).

Example. The patient weighs 70 kg. Plasma protein level is 5.0 g/100 ml. To raise to 7.0 g/100 ml. Deficit is 2 g/100 ml of plasma:

$$2 \times 7 \times 70 = 980 \text{ ml of plasma required}$$

$$2 \times 12 \times 70 = 1680 \text{ ml of whole blood required}$$

Rationale. One kilogram of body weight contains $0.05 \times 1000 = 50$ ml of plasma. To raise 1 g/100 ml, one needs 0.5 g of protein for 50 ml of plasma or 1 kg of body weight. We assumed a level of 7 g/100 ml in donor's plasma. Then 0.5 g is present in $0.5/7 \times 100 = 7.14$ ml. This rounds out to 7 ml of plasma. Since 60% of blood is plasma, in accordance with our assumptions, then $100/60 \times 7 = 12$ ml of whole blood, in round numbers. Thus 0.5 g of protein is contained in 12 ml of the donor's blood.

10.5.4 *Correction of Blood pH*

10.5.4.1 *Metabolic Acidosis*

To correct blood pH in metabolic acidosis calculate first the disparity between sodium and chloride levels. Administration of fluids to correct this disparity will often correct the condition in mild acidosis. If this correction is made in severe cases, the patient may still be in acidosis. In this case the nomogram of Figure 9.5 is used to find the base deficit. This is compared with the amount of sodium bicarbonate necessary to bring to normal pH, from the nomogram of Figure 10.2.[23] Figure 10.2 is a graphic representation of Table 9.4. Both these values are compared with the value obtained from the sodium-to-chloride disparity. If the latter value compares with that obtained from Figures 9.5 and 10.2, then this suggests that the acidosis is simply due to the disparity between

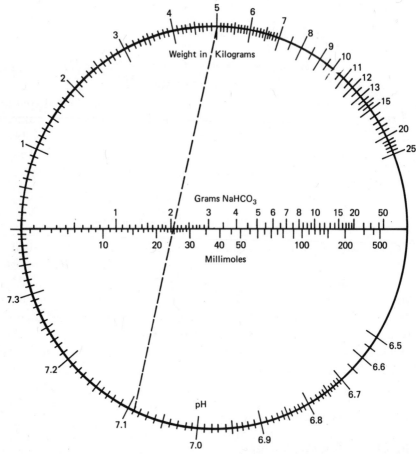

FIGURE 10.2 Nomogram to calculate amount of alkali required to bring to normal pH in acidosis. The nomogram will yield the amount of sodium bicarbonate or lactate required for 1 liter of extracellular fluid or 5 kg of body weight.[20]

Na and Cl. If the base deficit from Figure 9.5 is substantially greater, then this probably signifies an increased accumulation of organic acids such as lactic acid over that compensated for by hyperventilation.[24] If the result from Figure 10.2 is higher than that from Figure 9.5, then other factors are involved, such as respiratory problems.

In any of the above cases sodium bicarbonate is required if the pH is less than 7.28. If organic acids have accumulated in substantial amounts, added lactate will also not be metabolized. The pH of sodium lactate solutions is approximately 6.8 since they contain excess lactic acid over sodium for stability. If administered in severe acidosis, the lactate will not be metabolized and *actually aggravates the acidosis*.[25] However, sodium lactate is very effective in correcting a mild alkalosis.

Example: Metabolic acidosis. Patient weighs 70 kg; Na = 130, Cl = 110, K = 5.6 mEq/liter; CO_2 content is 9 mmol/liter; pH is 7.09; p_{CO_2} = 31 mmHg. Sodium deficit (see above) is 10 + 7 = 17 mEq/liter. Patient's requirement is 70/5 × 17 = 204 mEq of Na. $NaHCO_3$ is used because the pH is less than 7.25. From the nomogram of Figure 9.5, base deficit is 19 mEq for 5 kg of body weight or 228 mEq for 70 kg. This figure is close to that derived from the disparity between sodium and chloride ion.

From Figure 10.2, connecting the 7 kg mark to the pH mark of 7.09, we obtain 2.9 g of $NaHCO_3$ for 7 kg or 29 g for 70 kg. This corresponds to 341 mEq of $NaHCO_3$ required to bring to normal pH if respiration returns to the normal rate and depth. Subtracting 228 from 341, we note 113 mEq required above the base deficit indicated by Figure 9.5 to bring to normal pH.

An ampoule of $NaHCO_3$ contains 44.6 mEq (Table 10.11). This suggests five ampoules for the 228 mEq from base deficit and 7.6 ampoules for the total requirement. Since the patient may dispose of some of his deficit through the urine and by metabolism as correction fluids are supplied one necessarily corrects to normal pH cautiously, taking repeated measurements of pH, p_{CO_2}, p_{O_2}, and electrolytes during the process of repair. On the basis of earlier discussion, one brings to a pH somewhat less than normal (e.g., 7.35) to prevent alkalosis developing as the organic acids are removed and normal breathing is restored (see pp. 234–236).

It is to be noted that administration of sodium bicarbonate will at first only slowly correct the blood pH of the patient. The reason for this is that in the vicinity of pH 6.1, one reaches the flat part of the pH curve where maximum buffering activity exists (Figure 2.1). Therefore the plot of change in pH with the amount of bicarbonate administered is not a straight line. As one approaches pH 7.4, one reaches a point on the curve where small additions of $NaHCO_3$ will produce large changes in blood pH. Figure 10.3 plots the change in pH as one infuses the $NaHCO_3$ into the patient as calculated from Figure 10.2.

In explanation of Figure 10.3, let us assume that a particular patient showed a pH of 6.9 and calculations required that five ampoules of $NaHCO_3$ be administered (5 × 44.6 = 223 mEq). From Figure 10.3 it can be seen that after three ampoules had been administered, blood pH would reach only 7.14. After four ampoules, blood pH would be 7.24. Thus 80% of the calculated amount of $NaHCO_3$ will move the pH only 68% toward the normal. The last 20% will move it the rest of the way. Only one-half ampoule excess would bring the pH to 7.55. It is therefore apparent why caution must be used in administering $NaHCO_3$. Some prefer to bring to pH 7.3 with $NaHCO_3$ and continue

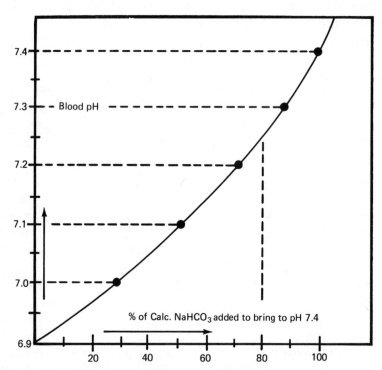

FIGURE 10.3   Theoretical plot of the percentage of required base administered vs. pH.

with sodium lactate for this reason. In many cases, if the patient is brought to pH 7.31–7.35, normal metabolism will be established and no more alkali need be administered. One then follows with maintenance fluids (Table 10.9 or Table 10.13) administered in accordance with the measured rate of urine output and the results of periodic blood analyses, until the patient's condition is stabilized.

In some patients, administration of $NaHCO_3$ will not move the pH toward normal when the full amount calculated from Figure 10.3 is administered, the pH remaining at values as low as 7.1. *This usually signifies that the condition is irreversible.* This occurs in some patients who have been in coma and severe metabolic acidosis for extended periods of time so as to destroy the metabolic regulatory mechanisms. Addition of alkali merely causes movement of organic acids of intermediate metabolism out of the tissues into the extracellular fluid spaces. Figure 10.3 then serves to advise as to an *upper limit* to the amount of alkali to be administered. If pH change does not move approximately in accordance with Figure 10.3, then the condition is probably irreversible.

10.5.4.2 *Metabolic Alkalosis*

As pointed out, this condition occurs commonly in patients who have been subjected to gastrointestinal drainage by nasogastric tube. An example will now be presented. It also occurs with excretion of excess chloride over sodium in the urine, the sodium retention being a result of damage to the central nervous system.[26]

Example. Patient weighs 70 kg; Na $= 136$, Cl. $= 80$, K $= 2.6$ mEq/liter; pH is 7.58; CO_2 content is 50 mmol/liter; $p_{CO_2} = 50$ mmHg. There is a sodium deficit of 4 mEq and a chloride deficit of 23 mEq. The net chloride deficit is 19 mEq/liter of extracellular fluid. From Figure 9.5, base excess is 22 mEq/liter, and from Table 9.4 chloride requirement is 18 mEq/liter. Since these figures are only approximations, we may say that they are in agreement. Note that when the fluids are set up, additional NaCl will be needed to correct to Na $= 140$ and Cl $= 103$ mEq/liter.

For a 70-kg man the chloride deficit is $70/5 \times 19 = 266$ mEq. Potassium deficit is $4.6 - 2.6 = 2.0$ mEq/liter. Total potassium deficit is $2 \times 70/5 = 28$ mEq.

The potassium administered as KCl would add 28 mEq of chloride ion and decrease the chloride deficit by that amount. Potassium chloride also acts to correct an alkalosis by excretion of excess potassium in the urine and retention of chloride to adjust the alkaline pH to normal.[27] The chloride deficit then becomes $266 - 28 = 238$ mEq.

The 28 mEq of KCl is added to the first liter of fluids administered. Since potassium loss within the cell is not being taken into consideration in the calculations and since some potassium will be excreted during fluid administration, much more potassium will be required. In addition, over the next 24 hr, with normal kidney function, one can expect an excretion of approximately 70 mEq of potassium in the urine (see Tables 10.8 and 10.9). It is also advisable not to add more than 30 mEq of K^+ to a liter of I.V. fluids unless it is to be administered very slowly, because of the potential hazard to cardiac muscle. Generally any repair or maintenance solution with potassium at higher than 5 mEq/liter should be administered with caution.

The total requirement then in this patient initially is 28 mEq of K^+ and 266 mEq of Cl plus $70/5 \times 4 = 56$ mEq of NaCl. From Table 10.10, the NaCl represents 364 ml of normal saline. Since one would administer approximately three liters in 24 hr (Table 10.9) to this patient, one might choose to put 20 mEq of KCl in the first liter. Five hundred milliliters of 2.14% NH_4Cl would supply 200 mEq (Table 10.10) of chloride ion or all of the chloride deficit. This is true since the net chloride deficit after KCl and NaCl are added is 266–28.

While there are many ways of approaching the particular problem under discussion, a conservative approach would be to use 300 ml of normal saline, 200 of 5% glucose in water, 20 mEq of KCl, and 500 ml of 2.14% NH_4Cl to make the first liter. Subsequent fluids could complete the pH and electrolyte repair during the first 24 hr.

Ammonium chloride is effective in promptly correcting an alkalosis for several reasons. The pH of NH_4Cl solution is 4.5, which is acidic relative to blood. It thus acts promptly on the condition being treated. The ammonium ion is disposed of in two ways. First, it is rapidly converted to the neutral urea and excreted as such. In addition, the ammonium ion is rapidly cleared by the kidney, serving to exchange for hydrogen ion in neutralizing the various anions of urine. With liver disease the rate of ammonia metabolism may be compromised.

10.5.4.3   *Respiratory Alkalosis and Acidosis*

As pointed out before, the blood pH cannot be corrected by using repair solutions in these conditions. One has to focus treatment on those measures that will improve ventilation of the lungs. However, these conditions may be accompanied by some metabolic disturbance resulting in metabolic acidosis or alkalosis. Fluid therapy may then be used to assist the patient by correcting that portion of the problem contributing to the abnormal blood pH. As a minimum, any significant variation of electrolyte levels from the normal should be corrected.

10.6 RECAPITULATION

The composition and volume of repair fluids are calculated on the basis of the blood pH, electrolyte, protein, and hematocrit values obtained in the laboratory. Priority is given to the correction of blood pH, since normal metabolism is pH dependent. For example, in severe acidosis and diabetic coma, insulin may not be effective unless blood pH is corrected first.

One needs to add to the repair solutions, maintenance fluids calculated on the basis of the normal intake for sodium, potassium, and chloride for the next 24 hr. To this must be added excessive losses from drainage fluids. For this reason, the volumes of drainage fluids and urine need to be measured and an aliquot analyzed for their electrolyte content.

For infants, it is recommended that after repair solutions have been administered, maintenance fluids be based on the volume and electrolyte content of human breast milk that the healthy infant would normally consume over a 24-hr period.

It is also advisable to prepare a solution regimen designed for the particular patient from basic solutions normally available, rather than depend upon special formulas.

Treatment with repair solutions should be conservative, monitored from time to time with data obtained from the laboratory. It is always easier to add to the fluids than to backtrack to correct errors.

10.7 SELECTED READING—FLUID AND ELECTROLYTE APPLICATIONS

Winters, R. W., Ed., *The Body Fluids in Pediatrics*, Little, Brown & Co., Boston, Massachusetts (1973).

Shoemaker, W. C. and Walker, W. F., *Fluid–Electrolyte Therapy in Acute Illness*, Yearbook, Chicago, Illinois (1970).

Bland, J. H., *Clinical Metabolism of Body Water and Electrolytes*, Saunders, Philadelphia, Pennsylvania (1963).

Goldberger, E., *A Primer of Water, Electrolyte and Acid Base Syndromes*, Lea and Febiger, Philadelphia, Pennsylvania (1970).

Smith, R., Ward, E. E., and Garrow, J. S., *Electrolyte Metabolism in Severe Infantile Malnutrition*, Pergamon, New York (1968).

Weisberg, H. F., *Water, Electrolyte and Acid Base Balance*, 2nd ed., Williams and Wilkins, Baltimore, Maryland (1964).

Snively, W. D., Jr. and Sweeney, M. J., *Fluid Balance Handbook for Practitioners*, C. C. Thomas, Springfield, Illinois (1956).

Statland, H., *Fluid and Electrolytes in Practice*, 3rd ed., Lippincott, Philadelphia, Pennsylvania (1963).

Wolf, A. V. and Crowder, N. A., *Introduction to Body Fluid Metabolism*, Williams and Wilkins, Baltimore, Maryland (1964).

Roberts, K. E., Parker, V., and Poppell, J. W., *Electrolyte Changes in Surgery*, C. C. Thomas, Springfield, Illinois (1958).

Wise, B. L., *Fluid and Electrolytes in Neurological Surgery*, C. C. Thomas, Springfield, Illinois (1965).

Soffer, A., Ed., *Potassium Therapy*, C. C. Thomas, Springfield, Illinois, (1968).

Bajusz, E., Ed., *Electrolytes and Cardiovascular Disease*, Williams and Wilkins, Baltimore, Maryland (1965).

Maxwell, M. H. and Kleeman, C. R., Ed., *Clinical Disorders of Fluid and Electrolyte Metabolism*, McGraw-Hill, New York (1962).

Dybkaer, R. and Jorgensen, K., *Quantities and Units in Clinical Chemistry*, Munksgaard, Copenhagen (1967).

Talbot, N. B., Richie, R. H., and Crawford, J. D., *Metabolic Homeostasis*, Harvard Univ. Press, Cambridge, Massachusetts (1959).

Elkington, J. R. and Danowski, T. S., *The Body Fluids*, Williams and Wilkins, Baltimore, Maryland (1955).

10.8 REFERENCES

1. Oliver, W. J., Graham, B. D., and Wilson, J. L., Lack of scientific validity of body surface as basis for parenteral fluid therapy, *J. Am. Med. Assoc.* **167**:1211–1218 (1958).

2. Gamble, J. L., *Chemical Anatomy, Physiology and Pathology of Extracellular Fluid*, 6th ed., Harvard Univ. Press, Cambridge, Massachusetts (1954).

3. Natelson, S., Pennial, R., Crawford, W. L., and Munsey, F. A., Non-casein protein to casein ratio of feeding formulas. Effect on blood component levels, *Am. J. Dis. Child.* **89**:656–668 (1955).

4. Talbot, N. B., Crawford, J. D., Kerrigan, G. A., Hillman, D., Bertucio, M., and Terry, M., Application of homeostatic principles to the management of nephritic patients, *New Eng. J. Med.* **255**:655–663 (1956).

5. Heeley, A. M. and Talbot, N. B., Insensible water losses per day by hospitalized infants and children, *Am. J. Dis. Child.* **90**:251–255 (1955).

6. DuBois, D., DuBois, E. F., Sawyer, M., and Stone, R. H., Measurement of surface area of man, *Arch. Intern. Med.* **15**:868–881 (1915); **17**:855–862 (1916).

7. Boyd, E., *Growth of Surface Area of Human Body*, Inst. of Child Welfare, Monograph Series 10, Univ. of Minnesota Press, Minneapolis, Minnesota (1935).

8. Talbot, N. B., Kerrigan, G. A., Crawford, J. D., Cochran, W., and Terry, W., Application of homeostatic principles to the practice of parenteral fluid therapy, *New Eng. J. Med.* **225**:865–862, 898–906 (1955).

9. Darrow, D. C., The physiologic basis for estimating requirements for parenteral fluids, *Pediat. Clin. N. Am.* **6**:29 (1959).

10. Benedict, F. G. and Talbot, F. B., *Metabolism and Growth from Birth to Puberty*, Publication #302, Carnegie Institition of Washington, (1921).

11. Canadian Council on Nutrition, *Can. Bull. Nutr.* **6**:1 (1964).

12. Edelman, I. S. and Leibman, J., Anatomy of body water and electrolytes, *Am. J. Med.* **27**:256–277 (1959).

13. Pincus, G., Thimann, K. V., and Astwood, E. B., Eds., *The Hormones*, Academic Press, New York (1964).

14. Hartwig, Q. L., Cotlar, A. M., Shelby, J. S., Atik, M., and Cohn, I., Jr., Tolerance to intravenously administered fat emulsions, *Surgery* **49**:308–312 (1961).

15. Elman, R., Pareira, M. D., Conrad, E. J., Weichselbaum, T. E., Moncrief, J. A., and Wren, C., The metabolism of fructose as related to the utilization of amino acids when both are given by intravenous infusion, *Ann. Surg.* **136**:635–642 (1952).

16. Smith, Harris L., and Etteldorf, James N., Comparison of parenteral fluid regimen in treatment of diarrhea, *J. Pediat.* **58**:1–16 (1961).

17. Talbot, N. B., Crawford, J. D., and Butler, A. M., Homeostatic limits to safe parenteral fluid therapy, *New Eng. J. Med.* **248**:1100–1108 (1953).

18. Nyhan, W. L. and Cooke, R. E., Symptomatic hyponatremia in acute infections of the central nervous system, *Pediatrics* **18**:604–613 (1956).

19. Weil, W. B. and Wallace, W. M., Hypertonic dehydration in infancy, *Pediatrics* **17**:171–183 (1956).

20. Natelson, S. and Alexander, M. O., Marked hypernatremia and hyperchloremia with damage to the central nervous system, *Arch. Intern. Med.* **96**:172–175 (1955).

21. Dudrick, S. J., Wilmore, D. W., Vars, H. M., and Rhoads, J. E., Can intravenous feeding as the sole means of nutrition support growth in the child and restore weight loss in the adult? An affirmative answer, *Ann. Surg.* **169**:974–984 (1969).

22. Heird, W. C., Driscoll, J. M., Jr., Schullinger, J. N., Grebin, B., and Winters, R. W., Intravenous alimentation in pediatric patients, *J. Pediat.* **80**:351–372 (1972).

22a. Filler, R. M. and Erakus, A. J., Case of the critically ill child: Intravenous alimentation, *Pediatrics* **46**:456–461 (1970).

23. Natelson, S. and Barbour, J. N., Equation and nomogram for approximation of alkali requirements in acidosis, *Am. J. Clin. Pathol.* **22**:426–439 (1952).

24. Waters, W. C. III, Hall, J. D., and Schwartz, W. B., Spontaneous lactic acidosis: the nature of the acid–base disturbance and consideration in diagnosis and management, *Am. J. Med.* **35**:781–793 (1963).

25. Natelson, S., Crawford, W. L., and Munsey, F. A., Correlation of clinical and chemical observations in the immature infant, Illinois Dept. of Public Health, Division of Preventive Medicine publication (1962), p. 117.

26. Natelson, S., Chronic alkalosis with damage to the central nervous system, *Clin. Chem.* **4**:32–41 (1958).

27. Schwartz, W. B., Van Ypersele-De Strittou, C., and Kassirer, J. P., Role of anions in metabolic alkalosis and potassium deficiency, *New Eng. J. Med.* **279**, 630–639 (1968).

The Kidney and Sweat Glands in Fluid and Electrolyte Balance

The Kidney

11.1 KIDNEY FUNCTION AND MAINTENANCE OF THE STEADY STATE

In Figure 4.2 the central nervous system is shown as controlling blood flow to the kidney and thus influencing kidney function. In man, impaired sympathetic nerve function leads to an exaggerated diuretic response to saline loading and impairment of sodium conservation in response to sodium restriction. Anesthesia increases saline diuresis.[1] The antidiuretic hormone (vasopressin) from the posterior pituitary stimulates reabsorption of water in the distal tubule. Thus, both directly and through humoral agents, the central nervous system modulates kidney function.

The objective of the central nervous system is to integrate the function of the kidney into the overall mechanism for the maintenance of constant fluid volume, osmotic pressure, electrolyte levels, and pH. These functions have been indicated in a general way in Figures 4.2, 5.3, and 6.1, which also relate the renin–angiotensin–aldosterone system to the feedback mechanism to the central nervous system. In this chapter the present understanding of how these functions are carried out will be explored.

11.2 KIDNEY STRUCTURE

The kidney contains some 1.25×10^6 functioning units, called *nephrons* (Figure 11.3). These include a highly branched and tightly coiled capillary network called the *glomerulus*.[3] This is fitted into the modified blind end of a uriniferous tubule called *Bowman's capsule*. The combination of both these structures constitutes the *Malpighian corpuscle*.[2-6] These corpuscles are found in the outer part (cortex) of the kidney (Figures 11.1 and 11.2).

The uriniferous tubule is itself divided into three major parts (Figure 11.3). The first division, the *proximal tubule*, leads from Bowman's capsule into the second narrowed segment, called the *loop of Henle*, which again widens as it becomes the *distal convoluted tubule*. Several distal tubules finally connect with a single large duct, the *collecting duct*. The collecting ducts eventually empty urine into the *renal pelvis* (Figure 11.3). The renal pelvis of each kidney is connected with the bladder by the ureters (Figure 11.1), which aid the transfer of urine by a peristaltic movement.[7] As the pressure in the urinary bladder exceeds a limiting value, a reflex contraction of the muscular bladder wall causes the urine to be excreted through the urethra to the outside. The voiding reflex action can be aided or inhibited by voluntary muscular controls.

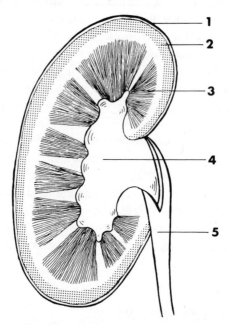

FIGURE 11.1 Saggital section of kidney showing the cortex containing the glomeruli, the renal pyramids composed of the collecting ducts and loops of Henle, the renal pelvis, and the ureter. 1. Renal capsule. 2. Renal cortex. 3. Renal pyramid (outer and inner medulla). 4. Renal pelvis. 5. Ureter.

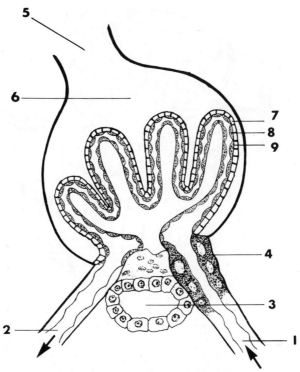

FIGURE 11.2   Malpighian (renal) corpuscle composed of the glomerulus fitted into Bowman's capsule. The juxtaglomerular cells (which secrete renin) are shown in the wall of the afferent arteriole where it touches the distal tubule. 1. Afferent arteriole. 2. Efferent arteriole. 3. Distal tubule. 4. Juxtaglomerular granular cells. 5. Proximal tubule. 6. Bowman's capsule. 7. Cell wall of Bowman's capsule. 8. Capillary endothelial cell wall. 9. Basement membrane of endothelial cell.

11.3   *BLOOD FLOW TO THE KIDNEY*

The total quantity of blood presented to the kidney from the renal artery for processing per minute is called the *renal blood flow* (*RBF*). The bulk of this blood is processed and modified during its passage through the kidney.

The total blood flow to the kidney is 525–625 ml/min/100 g of tissue. Seventy-five percent of this total (400–500 ml/min/100 g tissue) flows to the renal cortex, which contains the glomeruli. Approximately 120 ml/min/100 g of tissue flows to the outer medulla and

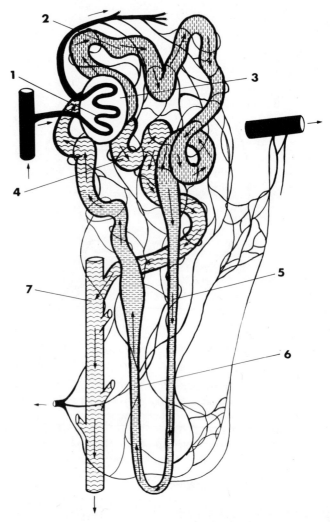

FIGURE 11.3   The structure of the nephron, composed of the glomerulus, Bowman's capsule, the renal tubules, and the collecting duct, is shown enveloped by a capillary network arising from the efferent arteriole. 1. Glomerulus. 2. Proximal tubule. 3. Bowman's capsule. 4. Distal tubule. 5. Descending limb of loop of Henle. 6. Ascending limb of loop of Henle. 7. Collecting duct.

25 ml/min/100 g of tissue to the inner medulla, for a total of 145 ml/min/100 g of tissue to the renal pyramids (medulla). For a 70-kg man the total volume is of the order of 1200 ml/min or 25% of his cardiac output. We are obviously dealing with a system of huge capacity.[8]

Normally, a small portion of the renal blood supply bypasses the afferent arteriole of the glomerulus and is shunted to venous drainage without being processed by the kidney. These *shunts* may increase significantly in certain disease states.[9,10]

11.3.1 Filtration to Bowman's Capsule (Figure 11.2)

The main renal artery bringing blood to the kidneys undergoes a series of divisions. Eventually blood enters a structure called the *glomerulus* via the afferent arteriole at a pressure which was long considered to be approximately 65–75 mmHg, or about 70% of the systemic arterial pressure, which is 90 mmHg.[11] This was at variance with measurements taken in capillaries in other parts of the body, where the hydrostatic pressure is only approximately 37 mmHg at the afferent end of an arteriole–venule system (see Figure 8.2).

Recent measurements have indicated that the value of 65–75 mmHg as originally determined was an overestimate and was obtained because the pressure was measured with the capillary system blocked. Measurements taken with modern instruments, which permit measurement during free flow through the capillaries, have demonstrated that this pressure in the arteriole is actually only 45 mmHg if the capillary is not blocked.[12] This can be seen in Table 11.1, comparing pressure values obtained with and without stopping the blood flow in the capillary.

TABLE 11.1

Net Filtration Pressure in the Glomerulus as a Result of Balance Between Hydrostatic and Osmotic Pressure on Both Sides of the Capillary Wall: Comparison of Stop-Flow and Continuous Flow Measurement

Pressure[a]	Stop-flow, mmHg		Continuous flow, mmHg	
	Afferent end	Efferent end	Afferent end	Efferent end
P_C	75	75	45	45
P_B	10	10	10	10
Π_C	20	35	20	35
Net	45	30	15	0

[a] The subscript C means within the capillar and B within Bowman's capsule.[12]

In Figure 8.2 it is shown that the net filtration in a capillary system, as in the glomerulus, is the result of the balance of forces between the hydrostatic pressures and the osmotic pressures on both sides of the capillary wall. The net hydrostatic pressure ΔP is equal to the hydrostatic pressure inside the capillary P_C minus the hydrostatic pressure in Bowman's capsule P_B:

$$\Delta P = P_C - P_B$$

Similarly, the net oncotic pressure is the colloid osmotic pressure inside the capillary Π_C minus the opposing colloid osmotic pressure in Bowman's capsule Π_B:

$$\Delta \Pi = \Pi_C - \Pi_B$$

The oncotic pressure is due to the proteins in the plasma, which are retained as the filtrate moves into Bowman's capsule. Salt, glucose, and urea concentrations are essentially the same on both sides of the membrane. Since normally only minute amounts of albumin pass the membrane, the oncotic pressure is therefore almost solely due to the plasma proteins, and is 20 mmHg.

The net filtration pressure is then $(P_C - P_B) - (\Pi_C - \Pi_B)$ or 15 mmHg, as indicated in Table 11.1, when the capillary is not blocked. This value is 45 mmHg when the capillary is blocked. Note that in Table 11.1 the colloid osmotic pressure in Bowman's capsule is ignored. This is because the protein concentration in the blood is 7 g/100 ml and is negligible in the normal filtrate.

From Table 11.1 it can be seen that a hydrostatic force of 45 mmHg is being opposed by a back hydrostatic pressure of 10 mmHg and an osmotic pressure of 20 mmHg, for a net pressure of 15 mmHg at the afferent end of the arteriole. This gradually diminishes to zero at the efferent end. The net result is a movement of filtrate into Bowman's capsule.

The membrane through which the fluid passes is approximately twice as permeable as the capillary wall in other parts of the body. This results in a high rate of glomerular ultrafiltration. This is proportional to the net pressure applied $(\Delta P - \Delta \Pi)$. We then have

$$\text{filtration rate [nl/cm}^2\text{/sec]} = K(\Delta P - \Delta \Pi)$$

The constant k is a measure of the permeability of the capillary wall, and is the rate when 1 mmHg of pressure is applied. For the rat glomerulus it is 41 nl/sec/cm^2/mmHg. The k value for other capillaries in the rat is only a fraction of this value.[13]

In the human the glomerular filtrate rate for all the glomeruli combined is 125 ml/min. The estimated area is 1.56 m^2 for the total filtering surface. From these figures one can calculate the volume/cm^2/sec/mmHg or the k value. There are 15,600 cm^2 in 1.56 m^2. Converting to microliters, we have $125,000/15,600 = 8.0$ μl/cm^2/min. Converting to nanoliters and seconds, we have $8000/60 = 133.3$ nl/cm^2/sec. From Table 11.1, the net driving force is 15 mmHg. Dividing by this number, we have $133.3/15 = 8.9$ nl/cm^2/sec/mmHg as the value of k. This is an underestimate since the pressure varies from 15 to 0, or possibly an average of 7.5 mmHg throughout the glomerulus. This would multiply the figure by a factor of 2 and yield 18 for the value of k, as compared to 41 measured in the rat.[13] This filtration rate would increase if the tubule were blocked due to the pressure increase. Others have reported measured values for k (called the "effective hydraulic permeability of the glomerular capillary wall") of the order of 3–6, which is closer to the range calculated here.[14]

11.3.2 *Glomerular Filtration Rate (GFR)*

Plasma fluid flows from the glomerular capillaries, passes across the capillary cell wall, through a thin basement membrane, and across the cell wall of Bowman's capsule. The total distance the fluid must travel in passing through these three barriers is of the order of 1 μm. The solution which appears in Bowman's capsule is essentially an ultrafiltrate of plasma, protein being the major plasma constituent, which, normally, is unable to pass the basement membrane in significant amounts. The glomerular capillary walls are constructed to provide a maximum of diffusion area. As pointed out, this is estimated to be 1.56 m^2 in the normal adult.[3] This considerable surface area supports the rapid accumulation of a large glomerular filtrate. The filtrate is formed at a rate of approximately 125 ml/min in the normal individual. This is called the *glomerular filtration rate (GFR)*.

A *more concentrated blood* with a higher protein level and higher oncotic pressure leaves the glomerulus via the efferent arteriole, 18–20% of the plasma water having been removed. The efferent arteriole almost immediately breaks up into a capillary network around the uriniferous tubules and provides their blood supply.[15] This is often called the "peritubular circulation." Re-entry of water into the vascular compartment from the tubules is facilitated by the extensive branching of this blood supply and its increased osmotic pressure (Figure 11.3).[16,17]

11.4 FUNCTION OF THE URINIFEROUS TUBULES (GENERAL SURVEY) (Figure 11.3)

The dimensions of the average tubule have been given as 20–50 μm in diameter and 50 mm in length. The various sections of the uriniferous tubules are specialized to perform different tasks as the ultrafiltrate is modified to form the urine, which passes into the renal pelvis.[18]

11.4.1 The Proximal Tubule

In the first section of the tubule, *the proximal tubule*, 85% of the glomerular filtrate is reabsorbed as water along with most of the physiologically and metabolically important substances, such as glucose, amino acids, phosphate, Na^+, K^+, and Cl^-. Active secretion of various substances, such as creatinine and phenolsulfonephthalein dye, also takes place in the proximal tubule.[18]

Since both salts and water are reabsorbed with equal facility, the glomerular filtrate remains iso-osmotic with respect to plasma in the proximal tubule.

11.4.2 The Loop of Henle

As the filtrate passes through the *loop of Henle* and to the upper limb, extrusion of Na^+ and Cl^- begins but little water is able to leave the tubule. The net result is to make the fluid entering the distal tubule hypo-osmotic and the *renal medulla*, into which the loops of Henle project, extremely hyperosmolar with respect to the plasma.[19,20]

11.4.3   *The Distal Convoluted Tubules*

These tubules receive the filtrate from the ascending portion of the loops of Henle, and pass through the renal medulla to connect with the collecting ducts. Under the influence of the antidiuretic hormone (ADH) the distal tubules reabsorb water.[21,22] The urine thus becomes concentrated as water passes out into the hyperosmolar renal medulla to be carried away by the tubular capillary network. The urine is acidified in the distal tubules and reaches its final pH of approximately 5.0.[23-25]

11.4.4   *The Collecting Ducts*

The urine passes from the distal tubules to the collecting ducts, where some further concentration takes place as water continues to be reabsorbed under the influence of the antidiuretic hormone. When the urine reaches the renal pelvis it is in its final volume and composition.

11.5   *MECHANISM FOR URINE FORMATION*

11.5.1   *Introduction*

A major objective of the kidney is to conserve base (excrete H^+ generated during metabolism) and excrete unneeded electrolytes and other anions, particularly phosphate, potassium, chloride, and nitrogenous wastes.[26-28] In order to do this, a certain amount of water must be eliminated. However, this must not be excessive and in order to conserve water it is therefore *necessary to excrete a urine of higher osmotic pressure than that which exists in the plasma.* The objective is also to conserve glucose and as much sodium as is required for normal metabolism. As pointed out earlier, the cation of our foods is mainly potassium. Therefore potassium is usually present in excess. Sodium, however, may be in short supply and needs to be conserved.

The osmotic pressure of plasma ranges, normally, from 285 to 305 mosmol/liter. It is not uncommon for urine osmotic pressure, however, to range from 400 to 1200 mosmol/liter and occasionally to rise to as high as 1450 mosmol/liter. To produce this urine, it is required that a system of osmotic multiplication exist.

The problem may require the concentration of the plasma ultra-filtrate from the glomerulus as much as fourfold.[29] The energy required is substantial and can be calculated from the Gibbs-Helmholtz equation $-\Delta F = RT \ln(C_1/C_2)$, where C_1/C_2 is the ratio of the concentrations in urine and in blood and may be equal to 4, $R = 2$ cal/mol, and T is the absolute temperature of the body ($\sim 310°K$). The natural logarithm of 4 is 1.4. Multiplying the factors $2 \times 310 \times 1.2$, we obtain 744 cal to move 1 mol of sodium against this gradient, or 0.744 cal/mEq.

The energy comes from the glycolytic process. In glycolysis in the kidney, ATP is generated from ADP and inorganic phosphate.[30,31] In the loop of Henle the cells lining the walls contain ATPase which hydrolyzes the ATP, releasing 10,000 cal/mol, forming ADP and inorganic phosphate, which is again utilized in the glycolytic process. This ATPase is activated by both Na and K.[32-35] The rate of ATP hydrolysis and thus the rate of glycolysis are indirectly controlled by the Na and K levels. The mechanisms which have been proposed as to how this results in extrusion of Na ion from the lumen of the ascending loop of Henle into the interstitial fluid will now be discussed.

Urine concentration results from the concerted action of at least four systems. These comprise:

1. The "osmotic multiplier" of the loop of Henle.
2. The countercurrent osmotic exchanger of the blood capillaries, which extend in a loop to the tip of the renal pyramid and back to the cortex.
3. The osmotic exchanger in the distal convoluted tubules and collecting ducts.
4. The effect due to an osmotic gradient set up by an increasing concentration of albumin and salts as one moves from the cortex to the tip of the renal pyramid.

11.5.2   *Proximal Tubule Reabsorption*

Filtration in the glomerulus results in an ultrafiltrate of the same osmotic pressure as the blood. As described, the capillaries that comprise the glomerulus continue as the efferent arteriole, which now forms a network around the proximal tubule (Figure 11.3). Since the blood has been concentrated relative to the ultrafiltrate, an opportunity is now given for the water that has filtered through to return to the blood stream. The membrane is not permeable to certain substances, such as creatinine, and is poorly permeable to urea. However, sodium, bicarbonate, and glucose move into the blood stream with the water. A major force is the elevated peritubular albumin concentration as compared to the filtrate, which is negligible. This colloid osmotic pressure accounts for at least 50% of the water that is reabsorbed.[16,36]

The loop of Henle can be divided into three sections. The thin descending limb, the thin acending limb, and medullary thick ascending limb (Figure 11.4). Active transport of sodium and chloride takes place only in the thick ascending limb. While the general thesis has been held that this pump is a sodium pump, chloride following passively, some claim that the pump is actually an electrogenic *chloride pump*, sodium following passively.[34]

The net result is to produce a fluid hypo-osmotic to the original plasma since salt has been removed to the interstitial spaces and carried off by the blood stream. For this reason some call the thick ascending limb of the loop of Henle, "the diluting tubule." It is also classified by some as part of the distal convoluted tubule.

11.5.2.1   *The Osmotic Multiplier (Loop of Henle)*[37]

The loops of Henle extend down from the kidney cortex to the medullary tip and return to the cortex. This is not true for all the loops of Henle but these are the ones under discussion (Figure 11.4). At first, the descending arm of the loop resembles, to some extent, the proximal tubule in that it is porous to water and electrolytes.[38-40] The fluid picks up substantial amounts of sodium and loses water on its downward path and the osmotic pressure equilibrates with that outside the lumen. At the medullary tip the osmotic pressure within the lumen is approximately the same as that in the interstitial fluid. After it turns at the medullary tip and ascends, it thickens and becomes impervious to water, but permits salt to be extruded. In this thick ascending segment, chloride is extruded by active transport and sodium moves with it.[44,45]

As the thick loop ascends, the opposing osmotic force, acting against the active transport mechanism, decreases progressively and the tendency to secrete sodium to the interstitial fluid is facilitated. *However, less sodium chloride is available for secretion as the fluid in the lumen becomes depleted and thus lesser amounts of salt are secreted as it ascends. A*

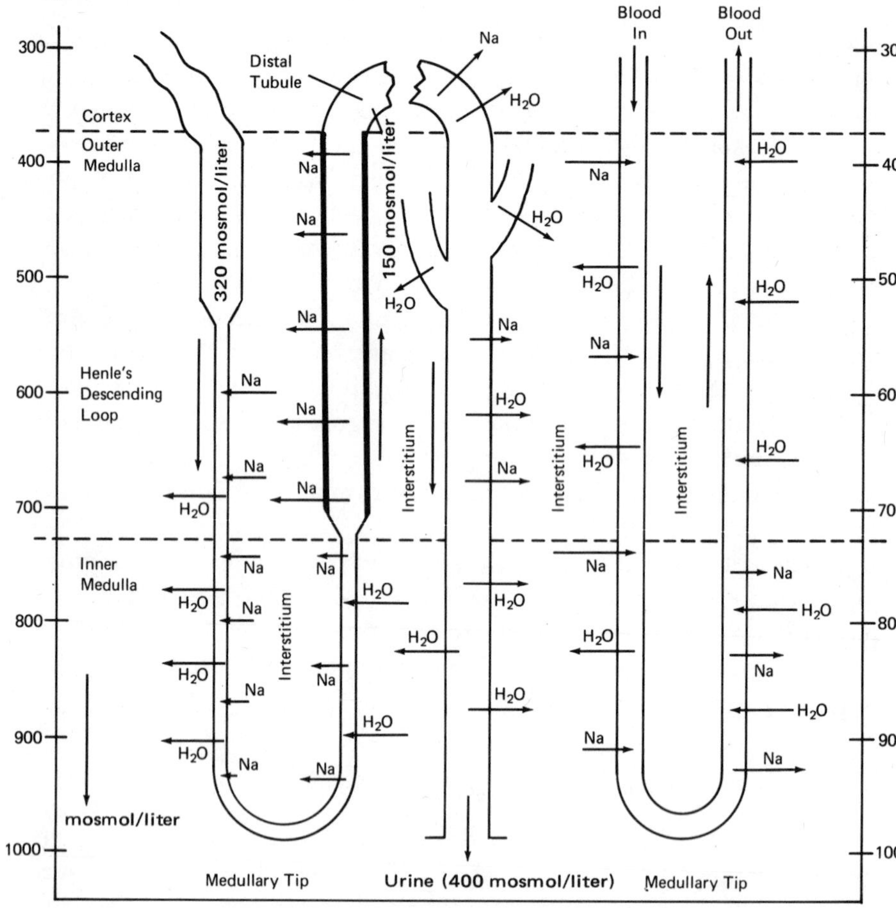

FIGURE 11.4   Schematic representation of mechanisms for recovering Na and H_2O and producing a hyperosmotic urine. Thin lines represent walls permeable to H_2O. Thick line in loop of Henle represents impermeable wall. Blood capillary reclaims both water and Na and accentuates osmotic gradient.

concentration gradient then exists as illustrated schematically by the osmolar scale in Figure 11.4. *Note that the closer to the medullary tip, the more salt is being pumped out. This establishes the initial concentration gradient.*

11.5.2.2.   *The Countercurrent Osmotic Exchanger (Vasa Recta)*[41–43]

Reclaimed water and sodium need to be carried away by the blood capillaries (vasa recta) in the medulla. The object is to perform this process efficiently, maintaining and intensifying the concentration gradient that is required to produce a concentrated urine. This is done by a branch from the efferent capillary, which leaves the glomerulus, moving down into the medullary tip and returning as indicated on the right side of Figure 11.4.

The renal tubules are separated by an interstitial space through which the vasa recta also run. This is also indicated in Figure 11.4. This interstitial space is shared in common by the descending and ascending limbs of the loop of Henle, the vasa recta, and the collecting tubules. The active extrusion of salt from the ascending loop of Henle initially supplies the energy to maintain a concentration gradient from the medulla to the medullary tip.[44]

As pointed out, a concentration gradient is started by the thick ascending limb of the loop of Henle since it becomes progressively depleted of salt as it moves up (Figure 11.4). The extruded NaCl moves to the interstitial space, creating an area there of high concentration of NaCl along the thick ascending limb.[45] The descending limb of the blood capillary is readily pervious to salt and water and moves through this hypertonic interstitial area. Water then moves out of the capillary and salt moves in. The blood moving rapidly in the descending loop of the capillary becomes progressively hypertonic and carries the salt forward. It thus moves the salt toward the medullary tip. As it moves to below the thick portion of the loop of Henle, it is at first hypo-osmotic with respect to its surroundings, then iso-osmotic, and finally hyper-osmotic as it turns the loop. Salt then moves out to its surroundings to raise the salt concentration to its highest level at the medullary tip; this is shown in Figure 11.4.

On its return trip toward the cortex the capillary tends to lose salt. However, the amount of sodium lost on the ascending loop is less than was gained on the descending loop. The water intake on the ascending loop is higher than that lost in the descent. The net result is a gain in water and Na by the blood. *The osmotic pressure of the blood remains the same, finally, but the volume increases.* On the other hand, the osmotic pressure in the loop of Henle decreases markedly to 150 mosmol/liter as indicated in Figure 11.4.

The capillaries that move down from the glomerulus into the medulla have muscular walls and a peristaltic action. *It is felt by some that one action of the antidiuretic hormone is to control this action and thus the flow through the vasa recta,* a slow flow allowing more time and thus a better opportunity to form and maintain the concentration gradient.[47]

From Figure 11.4 it is apparent that the collecting ducts and distal tubules are involved in the process. Since they pass through hyperosmotic regions, water in the distal tubule is extracted into the interstitial fluid by simple osmosis, in the presence of the antidiuretic hormone. This water is reclaimed by the blood capillaries and returned to the general circulation.

11.5.2.3 *The Osmotic Exchanger (Distal Tubules and Collecting Ducts)*[46]

When the fluid leaves the loop of Henle the osmotic pressure is only 150 mosmol/liter. The interstitial spaces immediately around it and the blood are at ~ 300 mosmol/liter. Thus water moves out to the interstices and into the capillaries by passive transfer and is carried away by the blood. This process is accelerated by the action of ADH, which prevents sodium return to the lumen of the distal tubule and, by stimulating active transfer, actually moves some sodium out.[47] By the time the ultrafiltrate reaches the collecting ducts the osmotic pressure is ~ 300 mosmol/liter. The collecting ducts now move toward the medullary tip, where the osmotic pressure is of the order of 1000–1200 mosmol/liter. Water diffuses out into the interstitial fluid. ADH is also active in stimulating this process. The net result is a urine approaching the osmotic pressure of the medullary tip.

11.5.2.4 *The Effect of Albumin Concentration in the Interstitium*[48–50]

As one progresses from the cortex to the medullary tip, the concentration of albumin increases significantly. As a result, a built-in osmotic gradient exists which facilitates the movement of water from the collecting tubules and helps in producing a concentrated urine. A concentration gradient also exists in the vasa recta for albumin, due to loss of water on the descent to the renal medulla. At the medullary tip the albumin is normally 1.5 times as concentrated as in the blood. Some believe that this albumin moves into the interstitial spaces and creates an albumin concentration gradient. Others feel that the salt gradient in the capillary removes water from the interstitial fluid, thus producing the gradient in the interstices. In any case, an albumin gradient does exist in the capillaries and in the interstices, and contributes to the maintenance of the osmotic gradient in the kidney.

Several investigators have taken issue with some of the concepts discussed above. They point out that many loops of Henle do not descend to the medullary tip in the human. In some species none of the loops descend. Some point out that under the microscope and electron microscope no difference, histologically, can be seen between many of the descending and ascending loops of Henle. It has also been pointed out that ADH seems to have a special action in urea transport across the membranes which is not being considered.[51,52]

In conclusion, the mechanisms discussed only partly explain the phenomenon of urine concentration by the kidney, but are useful in the understanding of renal physiology in relationship to the maintenance of constant pH and osmotic pressure in the blood.

11.6   *THE JUXTAGLOMERULAR APPARATUS*

In order to control the excretion of water and salts, a feedback system exists which monitors the composition of the fluid in the uriniferous tubules as it emerges from the thick ascending limb of Henle. The sodium content and osmotic pressure determine the subsequent behavior in the distal tubule and collecting ducts and eventually the composition of the urine.

The geography of this system is seen in Figures 11.2 and 11.3, and is shown more distinctly in Figure 11.5. Here a simplified schematic drawing indicates how the uriniferous tubules fold over to come eventually in contact with the blood capillaries as they enter and leave the glomerulus. This assembly is called the *juxtaglomerular* apparatus since it is adjacent to the glomerulus.

In Figure 11.5 the "granular cells" in the wall of the afferent capillary are seen. These cells produce renin and also the converting enzyme referred to in Figure 11.6. In response to changes in sodium ion and osmotic pressure in the lumen of the distal tubules, more or less renin is produced. When the fluid is hyperosmotic (elevated Na level) then water flows from the afferent capillary toward the fluid, resulting in reduced pressure in the capillary. This stimulates renin production so as to reclaim this water and sodium.[53–55]

The juxtaglomerular apparatus shown in Figure 11.2 is then the site of production of the enzyme *renin*. When released into the blood stream, this substance is capable of converting a plasma polypeptide called *angiotensinogen* to *angiotensin I*, a decapeptide. Angiotensinogen is

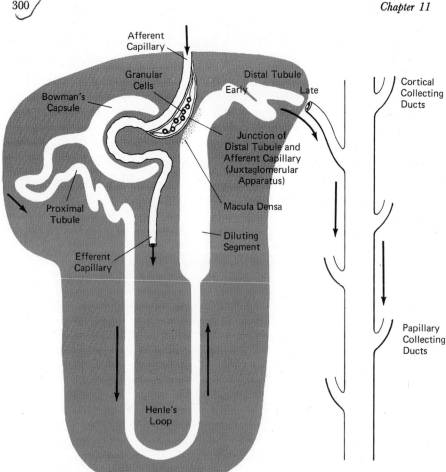

FIGURE 11.5 Schematic representation of the spatial relationship between the distal tubule and afferent glomerular capillary to produce the juxtaglomerular apparatus.

itself synthesized in the liver. A second enzyme normally present in plasma catalyzes the conversion of angiotensin I to *angiotensin II*, an octapeptide. Angiotensin II stimulates contraction of the smooth muscle fibers in arteriole walls, and thus has a direct hypertensive effect by increasing peripheral resistance. Angiotensin II is also a powerful stimulant of *aldosterone* secretion from the adrenal cortex.[53-56]

The action of aldosterone in the normal individual is to facilitate sodium resorption in the distal tubule by supporting ion exchange of sodium for hydrogen and potassium. This action provides the fine control on sodium excretion since 85% of the sodium present in the glomerular filtrate has already been reabsorbed in the proximal tubule.[57]

As pointed out above, increased osmotic pressure (elevated NaCl level) in the uriniferous fluid causes a decrease in intra-arteriolar pressure in the afferent arteriole. When intra-arteriolar pressure in the afferent arteriole is lowered, renin production is stimulated, which is secreted into the blood stream from the granular cells.[58-62] A rise in blood pressure follows because angiotensin II acts directly on systemic arteriolar smooth muscle. This causes an increase in filtration rate in the glomerulus.

An increase in glomerular filtrate tends to dilute the fluid in the tubules. Simultaneously, the action of the aldosterone whose secretion has been stimulated by angiotensin II causes increased reabsorption of sodium from the distal tubules in exchange for some potassium but mostly hydrogen and ammonium ion. The net result is a diluted urine.

In a similar manner, when a dilute fluid is presented in the distal tubules at the juxtaglomerular apparatus, the lowered sodium level reduces the amount of renin produced, resulting in decreased filtration and sodium reabsorption so as to produce a more concentrated urine.[63-67]

From the above, it is apparent that a servomechanism based on the feedback principle exists so as to regulate the volume and osmotic pressure of the urine formed.

Hypertension is commonly found in association with certain types of renal disease, notably arteriolar nephrosclerosis and advanced chronic glomerulonephritis. Also associated with hypertension, but to a lesser degree, are acute glomerulonephritis and possibly pyelonephritis. In glomerulonephritis and pyelonephritis nonspecific vascular damage occurs throughout the kidney as a consequence of inflammation and scarring. It is believed that an element of the hypertension associated with all these renal diseases is probably vascular in origin and can be explained by disturbance of the intra-arteriolar pressure control of the secretion of renin by the kidney.[73-76]

Excessive secretion of renin results in increased plasma levels of angiotensin II and excessive secretion of aldosterone which results in an abnormally large resorption of sodium and increased retention of water,[68,69] resulting in edema. The increase in extracellular fluid and plasma volume further elevates systemic blood pressure, aggravating

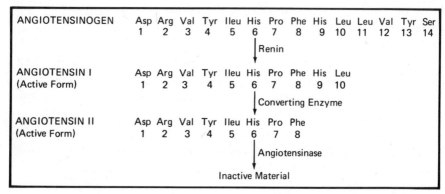

FIGURE 11.6   Hydrolysis of angiotensinogen to the active angiotensins.[56,77]

the condition.[69,70] Acutely induced hypertension in the perfused kidney augments sodium excretion.[72] It is apparently the secondary effect of aldosterone in the human that causes sodium and water retention.[71]

Figure 11.6 illustrates the conversion of angiotensinogen (equine or porcine) to the active angiotensins and their inactivation by angiotensinase, present in the kidney. For bovine angiotensins, the isoleucine in position 5 is replaced by valine. Human angiotensin resembles the porcine angiotensin.

11.7   RENAL POTASSIUM EXCRETION

The narrow range of potassium concentration in plasma and interstitial fluid is maintained by an interaction of both tubular reabsorption and tubular excretion at different levels of the nephron.

Usually, potassium excretion exceeds the glomerular filtration rate and therefore a tubular mechanism of potassium excretion must exist. By means of tracers it can readily be demonstrated that most of the potassium which appears in the urine is secreted in the tubules.[78]

Intake of sodium bicarbonate, resulting in increased excretion of sodium ion and an alkaline urine, results in an increased excretion of potassium ion. On the other hand, urinary potassium excretion is decreased in metabolic acidosis. For this reason it has been postulated that potassium ion exchanges for hydrogen ion in the distal tubular lumen.[79]

There are several factors which result in rapid urinary potassium loss. Simple dehydration due to decreased water intake or increased water loss, such as with diarrhea, results in large losses of potassium. Infusion of saline also results in potassium loss in the kidney. These can be explained by the hypothesis that secretion of potassium in the distal tubule exchanges for either hydrogen or sodium, in the latter case conserving sodium ion. Administration of sulfate, phosphate, ferrocyanide, or any of a number of poorly reabsorbable anions results in potassium secretion in the distal tubule. This is also true for cationic amino acids.[80] Potassium forms rather poorly dissociated complexes with these anions.

In the proximal tubule the concentration of potassium is the same as in the plasma, and apparently the proximal tubule does not actively secrete potassium.[81] During reabsorption of water some extra potassium moves back into the blood stream and the concentration is somewhat below the plasma level by about 5–10% when the fluid reaches the loop of Henle.[81]

The ascending limb of the loop of Henle, especially the thick part, is an important site of tubular potassium reabsorption. Thus, as NaCl is extruded, K^+ moves with it. This can be blocked with certain diuretics, such as Furosemide, Ethacrynic acid, Chlormeradrin, and Hydrochlorthiazide, all of which inhibit oxidative metabolism.[82,83]

It is at the distal tubule and less at the collecting ducts that most (75%) of the potassium is normally secreted.[84] The potassium is therefore secreted mainly in the cortex. As pointed out above, the amount secreted depends upon total potassium intake in the diet and the well being of the individual.

The mineralocorticoids (e.g., aldosterone) modify the renal tubular transport of potassium. Administration of the mineralocorticoids stimulates potassium secretion and increased tubular resorption of sodium. The two effects can be dissociated but in general, in the presence of adequate dietary potassium and sodium the mineralocorticoids cause sodium retention and potassium excretion in the distal tubule.[85]

The effect of potassium on hydrogen ion secretion takes place mainly in the distal convoluted tubule. Hypokalemia stimulates hydrogen ion secretion and administration of potassium salts raises the pH of the urine.[86] Thus potassium excretion is a significant factor in regulating urinary pH.

In summary, potassium is absorbed in the lumen of the proximal tubule and loop of Henle, most probably by active transport. In the distal tubule, where most of the potassium is excreted, it seems to be a passive flux in exchange for hydrogen and sodium ions.[87] Aldosterone stimulates potassium excretion and sodium retention.

11.8   *HYDROGEN ION EXCRETION*

The hydrogen ion excreted in the urine is the sum of the "titratable acidity" and the ammonium ion excreted. The sum of these two numbers minus any bicarbonate which may be present represent the total loss of hydrogen ion from the blood stream. The titratable acidity is obtained by titrating with 0.1 N NaOH to pH 7.4. Thus the titratable acidity represent the extent of acidification of the nonbicarbonate buffers such as phosphate and sulfate.

Blood pH is 7.4. Urine pH normally ranges from 4.8 to 5.3 but can drop as low as 4–4.5. This lower limit is set by the pH of NH_4Cl, which is 4.0. Any lower pH would hydrolyze the huge amounts of urea present, generating ammonium salts. Thus a pH gradient of approximately 2.5 pH units exists between blood and urine. How this is accomplished will now be discussed.

11.8.1   *Proximal Tubule*

The proximal tubule has only a small role in generating titratable acid. Most of the acidification takes place beyond this point, particularly in the distal tubule.

Approximately 80% of the filtered bicarbonate in the glomerulus is reabsorbed with the water in the proximal tubule.[88] In the distal tubules most of the other 20% of bicarbonate is recovered. In the proximal tubule reabsorption of the $NaHCO_3$ lowers the pH of the fluid in the lumen to 6.8–7.0. Proximal fluid bicarbonate level then drops to 8–10 mEq/liter as it enters Henle's loop, compared to 27 in the original filtrate.

By the time the distal tubules are reached, the pH has dropped to 6.3 and bicarbonate level is down to 2 mEq/liter, the rest being reabsorbed in the loop of Henle. Final acidification takes place in the distal tubules and the minimal pH has already been achieved when the fluid enters the collecting ducts.[86]

Since the proximal tubule filtrate is alkaline relative to urine, failure to reabsorb this filtrate along with sodium bicarbonate will result in losses of base from the blood to the urine and a resultant acidemia. This results in a condition called *proximal renal tubular acidosis* (proximal RTA). This may occur in infancy without apparent explained cause (idiopathic proximal RTA). In many of these cases, the condition is transient and improves spontaneously as the child matures.[86a] In the thirties, a substantial number of children in England developed this condition (Lightwood syndrome). Subsequently, the condition disappeared.[86a] Apparently, some change in environment or diet had taken place, which removed the cause of this condition. Other diseases associated with proximal RTA have been listed in the discussion of phosphate excretion (Section 7.8.2). In proximal RTA, urinary pH will not usually exceed 5.5 and titratable acidity and ammonia excretion will usually hover around the lower limit of normal. Nephrocalcinosis, polyuria, and hyperchloremic acidosis are common findings in this condition.

11.8.2 *Ammonia Excretion*

Seventy percent of the ammonia that filters through the glomeruli is not reabsorbed along with the water in the proximal tubule. The ammonia-containing fluid leaving the proximal tubule is partly alkalinized as a result of water resorption in the descending limb of the loop of Henle. The collecting duct, on the other hand, is acid. *Ammonia* (NH_3) *then diffuses from the descending loop of Henle through the interstitial fluid directly to the collecting ducts.*[89-91] Thus the distal convoluted tubules are bypassed. The amount of ammonia excreted is a direct function of the acidity in the collecting ducts and serves to prevent the urine from becoming so acidic that it damages the tissues.[92,93] In the normal adult, 20–70 mEq of NH_3 is excreted per 24 hr.

Ammonia excretion is lowered in metabolic and respiratory alkalosis and increased in metabolic and respiratory acidosis. In Addison's disease, hypercorticalism, and in renal disease, with damage to the distal tubules, ammonia excretion is decreased. High protein diets, starvation, and increased acid intake will all result in increased excretion of ammonia. Vegetarian diets or increased alkali intake will result in decreased ammonia excretion.

11.8.3 *Carbonic Anhydrase*

Carbonic anhydrase is intimately involved in hydrogen ion excretion in the urine. It is located in the inner wall of the proximal but not in the wall of the distal tubule. In both tubules, however, carbonic anhydrase is present in the cells. Inhibition of carbonic anhydrase with acetazolamide (Diamox) prevents the elaboration of an acid urine, titratable acidity dropping to zero. As blood p_{CO_2} rises, reabsorption from the proximal tubule is increased. This would be expected since as the blood is acidified, alkali would tend to be reabsorbed. This effect is markedly depressed if a carbonic anhydrase inhibitor (acetazolamide) is present.[95] Thus carbonic anhydrase is involved in the reabsorption of bicarbonate from the proximal tubule.[94] It is the presence of carbonic anhydrase in the inner wall which prevents the urine from becoming acid in the proximal tubule. This is illustrated in Figure 11.7.

11.8.4 *The Distal Tubule*

As pointed out earlier, the fluid reaches the distal tubule at pH 6.3 and final acidification takes place here. The mechanisms acting in the distal tubule are the exchange of K^+ and Na^+ of the fluid for H^+, and the liberation of ammonia in the lumen to buffer the acidity and in reabsorbing the Na^+ to exchange this for hydrogen with the phosphate system.

In summary, the regulatory capacity of the kidney with respect to the maintenance of constant pH in the blood and conservation of blood OH^- comprises four basic mechanisms:

1. Na^+ and K^+ are ion-exchanged with H^+; the latter is then excreted in combination with an anion in the urine.
2. The deamination of glutamine to form glutamic acid liberates NH_3 into the tubular lumen, where it can combine with H^+ to form the ammonium ion, which is excreted in combination with an anion.[96]
3. Carbonic anhydrase catalyzes the reaction $CO_2 + H_2O \rightleftharpoons$

$H_2CO_3 \rightleftharpoons H^+ + HCO_3^-$. The H^+ of the blood is exchanged for Na^+ in the tubular lumen, and $NaHCO_3$ is reabsorbed. H^+ in combination with an anion is thus excreted. This occurs mainly in the distal tubules.[97]

4. Electroneutrality is maintained by the shift of phosphate ion from mostly a dibasic phosphate in the blood to almost totally a monobasic phosphate in the urine.

In explanation of point 4, the pK' for the reaction

$$H_2PO_4^- \rightleftharpoons H^+ + HPO_4^{2-}$$

is 7.2 (Table 2.1) The equation for the phosphate buffer then becomes

$$pH = 7.2 + \log \frac{[HPO_4^{2-}]}{[H_2PO_4^-]}$$

and in the blood,

$$7.4 = 7.2 + \log \frac{[HPO_4^{2-}]}{[H_2PO_4^-]}$$

so that

$$\frac{[HPO_4^{2-}]}{[H_2PO_4^-]} = \text{antilog } 0.2 = 1.58$$

This implies that in the plasma, 61% of the phosphate is in the form of Na_2HPO_4 and 39% is as NaH_2PO_4. At the pH of the urine (approximately 5.0), we have

$$5.0 = 7.2 + \log \frac{[HPO_4^{2-}]}{[H_2PO_4^-]}$$

so that

$$-2.2 = \log \frac{[HPO_4^{2-}]}{[H_2PO_4^-]}$$

or

$$2.2 = \log \frac{[H_2PO_4^-]}{[HPO_4^{2-}]}$$

and

$$\frac{[H_2PO_4^-]}{[HPO_4^{2-}]} = \text{antilog } 2.2 = 165$$

The ratio of HPO_4^{2-} to $H_2PO_4^-$ becomes 1/165. This indicates that in urine, more than 99% of the sodium salt of phosphate is in the

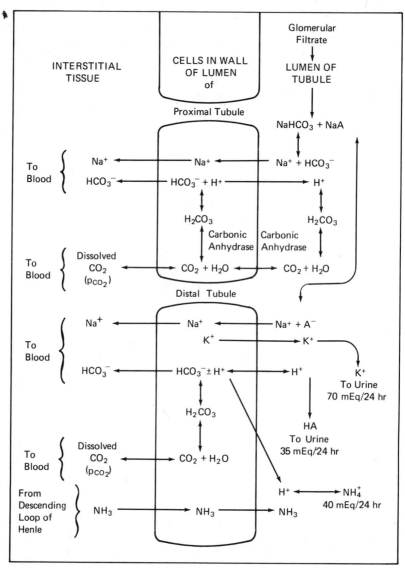

FIGURE 11.7 The various processes involved in the acidification of the urine. Hydrogen ion is excreted mainly in the distal tubule along with potassium and ammonium ion. Ammonia derives from ingested amino acids via glutamine formation. Sodium ion is reabsorbed in the distal tubules in exchange for the hydrogen ion.

form of NaH_2PO_4, while in the blood, $\frac{2}{3}$ of the sodium is in the form of NaH_2PO_4. Thus sodium is conserved, its place being taken by hydrogen ion when phosphate is excreted in the urine.

The mechanism of acidification of urine is summarized in Figure 11.7. In the proximal tubule, carbonic anhydrase is present in both the cell and wall of the lumen. $NaHCO_3$ is reabsorbed and H^+ is largely reclaimed, the urine remaining neutral in pH. Note that the blood p_{CO_2} level affects the process.

In the distal tubule there is no carbonic anhydrase in the wall of the lumen, but it is present in the cell. Sodium and some potassium ion in the lumen are exchanged for H^+ in the blood, which then combines with the anions in the lumen, such as phosphate. In the distal tubule K^+ exchanges for Na^+ and some H^+, resulting in a net excretion of K^+. Ammonia is derived from glutamine.[96] At least 80% of the NH_3 is in the form of glutamine before it is excreted. NH_3 from the descending loop of Henle diffuses to react with hydrogen ion in the distal tubule, $NH_3 + H^+ \rightleftharpoons NH_4^+$.[98] Some ammonium ion present in the interstital fluid also exchanges for Na^+ and H^+ by simple ion exchange.

Deficiency of the distal tubule to excrete adequate amounts of H^+ or NH_4^+ results in loss of base from the blood to the urine, with a resultant acidemia and a more alkaline urine pH ranging from 6.0 to 7.2. This condition is called distal renal tubule acidosis (distal RTA). The urine is more alkaline than that seen in proximal RTA and the condition is not a transient one, even when it occurs in infants and children.

Numerous pathological conditions which result in distal tubular damage can result in distal RTA. These include many of the conditions listed in Section 11.16 where the distal tubule is involved.

11.8.5 *Diuretics*

The objective of the use of diuretics is to influence the kidney so as to excrete water and salts which have been retained to the detriment of the patient's well being. Mercury salts and strong tea extracts from various sources (theophylline, caffeine, theobromine, etc.) were used for this purpose from ancient times. With an understanding of the function of the various components of the kidney, drugs are now synthesized with specific objectives in mind. These may be classified as follows:

Mercurials. These seem to act at the proximal tubule since the urine becomes more dilute when these drugs are administered.[98a] They seem to act in stimulating excretion of chloride and have been known to produce an hypochloremic alkalosis. Mercurials are toxic and produce necrotic changes in the tubules. Certain individuals are allergic to mercurials and in these cases administration of mercurials will cause a kidney shutdown. A typical mercurial is meralluride (Mercuhydrin) which is usually administered with theophylline.

$$\overset{\displaystyle OCH_3}{\underset{\displaystyle HOOC \cdot CH_2 \cdot CH_2 \cdot CO \cdot NH \cdot CO \cdot NH \cdot CH_2 \cdot CH \cdot CH_2 \cdot HgOH}{|}}$$

Carbonic Anhydrase Inhibitors. These cause an increase of serum bicarbonate levels stimulating excretion of sodium and potassium (see Table 6.5). Their action is in the proximal tubules.[98b] Hyperchloremic acidosis ensues and the action of the drug is inhibited. These compounds are of limited use today but with resistant edema they are effective in combination with drugs which act on the loop of Henle such as furosemide. Carbonic anhydrase inhibitors are usually sulfonamides. Typical is acetazolamide (Diamox).

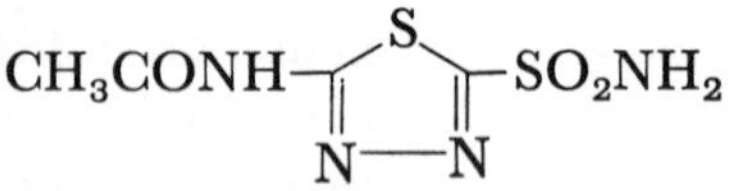

acetazolamide

The Thiazides. These diuretics are commonly used along with low salt intake because of their effectiveness and relatively low toxicity. They block sodium reabsorption in the cortical diluting segment of the distal tubule. Potassium is also lost and hypokalemia also often follows the use of these drugs.[98c] Titratable acidity and ammonium are increased somewhat in the urine. Since hypochloremia and alkalosis often develops during thiazide treatment, some refer to these compounds as "Chloruretic" sulfonamides. A typical thiazide is chlorothiazide (Diuril).

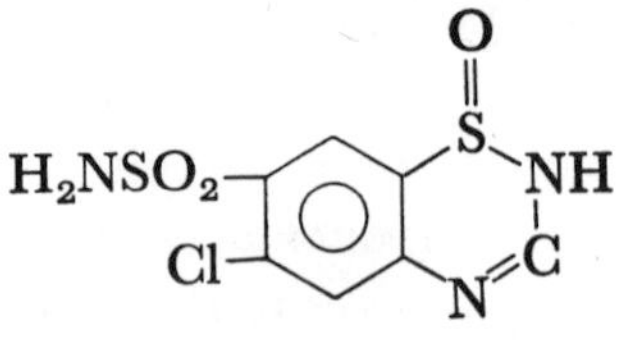

chlorothiazide

Potassium-Retaining Diuretics. Certain apparently unrelated compounds will cause a diuresis but not result in large losses of potassium with the sodium. The original rationale for these compounds was to prepare an aldosterone inhibitor. Since excessive formation of aldosterone results in potassium excretion and sodium retention, the inhibitor should cause a sodium diuresis and potassium retention. Spironolactone is a steroid which has this effect. Other unrelated compounds like amiloride (a guanidino derivative) and triamterene, a strongly basic polycyclic compound, also have this effect and are apparently not aldosterone inhibitors. All three drugs thus work on the K for Na exchange system in the distal tubules and prevent its action.[98d]

Sodium Pump Inhibitors in Loop of Henle. Since the ascending thick loop of Henle pumps sodium and chloride back into the blood, inhibition of this system should result in loss of salt and a diuresis. For this purpose, various compounds have been proposed and utilized. Two drugs used at present are furosemide and ethacrynic acid. They do not affect glomerular filtration rate and produce a huge diuresis in many patients but are ineffective in others.[98e] Ethacrynic acid has been reported to produce nausea, vomiting and abdominal pain. Both drugs seem to tend to cause hearing impairments which may not be reversible. Auditory acuity should be tested when administering these drugs in large doses. The formulas for these compounds are as follows:

furosemide

ethacrynic acid

11.9 *RENAL FUNCTION TESTS*

The immediate objective of the *kidney function tests* is to provide a tool through which the ability of the kidney to perform its regulatory and excretory activities can be investigated. The final goal is an evaluation of the functioning of the individual components of the kidney, as well as the total renal functional capacity. If it has been determined that a disease process exists, then the purpose of these tests will be to quantitate its progression or regression.

For screening purposes, blood urea nitrogen, creatinine, and non-protein nitrogen have been widely used on a routine basis. *They are not alternatives but complement each other, yielding different information.*

11.9.1   Urea Nitrogen

The molecular weight of urea is 60; its two nitrogens have a total molecular weight of 28. The factor $60/28 = 2.14$ will convert urea nitrogen levels to urea levels. For example, a urea N level of 10 mg/100 ml means 21.4 mg of urea/100 ml. In general, urea levels are listed as *urea nitrogen levels.*

Urea is synthesized in the liver, and with liver disease without kidney involvement serum urea nitrogen levels are very low, of the order of 5 mg/100 ml as compared to the normal of 10–18 mg/100 ml. Such low values are also observed in pregnant women.

Elevated blood urea nitrogen levels do not necessarily indicate impaired kidney function. Values as high as 60 mg/100 ml may be seen in diabetic coma with severe dehydration without kidney disease. Serum urea nitrogen in the plasma of children on high-protein milk formulas may reach levels as high as 25–30 mg/100 ml, as compared to 7–10 mg/100 ml when on breast milk. Thus the urea nitrogen level is influenced by protein intake.

The two major procedures for estimating urea nitrogen levels are with diacetyl[99] and with the blue color formed by alkaline phenol after urease converts the urea to ammonia.[100] Of these only the diacetyl method measures urea levels provided no substances (e.g., thiosemi-carbazide, phenazone) are added to increase sensitivity. These additives change the nature of the determination so that ammonia and other substances interfere.[100] Similarly, the alkaline phenol method is not specific. *As a result, in many laboratories the so-called urea nitrogen level is partly a nonprotein nitrogen level.*

11.9.2   Nonprotein Nitrogen

Nonprotein nitrogen (NPN) comprises all the nitrogenous components of the serum after the proteins have been removed. The major components are listed in Table 11.2.

TABLE 11.2

Approximate Distribution of Nonprotein Nitrogen Among Various Components in Serum from a Healthy Adult

Component	Nitrogen, mg/100 ml	Percent of total
Urea	13.0	62.74
Uric acid	1.7	8.20
Creatinine	0.25	1.20
Amino acids	0.70	3.37
Ammonia	0.07	0.34
Residual	5.0	24.13
Total	20.72	100.00

Normally, less than 28 mg of NPN/100 ml is found in the serum from fasting patients. This level rises markedly in dehydration from many causes and with kidney disease. A high protein intake will also raise NPN levels. This is true particularly in newborn infants, whose serum NPN levels do not exceed 22 mg/100 ml when on breast milk but rise to levels as high as 50–60 mg/100 ml when on high-protein formula.[101]

Since urea is made in the liver, NPN levels will rise more rapidly than urea levels in cases of severe uremia where there is extensive liver involvement in the disease process. *This is often masked when nonspecific methods for urea assay are employed.* Figure 11.8 illustrates the point. In case A, a liver problem which originally existed cleared up and at one time (12th day) rising urea levels were an indication of returning normal liver function. In case B, progressive increase in the disparity between urea and NPN levels indicates progressive deterioration of liver function.

11.9.3   *Creatinine*

Creatinine is formed by transamidination from arginine or canavanine to glycine to form guanidinoacetic acid. This is then methylated in the liver and cyclizes to form creatinine. Creatinine is an end point of metabolism and then appears in the urine.

$$
\begin{array}{l}
NH_2 \\
| \\
C{=}NH \\
| \\
HN \\
\cdot \\
(CH_2)_3 \\
\cdot \\
CH(NH_2)COOH
\end{array}
\quad
\begin{array}{l}
NH_2 \\
| \\
+ \; CH_2 \cdot COOH
\end{array}
\;\rightarrow\;
\begin{array}{l}
NH_2 \\
| \\
C{=}NH \\
| \\
NH \\
| \\
CH_2COOH
\end{array}
\quad
\begin{array}{l}
NH_2 \\
| \\
+ \; (CH_2)_3 \\
\quad\;\; CHNH_2 \cdot COOH
\end{array}
$$

arginine   glycine   guanidinoacetic   ornithine
                                                        acid

$$
\xrightarrow[\text{in liver}]{\text{methylation}}
\quad
\begin{array}{c}
HN \quad CH_3 \\
\| \quad\; | \\
H_2N \cdot C \cdot N \cdot CH_2{-}COOH \\
\\
\text{creatine}
\end{array}
\quad
\xrightarrow[\substack{\text{closure} \\ -H_2O}]{\text{ring}}
\quad
\begin{array}{c}
NH \quad CH_3 \\
\| \quad\;\; \diagup \\
C{-}N \\
| \quad\; | \\
HN \quad CH_2 \\
\diagdown \diagup \\
C \\
\| \\
O
\end{array}
$$

creatinine

Excreted along with the creatinine in the urine are some creatine and guanidinoacetic acid. In males, urine contains 20–28 mg of creatinine/kg/24 hr. Thus, a 70-kg man excretes ~ 1.63 g of creatinine/day. The urinary excretion levels of creatinine, creatine, guanidinoacetic acid (GAA), and guanidinosuccinic acid (GSA) are listed in Table 11.3. Normally, creatinine concentration is less than 0.9 mg/100 ml in serum. With damage to the kidney, creatinine levels rise to as high as 10–20 mg/100 ml. Levels higher than 2 mg/100 ml usually indicate kidney involvement in the disease process.[102] Serum creatinine levels correlate more closely with disease of the kidney than urea or NPN levels. In dehydration without kidney involvement, elevated urea or NPN levels are not accompanied by correspondingly elevated creatinine levels.

In summary, in screening patients for kidney disease, it is wise to accompany urea or NPN determination with assay of serum creatinine levels.

11.9.4   *Other Guanidino Compounds*

Guanidinoacetic acid (GAA) excretion is reduced markedly with kidney disease. From Table 11.3, the normal male excretes 0.6–0.9 mg of GAA/kg/24 hr in the urine. The female excretes more, the normal range being 1–1.4 mg/kg/24 hr. This is markedly reduced in uremia,

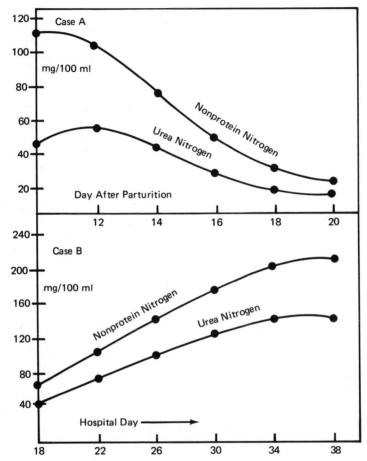

FIGURE 11.8 Disparity between urea and nonprotein nitrogen level in the plasma when kidney disease is associated with liver disease. Case A depicts serial analyses of blood from a patient recovering from toxemia of pregnancy associated with renal failure. Case B is a terminal case of uremia with progressive deterioration of liver function.

where values do not exceed 0.2 mg/kg/24 hr.[103,104] On the other hand, with the disappearance of GAA from the urine in kidney disease, excretion of another substance, guanidinosuccinic acid, is increased.[103,105,106] Normally, approximately 0.18 mg of GSA/kg in males and 0.25 mg of GSA/kg in females is excreted in 24 hr. In uremia this value rises markedly to levels ranging from 0.33 to 2.33 mg/kg/24 hr. This is an increase by a factor as high as 10. For this reason the

TABLE 11.3

Excretion of Creatinine, Creatine, Guanidinoacetic Acid (GAA), and Guanidinosuccinic Acid (GSA) in the Adult per 24 hr: Mean Values of 17 Females and 10 Males[103]

Substance	Male		Female	
	Mean	SD ±	Mean	SD ±
Creatinine				
mg/dl	149.9	64.4	94.8	37.5
mg/24 hr	1630.8	497.7	1142.8	367.8
mg/24 hr/kg	24.7	8.1	22.8	5.7
Creatine				
mg/dl	4.65	3.88	8.66	4.11
mg/24 hr	54.7	5.16	98.8	38.9
mg/24 hr/kg	0.83	0.80	1.98	0.71
GAA				
mg/dl	4.45	1.98	5.21	2.08
mg/24 hr	46.9	14.4	60.8	13.9
mg/24 hr/kg	0.71	0.25	1.22	0.22
GSA				
mg/dl	1.03	0.52	1.07	0.62
mg/24 hr	11.9	5.8	12.8	6.4
mg/24 hr/kg	0.18	0.10	0.25	0.10

GSA/GAA ratio has been proposed as a test for kidney function. The normal ratio does not exceed 0.3, while in the uremic the ratio will be at least 2 and go as high as 24. The GSA/GAA ratio is a very sensitive test for kidney function but the assay for these substances presents some difficulty at present.[103,107]

11.9.5 *Proteins in Urine*

Urinary proteins resemble plasma proteins, although normally proteins of molecular weight greater than 200,000 do not appear in the urine. The concentration of protein in the urine does not exceed 40–80 mg/100 ml. The proteins are concentrated by filtration through plastic porous filters (Amicon, Sartorius, Millipore) in the ordinary laboratory centrifuge or by concentrating by dialysis (Amicon). On electrophoresis, some albumin is seen, but the pattern is less distinct than that from plasma for the other fractions. Some protein is present at the albumin, α_1, α_2, β, and especially the γ sites.

In kidney disease involving the glomerulus, plasma proteins leak out. In addition, the glycoprotein of the basement membrane is found in the urine. This may be present in such a high concentration that it forms casts of the tubules which stain with metachromatic stains for polysaccharides. A glycoprotein normally present in urine in minute amounts is the *Tamm Horsfall* protein (uromucoid), which precipitates with 0.58 M NaCl solution.[107a] This glycoprotein is rich (9% in weight) in sialic acid (acetyl or glycyl neuraminic acid). In the hemagglutination reaction, neuraminic acid from the wall of the erythrocyte is hydrolyzed off, causing the cells to lose their charge and clump. This test is positive in certain virus diseases, such as influenza and the mumps, because these viruses produce a neuraminidase that hydrolyzes off the neuraminic acid from the cell. The Tamm Horsfall protein inhibits this test because it contains large amounts of neuraminic acid. This protein originates in the kidney and seems to be a fragment of the basement membrane of the glomerulus.[107b,107c]

In the myelomas an abnormal protein is often found in the urine, in which case it usually consists of one of the light chains (κ or λ). These are called Bence-Jones protein and have the peculiar property of precipitating on heating at 70°C after acidification of the urine with acetic acid, but redissolving at 100°C. This is distinct from the behavior of albumin, which precipitates on heating and does not redissolve.

Proteins are assayed in urine by precipitation with sulfosalicylic acid, determination with biuret, or with their reaction with commercially available dyes such as bromcresol green on paper.[167]

11.10 PROCEDURES FOR THE RENAL FUNCTION TESTS

11.10.1 *Introduction*

The procedures detailed below describe the techniques used in carrying out renal function tests in order to obtain the specimens to be sent to the laboratory for analysis. While variations of these procedures are not uncommon, the procedures described will serve the purpose for which they are designed. The procedures are described for the "normal" adult based on 1.73 m² of surface area. If the patient varies substantially from this figure, the volumes administered and the interpretation of the data need to be corrected for this variation.

11.10.2   *Renal Blood Flow (RBF) and Renal Plasma Flow (RPF)*

To measure renal blood flow, we need a substance whose plasma level we know and will be completely removed by the kidney and transferred to the urine. The amount that appears in the urine tells us how much reached the kidney per unit time and thus the *renal plasma flow (RPF)* and from the hematocrit value the *renal blood flow (RBF)*. Diodrast, ^{125}I-hippuran (*o*-iodohippuric acid), and *p*-aminohippuric acid have been used for this purpose.[108–112]

When *p*-aminohippuric acid (PAH) is given intravenously in quantities sufficient to cause a plasma concentration of 5–10 mg %, the kidneys respond both by glomerular filtration and by tubular secretion to remove 91–93.5% of this material from the plasma which flows through them.[111,113] If a constant plasma level of PAH is maintained, the determination of the amount excreted in the urine per unit time enables us to calculate the volume of blood passing through the kidneys during that time.

The patient drinks four glasses of water during the hour preceding the test in order to ensure adequate hydration. A urinary catheter is then inserted, and an initial specimen of both urine and blood is collected as a blank. A priming dose of 5 ml of a 4% solution of PAH is now given intravenously, followed by the maintenance solution of 4 ml/min of a 10 mg % solution of PAH for the duration of the test. At 30 min the collection of the first of three urine specimens is begun, each containing the total volume of urine passed during a 20-min interval. At the end of each collection period the bladder is irrigated with 20 ml of saline and this wash is added to the urine from the collection. Midway through each collection period a blood specimen is taken. The urines and bloods are now analyzed for PAH by diazotization and coupling to *N*(1-naphthyl)ethylene diamine, using PAH as the standard.[114,115] Table 11.4 illustrates typical data obtained in this test. The method of calculation is as follows.

1. Plasma concn. of PAH (mg/ml) × renal plasma flow (RPF) (ml/min) = urinary excretion (mg/min).
2. $0.06 \times RPF = 42$; $RPF = 42/0.06 = 700$ ml/min.
3. This must be corrected because the kidney removes only 91% of the PAH from the RPF. Thus we have $RPF = 700/0.91 = 769$ ml/min.
4. To calculate *renal blood flow* (RBF), we must correct for the red cell volume. Assuming a hematocrit value of 45%, the plasma will represent 55% of the total blood volume; then $RBF = 769 \times 100/55 = 1398$ ml/min.

The method of calculation described above is designed to show the principle of the procedure. The above result can be obtained by substituting in a formula. What we did was to multiply urine volume per minute by urine concentration and then divide by plasma concentration:

$$RPF = \frac{\text{urine volume}}{\text{time (min)}} \times \frac{\text{urine concentration}}{\text{serum or plasma concentration}}$$

Substituting the values for specimen 1 in Table 11.4, we find $RPF = (100/20)(800/5) = 800$ ml plasma/min; applying the correction as in step 3 above yields $800 \div 0.91 = 879$, and so we have

$$RBF = RPF \times \frac{100}{100 - \text{hematocrit}} = 879 \times \frac{100}{55} = 1598$$

Calculation for specimens 2 and 3 in Table 11.4 is done in a like manner. The mean of three tests is usually accepted as the true value and is the result obtained above (1389 ml/min).

TABLE 11.4

Renal Plasma Flow (PAH)

Specimen	Plasma concn., mg/100 ml	Urine concn., mg/100 ml	Urine volume	Total excreted, mg	Time, min	mg/min
1	5	800	100	800	20	40
2	7	912	90	820	20	41
3	6	1126	80	900	20	45
Mean	6	—	—	—	—	42

11.11   *GLOMERULAR FILTRATION RATE (GFR)*
(CREATININE AND UREA CLEARANCE)

The objective of the test is to determine the volume of fluid filtered per minute through all the glomeruli of the kidneys. This result in milliliters is called the glomerular filtration rate (GFR). Since the urine collected has been concentrated when passing through the tubules, due to water absorption, it is necessary to use a tracer (urea, creatinine, mannitol, inulin, thiosulfate, etc.) to calculate the GFR. The substance used as the tracer should not be reabsorbed or secreted by the tubules. This is true for mannitol and inulin.[116–120]

Inulin and mannitol are polysaccharides which when given intravenously can only enter the urine by passing with the glomerular filtrate into Bowman's capsule. Their secretion by the renal tubules is negligible.

The procedure for carrying out the test is similar to that described above for estimating renal plasma flow. The priming solution is 30 ml of a 10% inulin solution, which raises the plasma level to 100 mg %. The maintenance dosage is 4 ml/min of a 1.4% inulin solution. Inulin may then be measured in the blood and urine by the resorcinol method.[116,117] Calculations are as described below for a creatinine clearance.

11.11.1   *Clearance*

Because of the inconvenience in measuring the true GFR by inulin excretion, the concept of *clearance* (C) has been introduced. A clearance expresses the ability of the kidney to remove a specific substance from the blood per unit time. Though in common usage, clearance and GFR have become almost synonymous, their differences should be understood. Urea and creatinine are substances commonly used to measure clearance. Approximately 30–40% of the urea filtered is reabsorbed in the tubules. Thus a GFR determined by urea excretion is low or approximately 75 ml/min, in contrast to the GFR of 125 ml/min given by inulin excretion in the same individual. The value of 75 is not a true GFR and it is therefore called a "clearance." The same reasoning applies to creatinine, which is variably secreted by the tubules. A creatinine clearance may vary from 75 to 125% of the GFR as determined with inulin.[121-124] The creatinine clearance is taken, however, as 125 ml/min as a mean for the normal.[125]

Creatinine or urea is most conveniently used as the tracer since the blood level of these substances remains constant during the test. For creatinine, the clearance is constant at all urine outputs provided the bladder is emptied completely. For urea this is true only above a urine excretion of 2.2 ml/min. When doing a urea clearance with smaller outputs, one uses the standard clearance C_s calculated below, as against the maximum clearance C_m when excretion is more than 2.2 ml/min for urea.[126]

Let us assume that serum creatinine concentration is 1 mg/100 ml. At the end of 60 min, 120 ml of urine is obtained with a creatinine concentration of 55 mg/100 ml. It is apparent that the urine in passing the tubules has been concentrated 55-fold over the original concentration of the plasma ultrafiltrate. Therefore not 120 ml but $120 \times 55 = 6600$ ml has been filtered in 60 min, or 110 ml/minute. Now this might be normal for a particular individual, but what if one were dealing with a 50-lb child? It is therefore necessary to correct for the capacity of the kidney depending upon the filtering area available. This has been shown to be related to the surface area of the body. The normal surface area is considered to be 1.73 m². For this area a glomerular filtration rate of 125 ml/min is considered normal. Actually these values were determined 40–50 yr ago, and such an individual would be considered small by present standards (see Tables A.1 and A.2 in the Appendix). This does not affect the results, since correction factors would be adjusted to give the same percentage of normal. Tables 11.6 and 11.7 divide the surface area into 1.73 (the normal) to obtain a correction factor. Thus, referring to Table 11.6, the correction factor for an individual weighing only 55 lb and 4 ft, 5 in. in height is 1.75. If a GFR of 58.3 ml/min is obtained, one multiplies this by 1.75 and obtains approximately 102 ml/min, which is then reported as the GFR corrected to normal.

To obtain the GFR or clearance when inulin, mannitol, or creatinine is used as the tracer, one can see that we divide the urine concentration by the plasma concentration to estimate the degree of urine concentration. This result is multiplied by the number of milliliters of urine obtained per minute and the correction factor f for variance in height and weight:

$$\text{GFR} = \frac{\text{urine concn.}}{\text{plasma concn.}} \times \frac{\text{urine volume}}{\text{time (min)}} \times f = \frac{U}{P} \times \frac{V}{T} \times f$$

The result is then compared to the normal by dividing by 125 and multiplying by 100 to obtain a result in percent. Thus for the creatinine clearance described above the result is $(102/125) \times 100 = 81.6\%$ of normal, assuming a factor f of 1.0.

11.11.2 *Procedure (Creatinine or Urea Clearance)*

The patient voids and drinks two glasses of water. Collect a urine specimen 60 min later. Patient drinks a third glass of water. Collect a urine specimen 60 min later. Collect blood at the end of the first hour in between the duplicate tests. Obtain serum and urine urea levels. Assume the serum urea level is constant for both tests and obtain the urine urea level on each urine specimen. Calculate the clearance for each specimen separately. They should check each other if the test was done correctly. For creatinine the clearance is constant at all urine outputs provided the bladder is emptied completely. For urea this is true only above a urine excretion of 2.2 ml/min for the normal. If the volume of urine per minute times the correction factor for height and weight is less than 2.2 ml, then the values are calculated differently. Van Slyke corrects for this variance by taking the square root of the corrected volume per minute in the calculations, to arrive at a "standard clearance." The normal value for a standard clearance is 54 ml/min.[126,127]

The rationale for calculating the *standard clearance* is based upon the fact that if one plots the clearance for creatinine as ordinate against milliliters of urine excreted per minute, one obtains a straight line parallel to the x axis. This indicates that the clearance is a constant down to very low volumes. *It is important to keep in mind, however, that if in 1 hr one obtains a urine volume substantially less than 20 ml, then a serious error has been introduced, since even if only 5 ml of urine is retained in the bladder it would introduce a 25% error.* On the other hand, if such a plot is made for urea, a straight line is obtained down to 2.2 ml/min. The clearance then falls off rapidly as the volume excreted decreases. Van Slyke showed that this could be corrected for by calculating the standard clearance[126] as described above.

11.12 CALCULATIONS

$$C_m = \frac{\text{urine urea}}{\text{blood urea}} \times \text{correction factor} \times \frac{\text{total urine volume}}{\text{time}}$$

$$= \frac{U}{B} \times f \times \frac{V}{T}; \qquad \% \text{ of normal} = \%C_m = \frac{C_m}{75} \times 100$$

$$C_s = \frac{\text{urine urea}}{\text{blood urea}} \left[\text{correction factor} \times \frac{\text{total urine volume}}{\text{time}} \right]^{1/2}$$

$$= \frac{U}{B} \times \left(f \times \frac{V}{T} \right)^{1/2}; \qquad \% \text{ of normal} = \%C_s = \frac{C_s}{54} \times 100$$

A value of 54 ml/min is normal for a standard urea clearance.

Example 1. Maximum urea clearance. Patient has a serum urea B of 10 mg/100 ml. Urine area U is 120 mg/100 ml. Total volume of urine excreted in 1 hr ($T = 60$ min) is 360 ml. Patient weighs 154 lb and is 5 ft, 3 in. tall. Factor f from Table 11.6 is 100:

$$C_m = \frac{120}{10} \times 1.00 \times \frac{360}{60} = 72 \times \frac{\text{ml}}{\text{min}}$$

$$\%C_m = \frac{72}{75} \times 100 = 96\% \text{ of normal}$$

Example 2. Standard urea clearance. Patient has a serum urea of 10 mg/100 ml; urine urea is 520 mg/100 ml; 60 ml of urine was excreted in 60 min. Patient's height is 5 ft, 3 in., weight is 154 lb:

$$C_s = \frac{520}{10} \left(1.00 \times \frac{60}{60} \right)^{1/2} = 52 \text{ ml/min}$$

$$\%C_s = \frac{52}{54} \times 100 = 96\%$$

Example 3. Creatinine clearance. Patient has a serum creatinine of 1.00 mg/100 ml. Urine creatinine is 25 mg/100 ml. Total volume of

TABLE 11.5

Presentation of the Results of Clearance Tests

Clearance type	Time, min	Urine volume	Serum concn., mg/100 ml	Urine concn., mg/100 ml	Clearance	Percent of normal[a]
Urea C_m	60	120	10	360	72	96
Urea C_s	60	60	10	520	52	96
Creatinine[b]	60	80	1.0	90	120	96

[a] Factor $f = 1.00$.
[b] Creatinine clearance is always calculated as a maximum clearance.

urine excreted in 60 min is 240 ml. Patient is 5 ft, 3 in. tall and weighs 154 lb:

$$C = \frac{25}{1} \times 1.00 \times \frac{240}{60} = 100 \text{ ml/min}$$

$$\%C = \frac{100}{125} \times 100 = 80\%$$

The data for examples 1–3 would be presented as in Table 11.5 assuming they are from the same individual.

While the clearance tests are designed to evaluate glomerular function, some insight into tubular function can also be gleaned from these studies. The ratio of urine concentration to serum concentration of urea and creatinine is a measure of the ability of the tubules to reabsorb water. Generally, values in excess of 30:1 are obtained in the urea clearance and 50:1 in the creatinine clearance. Ratios substantially lower than these figures suggest tubular damage, provided the test has been performed correctly and excessive amounts of water were not taken by the patient during the test.

Although the creatinine clearance is to be preferred because the results are closer to the true glomerular filtration rate, even with reduced urine output, it suffers from relative inaccuracy in the methodology for creatinine assay in serum. Urea assays can be performed more accurately but the necessity for calculating standard clearances in many cases with low urine output and the greater influence of tubular reabsorption on the values obtained makes this test less specific as a measure of glomerular function. In any case the creatinine clearance is presently the method of choice.

TABLE 11.6

Correction Factors to Normal Surface Area for Adults (1.73/A) from Equation, Surface Area $= W^{0.425} H^{0.725} \times 0.007184$

Height (top line: ft, in.; bottom line: cm)

Weight		3 11	4 1	4 3	4 5	4 7	4 9	4 11	5 1	5 3	5 5	5 7	5 9	5 11	6 1	6 3	6 5	6 7	6 9
lb	kg	120	125	130	135	140	145	150	155	160	165	170	175	180	185	190	195	200	205
33	15	2.37	2.30	2.24	2.18	x	x	x	x	x	x	x	x	x	x	x	x	x	x
44	20	2.10	2.04	1.98	1.93	1.88	x	x	x	x	x	x	x	x	x	x	x	x	x
55	25	1.91	1.85	1.80	1.75	1.71	1.66	x	x	x	x	x	x	x	x	x	x	x	x
66	30	1.76	1.71	1.67	1.62	1.58	1.54	1.50	1.47	x	x	x	x	x	x	x	x	x	x
77	35	1.65	1.60	1.56	1.52	1.48	1.44	1.41	1.37	1.34	1.31	x	x	x	x	x	x	x	x
88	40	1.56	1.52	1.47	1.43	1.40	1.36	1.33	1.30	1.27	1.24	1.21	x	x	x	x	x	x	x
99	45	1.49	1.44	1.40	1.36	1.33	1.30	1.26	1.23	1.21	1.18	1.15	1.13	x	x	x	x	x	x
110	50	1.42	1.38	1.34	1.30	1.27	1.24	1.21	1.18	1.15	1.13	1.10	1.08	1.06	1.03	x	x	x	x
121	55	x	1.32	1.29	1.25	1.22	1.19	1.16	1.13	1.11	1.08	1.06	1.04	1.02	1.00	x	x	x	x
132	60	x	1.28	1.24	1.21	1.18	1.15	1.12	1.09	1.07	1.04	1.02	1.00	0.98	0.96	0.94	0.92	x	x
143	65	x	x	1.20	1.17	1.14	1.11	1.08	1.06	1.03	1.01	0.99	0.97	0.95	0.93	0.91	0.89	x	x
154	70	x	x	1.16	1.13	1.10	1.07	1.05	1.02	1.00	0.98	0.96	0.94	0.92	0.90	0.88	0.87	0.85	x
165	75	x	x	x	1.10	1.07	1.04	1.02	0.99	0.97	0.95	0.93	0.91	0.89	0.87	0.86	0.84	0.83	0.81
176	80	x	x	x	x	1.04	1.01	0.99	0.97	0.94	0.92	0.90	0.89	0.87	0.85	0.83	0.82	0.80	0.79
187	85	x	x	x	x	x	0.99	0.96	0.94	0.92	0.90	0.88	0.86	0.85	0.83	0.81	0.80	0.78	0.77
198	90	x	x	x	x	x	x	0.94	0.92	0.90	0.88	0.86	0.84	0.83	0.81	0.79	0.78	0.76	0.75
209	95	x	x	x	x	x	x	0.92	0.90	0.88	0.86	0.84	0.82	0.81	0.79	0.78	0.76	0.75	0.73
220	100	x	x	x	x	x	x	x	0.88	0.86	0.84	0.82	0.81	0.79	0.77	0.76	0.74	0.73	0.72
231	105	x	x	x	x	x	x	x	x	0.84	0.82	0.81	0.79	0.77	0.76	0.74	0.73	0.72	0.70
242	110	x	x	x	x	x	x	x	x	0.83	0.81	0.79	0.77	0.76	0.74	0.73	0.72	0.70	0.69
253	115	x	x	x	x	x	x	x	x	x	0.79	0.78	0.76	0.74	0.73	0.72	0.70	0.69	0.68
264	120	x	x	x	x	x	x	x	x	x	0.78	0.76	0.75	0.73	0.72	0.70	0.69	0.68	0.66
275	125	x	x	x	x	x	x	x	x	x	0.76	0.75	0.73	0.72	0.70	0.69	0.68	0.66	0.65
286	130	x	x	x	x	x	x	x	x	x	x	0.74	0.72	0.71	0.69	0.68	0.67	0.65	0.64
297	135	x	x	x	x	x	x	x	x	x	x	0.72	0.71	0.69	0.68	0.67	0.66	0.64	0.63
308	140	x	x	x	x	x	x	x	x	x	x	0.71	0.70	0.68	0.67	0.66	0.65	0.63	0.62
319	145	x	x	x	x	x	x	x	x	x	x	0.70	0.69	0.67	0.66	0.65	0.64	0.62	0.61
330	150	x	x	x	x	x	x	x	x	x	x	0.69	0.68	0.66	0.65	0.64	0.63	0.61	0.60

TABLE 11.7

Correction Factors to Normal Surface Area for Infants and Children (1.73/A) from Equation, Surface Area $= W^{0.425}\, H^{0.725} \times 0.07184$

Weight — Height (top line: in.; bottom line: cm)

lb	kg	7.9	9.8	11.8	13.8	15.7	17.7	19.7	21.7	23.6	25.5	27.6	29.5	31.5	33.5	35.4	37.4	39.4	41.3	43.3	45.3	47.2
(cm)		20	25	30	35	40	45	50	55	60	65	70	75	80	84	90	95	100	105	110	115	120
2.2	1.0	x	23.4	20.5	18.3	16.6	15.3	14.1	13.2	12.4	x	x	x	x	x	x	x	x	x	x	x	x
3.3	1.5	x	19.7	17.2	15.4	14.0	12.8	11.9	11.1	10.4	9.83	x	x	x	x	x	x	x	x	x	x	x
4.4	2.0	x	17.4	15.2	13.6	12.4	11.4	10.5	9.82	9.22	8.70	8.24	x	x	x	x	x	x	x	x	x	x
5.5	2.5	x	15.8	13.9	12.4	11.3	10.3	9.57	8.93	8.38	7.91	7.50	7.13	x	x	x	x	x	x	x	x	x
6.6	3.0	x	14.6	12.8	11.5	10.4	9.56	8.86	8.26	7.76	7.32	6.94	6.60	6.30	x	x	x	x	x	x	x	x
7.7	3.5	x	13.7	12.0	10.7	9.75	8.95	8.29	7.74	7.27	6.86	6.50	6.18	5.90	5.65	x	x	x	x	x	x	x
8.8	4.0	x	13.0	11.4	10.2	9.21	8.46	7.84	7.31	6.87	6.48	6.14	5.84	5.57	5.33	5.12	x	x	x	x	x	x
9.9	4.5	x	x	10.8	9.65	8.76	8.05	7.45	6.96	6.53	6.16	5.84	5.56	5.30	5.07	4.87	4.68	x	x	x	x	x
11.0	5.0	x	x	10.3	9.23	8.38	7.69	7.13	6.65	6.24	5.89	5.58	5.31	5.07	4.85	4.65	4.48	4.31	x	x	x	x
12.1	5.5	x	x	x	8.86	8.05	7.39	6.84	6.39	6.00	5.66	5.36	5.10	4.87	4.66	4.47	4.30	4.14	3.99	x	x	x
13.2	6.0	x	x	x	8.54	7.75	7.12	6.60	6.16	5.78	5.45	5.17	4.92	4.69	4.49	4.31	4.14	3.99	3.84	3.68	x	x
15.4	7.0	x	x	x	x	7.26	6.67	6.18	5.77	5.41	5.11	4.84	4.60	4.39	4.21	4.03	3.88	3.74	3.60	3.46	3.26	x
17.6	8.0	x	x	x	x	6.86	6.30	5.84	5.45	5.11	4.83	4.57	4.35	4.15	3.97	3.81	3.67	3.53	3.39	3.26	3.21	3.09
19.8	9.0	x	x	x	x	x	5.99	5.55	5.18	4.86	4.59	4.35	4.14	3.95	3.78	3.63	3.49	3.36	3.34	3.21	3.04	2.93
22.0	10	x	x	x	x	x	5.73	5.31	4.95	4.65	4.39	4.16	3.96	3.78	3.61	3.47	3.33	3.21	3.09	2.97	2.87	2.79
24.2	11	x	x	x	x	x	x	5.10	4.76	4.47	4.22	3.99	3.80	3.63	3.47	3.33	3.20	3.09	2.97	2.87	2.78	2.70
26.4	12	x	x	x	x	x	x	4.91	4.59	4.30	4.06	3.85	3.66	3.49	3.34	3.21	3.09	2.97	2.87	2.78	2.66	2.58
28.6	13	x	x	x	x	x	x	x	4.43	4.16	3.93	3.72	3.54	3.38	3.23	3.13	3.01	2.89	2.78	2.70	2.58	2.51
30.8	14	x	x	x	x	x	x	x	x	4.03	3.80	3.61	3.43	3.27	3.13	3.01	2.89	2.78	2.70	2.58	2.51	2.44
33.0	15	x	x	x	x	x	x	x	x	x	3.69	3.50	3.33	3.18	3.04	2.92	2.83	2.70	2.62	2.51	2.44	2.40
35.2	16	x	x	x	x	x	x	x	x	x	3.61	3.43	3.27	3.10	2.92	2.84	2.74	2.66	2.58	2.44	2.40	2.31
37.4	17	x	x	x	x	x	x	x	x	x	x	3.33	3.15	2.98	2.88	2.75	2.66	2.58	2.51	2.40	2.31	2.25
39.6	18	x	x	x	x	x	x	x	x	x	x	3.20	3.09	2.93	2.79	2.66	2.58	2.54	2.44	2.34	2.28	2.20
41.8	19	x	x	x	x	x	x	x	x	x	x	x	2.98	2.88	2.75	2.62	2.54	2.47	2.37	2.28	2.22	2.14
44.0	20	x	x	x	x	x	x	x	x	x	x	x	2.93	2.79	2.70	2.58	2.48	2.39	2.31	2.23	2.16	2.10
46.2	21	x	x	x	x	x	x	x	x	x	x	x	x	2.75	2.62	2.51	2.43	2.34	2.25	2.16	2.11	2.04
48.4	22	x	x	x	x	x	x	x	x	x	x	x	x	x	2.58	2.47	2.37	2.28	2.19	2.11	2.04	2.03
50.6	23	x	x	x	x	x	x	x	x	x	x	x	x	x	x	x	2.34	2.22	2.16	2.08	2.01	1.94
52.8	24	x	x	x	x	x	x	x	x	x	x	x	x	x	x	x	2.31	2.19	2.14	2.06	1.99	1.92
55.0	25	x	x	x	x	x	x	x	x	x	x	x	x	x	x	x	2.26	2.18	2.10	2.03	1.97	1.91

The factors *f* for correcting for height and weight in the clearance tests are shown in Tables 11.6 and 11.7. One uses these numbers to multiply the number of milliliters cleared per minute to bring to normal equivalence.

11.13 THE PROXIMAL TUBULE: PROCEDURES FOR MEASURING ITS FUNCTION

Tests for proximal tubular activity are based upon its capacity to reabsorb and secrete certain substances. An estimation of maximal proximal tubular activity (T_m) is also a good indication of the functioning mass of the kidney. A substance first is chosen that is secreted by the tubules. Diodrast and *p*-aminohippuric acid are two such substances.[128,129]

Generally, *p*-aminohippurate (PAH) is chosen because it is easier to assay than Diodrast. In measuring plasma flow to the kidney, a low concentration of PAH is used so as not to overload the kidney. In this test, one attempts to completely remove all of the PAH that reaches the kidney, so that the blood is completely or almost completely cleared. The objective is to measure blood flow to the kidney using PAH as a tracer.

In contrast, to measure the efficiency of the tubule in secreting the PAH or Diodrast, one overloads the kidney. In this way the number of milligrams of PAH or Diodrast excreted per minute is a measure of maximum excretory capacity of the kidney and thus a measure of tubular function. It is to be noted that in this case a number will be obtained which is characteristic of the substance employed. For example, compare the amount excreted by the glomerulus plus the amount secreted by the tubules for Diodrast to that for PAH in Table

TABLE 11.8

Typical Results Obtained for Total PAH Excreted Through the Glomerulus and Tubule in the T_m Test with PAH

No.	Plasma concn., mg/100 ml	Urine concn. (20 min), mg/100 ml	Urine volume	Total PAH, mg	PAH mg/20 (mg/min)
1	30	1600	150	2400	120
2	34	1857	140	2600	130
3	34	1700	150	2550	127

11.10. If another substance were used, a different number would be obtained. Once the normal is established and correction is made for the surface area of the patient, the number becomes meaningful.

11.13.1   *Tubular Maximum Excretory Capacity* (T_m)

The glomerular filtration rate (GFR) is estimated by inulin excretion or creatinine clearance as previously described. The patient is then prepared as for the determination of renal blood flow as above, and specimens of blood and urine are collected as described with this test. The priming dose is 10 ml of a 10% solution of p-aminohippuric acid given intravenously, followed by a maintenance dose of 4 ml/min of a 2.5% solution. The plasma level of PAH is now maintained between 30 and 40 mg %, easily exceeding the level that can be completely cleared from the plasma by tubular secretion. Tubular maximum excretory capacity for PAH (T_mPAH) is calculated from the data obtained such as that in Table 11.8 and when done separately on each pair of urine and blood collections, the results should check with each other.

If the glomerular filtration rate is 125 ml/min, then the PAH appearing in the urine due to glomerular filtration alone is 125×0.30 or 37.5 mg/min. Subtraction from the total urinary excretion per minute, as measured in sample 1 yields, $120 - 37.5 = 82.5$ mg/min or the quantity of PAH excreted by the tubules per minute in the experiment.

The figure of 82.5 mg/min is the T_mPAH since the tubules are overloaded and hence working at maximal capacity. The normal T_mPAH in an adult is 77 mg/min.

11.13.2   *Tubular Maximum Reabsorption Capacity* (T_m)

When the plasma concentration of a substance that is freely filtered in the glomerulus and reabsorbed in the proximal tubule is raised, a point will be reached at which the substance can no longer be totally reabsorbed and some will begin to appear in the urine. The maximum quantity that the tubules can reabsorb is measured by subtracting the amount of the substance appearing in the urine per minute from the amount per minute entering the glomerular filtrate. Glucose is the substance commonly used in the test. The maximum reabsorptive capacity of the tubules for it is 320 mg/min in the average individual. The symbol used to indicate this determination is $T_m$glucose.[130]

Let us assume that a patient is maintained at a glucose level of 500 mg/100 ml by continuous infusion of glucose solution. At this level glucose will appear in the urine since the kidney's ability to reabsorb the glucose is exceeded. Let us now assume that in 1 hr one obtains 240 ml (4 ml/min) of urine containing glucose (5 g/100 ml). Having determined the glomerular filtration by inulin, mannitol, or creatinine clearance as 100 ml/min in this patient, then 500 mg of glucose was filtered in 1 min. But 4 ml of a 5000 mg % glucose solution contains 200 mg of glucose. Therefore $500 - 200 = 300$ mg of glucose must have been reabsorbed.

Since it is routine for patients to receive I.V. glucose solutions, some prefer this test. If performed, the glucose in both urine and plasma must be determined by the glucose oxidase or hexokinase methods after precipitation of any interfering substances with zinc hydroxide,[131] to avoid interference from other reducing substances with the values obtained. This also means that the patient must not be receiving other medication since these may interfere with the enzymatic glucose procedure if the drug or its metabolites are not removed with zinc hydroxide.

11.13.3 *Phenolsulfonephthalein (PSP) Dye Excretion Test*

A more convenient test of proximal tubular function is the *phenolsulfonephthalein (PSP) dye excretion test.*[132–134] The patient empties his bladder and drinks four glasses of H_2O. Then 6 mg of PSP dye in 1 ml of H_2O is injected intravenously and urine specimens are collected at 15, 30, 60, and 120 min. Care is taken to see that the bladder is completely emptied at each collection. The dye is filtered in the glomerulus, but is primarily eliminated via secretion in the proximal tubule. Normal subjects generally excrete 35% of the initial dose at 15 min and 40–50% by the first hour, and at 2 hr 70% of the dye has entered the urine. This test measures renal blood flow and tubular function when the glomerular filtration rate is normal.

Some PSP is excreted by way of the liver. With liver disease, the amount of PSP conjugated and excreted into the bile ducts is reduced.

Thus more PSP is presented to the kidney and may enter the urine, leading to an overestimation of the tubular function.

11.13.4   *Tubular Reabsorption of Phosphate (TRP)*

This test measures the reabsorption of phosphate in the proximal tubule and is an indirect measure of parathyroid function and the function of the kidney in converting 25-hydroxycholecalciferol to 1,25-dihydroxycholecalciferol. The details of this test are described in Section 7.7 in the discussion of mechanisms for maintaining constant phosphate levels. Low values suggest hyperparathyroidism.

11.14   *DISTAL TUBULE: PROCEDURES FOR MEASURING ITS FUNCTION*

It is difficult to separate clearly the activity of the loop of Henle from that of the distal tubule since the distal tubule acts on the fluid from the loop of Henle to concentrate it.[111] As pointed out above, many consider the thick ascending limb of the loop of Henle as part of the distal tubule and call it the "diluting segment of the distal tubule."

The activity of the distal tubule is measured by its ability to produce a concentrated urine. This is measured by the specific gravity at first on a casual specimen, usually a first morning specimen. If this value is below 1.010, the urine is said to be *hyposthenic* and one proceeds with one of the "urine concentration tests."[135]

With the introduction of osmometers, which can measure osmotic pressure directly and rapidly, the more informative urinary osmotic pressure can be determined.[136] This determination is unaffected by protein in the urine since protein contributes very little to osmotic pressure but substantially to specific gravity. By assaying for Na, K, Cl, and urine urea on the same 24-hr specimen, the osmotic pressure can be interpreted on a rational basis. The results are reported as in Table 11.9.

Observed osmotic pressure is the result measured on the osmometer. The calculated value assumes (only true approximately) that the Na and K salts are ionized 100%, each yielding two ions.

To convert urea N/100 ml to mosmol/liter, we multiply by 0.357.

TABLE 11.9

Typical Report on Urine Osmotic Pressure

Date ____	Specimen: Urine	24 hr vol. 1750 ml
Na	80 mEq/liter	160 mosmol/liter
K	40 mEq/liter	80 mosmol/liter
Cl	86 mEq/liter	— mosmol/liter
Urea	600 mg %	214 mosmol/liter
Osmolality	Observed. 552	Calculated. 454 2(Na + K) + urea

Thus $600 \times 0.357 = 214$ mosmol/liter. The factor is derived from the fact that we first multiply by 10 to convert to one liter, then multiply by the ratio of the formula weight of urea to that of the two nitrogens (60/28), and then divide by 60 to get the number of mmol of urea per liter. This is the same as multiplying the urea N/100 ml by 10/28, which equals 0.357.

Glucose has a molecular weight of 180, and 18 g/liter or a 1.8% glucose level in the urine would contribute 100 mosmol to the osmotic pressure. This introduces a significant error in the urine of a diabetic. The difference between the observed and calculated value above $552 - 454 = 98$ mEq would be equivalent to a 1.8% glucose level in the urine. The glucose test in urine with glucose oxidase paper is usually negative below 100 mg % of glucose and this would register a positive value in the test. Where a marked disparity exists between the observed and calculated osmotic pressure one should test the urine with glucose oxidase paper. Many other substances, such as ammonia, phosphates, and amino acids, contribute to urinary osmotic pressure. For this reason the observed value will always be higher than the calculated value.

Protein, on the other hand, offers little interference. For example, a 5% protein level (50 g/liter) would be approximately 1/1000 normal, since a formula weight for urinary protein will average approximately 50,000. This would then account for only 1 mEq, which can be ignored.

If the osmotic pressure of a casual morning specimen is significantly below 400 mosmol, then one performs a concentration test such as the one recommended below.

A further test for tubular function is to test the ability of the tubule to acidify the urine. This also requires an ammonia determination. This test is also described below.

11.14.1 *The Urine Concentration Test*

At 6 P.M. the evening preceding the day of the test the patient's dinner should consist of a high protein meal ingested with less than 200 ml of water. No further food or water is taken before the completion of the test. All urine passed during the night is discarded, but the bladder is emptied at 8 A.M., 9 A.M., and 10 A.M., the next morning and the specific gravity of these urine specimens measured. In a normal individual at least one specific gravity measurement will exceed 1.025. With progressively severe distal tubular damage, the specific gravity begins to approach that of the glomerular filtrate passing into Bowman's capsule. This diminished concentrating ability is called *hyposthenuria* or *isosthenuria* when the urine specific gravity is, respectively, lower than or equal to that of the protein-free glomerular filtrate, which is 1.010.

11.14.2 *Titratable Acidity*

A 24-hr urine sample is collected under toluene (10 ml for the 24-hr collection). A 5-ml aliquot is added to a beaker fitted with a magnetic stirrer. One gram of potassium oxalate is added and the stirrer is started. A combination glass electrode is inserted and the pH recorded. NaOH (0.01 N, 10 μEq/ml) is added until the pH is exactly 7.4.

$$\frac{24 \text{ hr vol}}{5} \times \frac{\text{ml NaOH} \times 10}{1000} = \frac{\text{mEq}}{24 \text{ hr}} \text{ titratable acidity}$$

Normal values range from 0.17 to 0.50 mEq/kg/24 hr or 12 to 35 mEq/24 hr for a 70-kg individual.

Now 2 ml of 30% formaldehyde is added to the above solution and it is stirred for 1 min. Formaldehyde combines with ammonia and releases the acid excreted as the ammonium salt. The solution becomes acid. Now one again titrates to pH 7.4 with 0.01 N NaOH.

$$\frac{24 \text{ hr vol}}{5} \times \frac{\text{ml NaOH} \times 10}{1000} = \frac{\text{mEq NH}_4^+}{24 \text{ hr}}$$

Normal NH_4^+ excretion is 0.42–1.0 mEq/kg/24 hr. For a 70-kg man this value is 30–70 mEq/24 hr.

11.14.3 *Renal Acid Excretion Test*

NH_4Cl, 100 mg/kg, is given orally to begin the test, and its administration should be complete by 1 hr. At 2 hr the bladder is emptied and the urine discarded. Now all urine passed in the next 6 hr is collected. The pH and total ammonia of the pooled urine specimens are now measured. Normal values are urine pH 5.3 or lower, with ammonia excreted at the rate of 30–90 mEq/6 hr. With distal tubular malfunction the ammonia excretion is low and the pH ranges from 5.7 to 7.0.

11.15 *OTHER KIDNEY FUNCTION TESTS*

Several additional kidney function tests proposed are, in effect, recalculations of the results obtained in the tests discussed above in attempts to localize defects. A few examples will illustrate this point.

The *filtration fraction* is the ratio of the GFR (inulin or creatinine clearance) divided by the renal plasma flow [*p*-aminohippurate (PAH) clearance] multiplied by 100 to yield the result in percent. What is measured here is the percentage of blood filtered as compared to that presented to the kidney. For example, let us assume that we have a reduced glomerular filtration rate. The defect may not be in the glomeruli, which may be functioning normally, but in a reduced flow to the kidney incident to renovascular disease. In this case, the value of the fraction should be high. On the other hand, if the blood flow is normal and the glomeruli are at fault, then a low ratio and percentage would be found. Normally 18–20% of the plasma should be filtered. In acute and chronic glomerulonephritis values of the order of 15 will be obtained. In early pyelonephritis without glomerular involvement normal values will be obtained. In arteriolar nephrosclerosis, high values, of the order of 22–25, will be obtained since the defect is in the renal blood flow and not in the glomeruli. In actual practice, however, if glomerular damage is severe, the renal plasma flow, as measured with

PAH, will be reduced as well. In nephrosclerosis significant glomerular destruction commonly occurs. Thus these ratios should be interpreted with caution.

A comparison of the maximum tubular excretion (T_mPAH) or the maximum tubular reabsorption ($T_m$glucose) with the plasma flow to the kidney will distinguish between a decreased tubular excretion or absorption due to tubular defect, as in pyelonephritis, and that due to decreased blood flow, as in arteriolar nephrosclerosis. The ratio of renal plasma flow to the T_mPAH is called the *plasma flow per unit excretory mass*. Similarly, the ratio of renal plasma flow to the $T_m$glucose is the *plasma flow per unit reabsorption mass*. For *p*-aminohippurate the normal ratio is 9. In chronic glomerulonephritis, the ratio will be high, reaching values of 20. This reflects the decreased blood flow to the tubules secondary to glomerular capillary destruction, which results in decreased tubular secretion. Normally, the ratio of renal plasma flow to the $T_m$glucose will be 2.0. Increased and reduced ratios in this test will parallel the same variation in the ratio of RPF to T_mPAH.

The ratio of the glomerular filtration rate to that of the maximum tubular excretion rate is sometimes calculated to compare glomerular with tubular function in the same individual. This is called the *filtration rate per unit functioning nephron*. This ratio is normally 1.7 for GFR/T_mPAH. Reduced values are obtained with acute and chronic glomerulonephritis, suggesting that the glomerulus is relatively more inefficient than the tubule function in this condition. In early pyelonephritis, elevated values will be obtained, indicating more tubule involvement in the disease.

A summary of the kidney function tests and normal values is given in Table 11.10 for the more sophisticated tests and in Table 11.11 for the usual screening tests for kidney functions.

The levels of the various components found in the urine of healthy adults are shown in Table 11.12.

11.16   *LABORATORY FINDINGS IN VARIOUS RENAL DISEASES*

The application of the laboratory findings to various types of kidney disease aids in the diagnosis and in following the course of the disease process. Since a defect in one section of the kidney eventually

TABLE 11.10

Summary of the Kidney Function Tests Employed, with Normal Values

Test	Values for mean "normal" individual		Function intended to be measured
	Male	Female	
Glomerular Filtration rate (GFR), ml/min (inulin clearance)	130 ± 20	115 ± 15	Glomerular function
Renal plasma flow (RPF)	700 ± 135	600 ± 100	Rate of blood flow through the kidney
Filtration fraction, %	18–20	18–20	Percentage of plasma filtered through the glomerulus of the plasma reaching the kidney; distinguishes between abnormal GFR due to glomerular damage and that due to decreased plasma flow
Maximum tubular excretion			Tubular excretory capacity and thus tubular function
T_mPAH, mg/min	75 ± 13	70 ± 10	
$T_m$diodrast mg/min	55 ± 15	50 ± 15	
Maximum tubular reabsorption $T_m$glucose, mg/min	375 ± 80	300 ± 55	Tubular absorptive capacity for glucose and therefore tubular function
Plasma flow per unit of excretory mass			The ratio of plasma flow to tubular excretory capacity; this explains decreased tubular excretory capacity on the basis of either tubular damage or deficiency in blood flow
RPF/T_mPAH, ml/mg	9 ± 1.5	8 ± 1.5	
RPF/$T_m$diodrast, ml/mg	13 ± 2.3	12 ± 2.0	
Plasma flow per unit of reabsorptive mass RPF/$T_m$glucose, ml/mg	2 ± 0.4	2 ± 0.5	Ratio of plasma flow to tubular absortive capacity for glucose; this distinguishes between tubular and blood flow deficiency with tubular damage
Filtration rate per unit functioning nephron GFR/T_mPAH, mg/ml	1.7 ± 0.4	1.5 ± 0.3	Ratio of glomerular to tubular function, to evaluate relative extent of damage of tubules as compared to glomeruli

TABLE 11.11

Summary of the Screening Kidney Function Tests

Test	Normal values (males and females)	Function intended to be measured
Titratable acidity	0.17–0.50 mEq/kg/24 hr	Distal tubular
Acid loading test (NH_4Cl)	30–90 mEq/6 hr/1.73 m²	Distal tubular
Urine concentration test (osmolality test)	Morning urine should show sp. gr. < 1.25 and osmotic pressure > 600 mosmol/liter	Distal tubular
Creatinine clearance	125 ml ± 20%	Glomerular
Urea clearance (maximum)	75 ml ± 20%	Glomerular
Urea clearance (standard)	54 ml ± 20%	Glomerular
Phosphate tubular reabsorption test (PTR)	89–95% phosphate reabsorbed	Parathyroid and proximal tubular
Phenolsulfonephthalein test (PSP)	Of 6 mg PSP administered there is excreted: 15 min, 35%; 1 hr, 40–50%; 2 hr, 70%	Proximal tubular
Ammonia excretion	0.42–1.0 mEq/kg/24 hr	Distal tubule

results in impairment of the other parts, there is extensive overlap and the primary cause of the disease may not be apparent from the laboratory findings.

11.16.1 *Acute Glomerulonephritis*[137–142]

Acute glomerulonephritis is an inflammatory condition resulting from an antigen–antibody response to bacterial products, resulting in diffuse and generalized proliferation of the endothelium and mesangial cells of the basement membrane. This results in impaired permeability of the basement membrane of the glomerulus. The tubules may not be involved. Hypertension is not a constant finding. The diffuse type of this condition is attributed to streptococcal infection. Other bacterial infections may cause focal type of acute glomerulonephritis.

TABLE 11.12

Range for Certain Factors Found in the Urine of the Normal Human for Urine Output of 70-kg Individual per 24 hr

Factor	Range	Factor	Range
Osmotic pressure	400–1300 mosmol/liter	17-Keto steroids (adult)	
		Female	6–15 mg
Freezing point depression	0.9–2.5°C	Male	10–20 mg
		Lactic acid	140–350 mg
pH	4.8–6.0	Lead	< 40 μg
Specific gravity	1.015–1.040	Magnesium	30–170 mg
Volume	600–2000 ml	Manganese	7–98 μg
		Metanephrines	0.3–0.9 mg
Albumin	49–112 mg/100 ml	Nickel	140–280 μg
Aldosterone	1.0–9 μg	Nitrogen, total	10–20 g
Amino acid nitrogen	50–200 mg	Oxalic acid	20–50 mg
Ammonia nitrogen	200–900 mg	Phenols, total	14–42 mg
Arsenic	< 90 μg	Phosphorus	
Ascorbic acid (as dehydro ascorbic acid)	14–21 mg	Inorganic	300–1000 mg
		Organic	0.6–13 mg
		Potassium	50–150 mEq
Bicarbonate	0.6–12 mEq	Pregnandiol (female)	
Bromide	0.84–7.7 mg	Foll. phase	0.1–1.2 mg
Calcium	42–300 mg	Luteal phase	1.2–9.5 mg
Catecholamines	30–90 μg	Postmenopausal	0.1–0.9 mg
Chlorine	40–250 mEq	Pregnandiol (male)	0.1–0.9 mg
Citric acid	400–1000 mg	Pregnantriol (female)	0.2–3 mg
Copper	10–30 μg	Postmenopausal	0.4–1.5 mg
Coproporphyrin	< 20 μg	Pregnantriol (male)	0.4–2.4 mg
Creatine (female)	50–230 mg	Protein	2–100 mg
Creatine (male)	30–150 mg	Riboflavin	140–800 μg
Creatinine	See Table 11.3	Selenium	< 140 μg
Cystine	30–70 mg	Silicon	4.2–14 mg
Estrogens, total (female)		Sodium	80–225 mEq
		Sulfur	
Foll. phase	5–25 μg	Total	360–1400 mg
Luteal phase	2–15 μg	Inorganic	240–1120 mg
Ovulatory peak	20–45 μg	Thiamine	90–500 μg
Postmenopausal	4–10 μg	Tin	9–18 μg
Estrogens, total (male)	10–25 μg	Urea nitrogen	6–16 g
Fluorine	0.5–5.0 mg	Uric acid	250–750 mg
Guanidinoacetic acid	40–90 mg	Urobilinogen	0.5–6 mg
Hydroxy indole acetic acid	< 17 mg	Uroporphyrin	< 20 μg
Iodine	7.0–140 μg	Zinc	110–450 μg
Iron	100–500 μg		

It has been demonstrated that the composition of the polysaccharide coat of the initially invading streptococcus resembles, immunologically, the basement membrane of the glomerulus. For this reason it is postulated that antibodies against streptococcus infection also attack the basement membrane, resulting in its destruction and the appearance of mucoproteins in the urine in large quantities, such as the Tamm Horsfall protein.[107a–107c]

Laboratory findings:

(a) Reduction in glomerular filtration rate, and urea and creatinine clearance observed is only 50% of those of the patients in the early stages of the disease.

(b) PSP excretion often normal.

(c) Protein levels commonly elevated in urine.

(d) Hematuria and oliguria not common but serious when they do occur.

(e) Ketone bodies found in urine secondary to dehydration due to vomiting and food refusal.

(f) ASO titer may be elevated.

(g) Red blood cell casts often present in urinary sediment.

11.16.2 *Chronic Glomerulonephritis*[140–142]

Chronic glomerulonephritis may follow from acute glomerular nephritis or certain systemic diseases, often of unknown origin. The disease is characterized by numbers of glomeruli and tubules becoming atrophic, showing hyalinization and sclerosis. There are often prominent secondary vascular changes such as hyalinization of small arterioles. Many arterioles are completely destroyed.

Laboratory findings:

(a) Reduction in GFR and urea and creatinine clearance.

(b) Elevation of serum NPN, urea, and creatinine levels.

(c) PSP excretion lowered with tubule damage.

(d) Urine volume increased with impairment of concentrating mechanism. Specific gravity fixed between 1.008 and 1.012 (isosthenuria).

(e) Persistent but variable proteinuria.

(f)   Intermittent hematuria.

(g) Blood hemoglobin level lowered with a normochromic normo-cytic anemia.

(h) Low plasma protein and high lipid levels (see the nephrotic syndrome).

11.16.3   *The Nephrotic Syndrome*[138–149]

The nephrotic syndrome is a clinical rather than a pathological entity, including several unrelated disease processes all characterized by a glomerular lesion which permits protracted leakage of protein into the urine. The protein leakage is usually limited to the smaller protein molecules. Thus usually approximately 75% of the urine protein is albumin. Hypertension is usually absent.

Laboratory findings:

(a)   Low plasma total protein and inverted A/G ratio.

(b) Serum electrophoresis shows low albumin and marked elevated α_2 and β globulins.

(c)   Total lipids and cholesterol are markedly elevated.

(d) NPN and urea nitrogen are often normal or even low.

(e)   Urine protein levels may be above 5 g/24 hr.

(f)   Serum calcium levels are low.

(g) PSP test is usually normal.

(h) Concentration and dilution tests are usually normal.

11.16.4   *Lipoid Nephrosis*[141–148]

Lipoid nephrosis occurs commonly in children. It is characterized by marked edema and fusion of the foot processes of the visceral epithelial cells of the glomeruli as seen with the electron microscope; most commonly it is classified under the nephrotic syndrome.

Laboratory findings: Chemical findings are as above under nephrotic syndrome. Cholesterol levels are above 1000 mg/100 ml, with total lipids as high as 4000 mg/100 ml. Urine protein electrophoresis patterns are complementary to the serum patterns, showing low or absent α_2

and β fractions, high albumin fraction, and some α_1 and γ globulins. Calcium levels are of the order of 5 mg/100 ml. Urea levels are generally low or normal.

11.16.5   *Diabetic Glomerulosclerosis*[145-150]

There are two types of diabetic glomerulosclerosis, the nodular, also called Kimmelstiel–Wilson disease, and the diffuse type. In the nodular type, the kidney lesions show a thickening of the basement membrane of the glomerulus and deposition of hyaline material in the intercapillary mesangeal areas. There is usually considerable renal arteriosclerosis and hyalinization of the efferent glomerular arterioles. In the diffuse type of diabetic glomerulosclerosis there is general thickening of the basement membrane and deposition of hyaline material in the intercapillary cells. With the diffuse type the clinical picture of the nephrotic syndrome is seen more often than with the nodular type.

Laboratory findings:

(a)   Diabetic type glucose tolerance curve.
(b)   Urea and creatinine clearance values are low with elevated blood NPN, urea, and creatinine levels.
(c)   Protein found in the urine. Low plasma protein levels.
(d)   Cholesterol and total lipids elevated (see the nephrotic syndrome).
(e)   Renal plasma and blood flow are decreased.

11.16.6   *Disseminated Lupus Erythematosus*[151-152]

Disseminated lupus erythematosus is sometimes classified as a collagen disease and affects the glomerulus. There is a proliferation of intracapillary and extracapillary cells with local necrosis of the glomerular tuft and fibrin deposition in the glomerular capillaries. The disease is apparently the result of antibodies having been formed in the patient against DNA both single- and double-stranded. Thus it is apparently an autoimmune disease. Circulating antibodies to DNA and RNA have been demonstrated.[154,155]

Laboratory findings:

(a) L.E. preparation positive in 70% of the cases.
(b) Antinuclear–antibody test positive in 95% of the cases.
(c) False positive VDRL test for syphilis often seen.
(d) Cephalin flocculation test positive.
(e) Urea and creatinine clearance depressed.
(f) In later stages PSP excretion depressed.
(g) Later stages may resemble the nephrotic syndrome.
(h) The fluorescent antibody test and complement fixation test are positive in 98% of the cases.
(i) Plasma antibodies against DNA positive.

11.16.7 *Arteriolar Nephrosclerosis*[156, 157]

Arteriolar nephrosclerosis is associated with benign or malignant essential hypertension and is often called hypertensive kidney disease. A severe thickening with partial occlusion of small preglomerular arterioles with sclerosis of the arterioles is seen in this condition.

Laboratory findings:

(a) Renal blood flow is markedly reduced to a greater degree than the glomerular filtration rate.
(b) Reduced urea and creatinine clearance.
(c) Excretion of PSP dye is prolonged.
(d) Elevated urea, NPN, and creatinine levels.
(e) Later stages resemble chronic glomerulonephritis.

11.16.8 *Pyelonephritis*[158–160]

The origin of this disease is bacterial infection of the kidney, most often by the infection ascending the tubules from the bladder. Occasionally the infection descends from the glomeruli, after generalized infection of the blood stream. The isolation of bacterial pathogens from the urine in the acute state is a specific laboratory finding.

Laboratory findings:

(a) PSP excretion decreased.
(b) Loss of ability to concentrate urine.

(c)   Decrease in maximum tubular excretion T_mPAH and tubular resorption $T_m$glucose.

(d)   Pathogenic bacteria isolated from the urine.

11.16.9   *Amyloidosis*[161–163]

A deposition of amyloid, a protein–polysaccharide complex, occurs chiefly between the glomerular capillary endothelial cells and the basement membrane. Degenerative fatty changes are also usually present in the tubules.

Laboratory findings: The congo red dye retention test is useful in the diagnosis of amyloidosis, but it has been discouraged lately because of side effects and nonspecificity. The nephrotic syndrome, with low plasma and high urinary protein and high serum lipids, may develop. Electrophoresis patterns may resemble that in the nephrotic syndrome but may also show a monoclonal gamopathy. Occasionally light chains are found in the urine, and a positive Bence-Jones test may be found.

11.16.10   *Neoplastic Disease of the Urinary Tract*[164]

Tumors may originate in the kidney, bladder, and prostate, or invade these organs from distant foci.

Laboratory findings: Carcinomas of the kidney are clinically silent in as many as 50% of the cases. Late in the disease, hematuria is usually present. In most neoplastic disease of the urinary tract, elevations in the urinary lactic dehydrogenase levels are observed. With tumors of prostatic origin, acid phosphatase is elevated, especially in the advanced stages of the disease. Alkaline phosphatase may also be elevated if metastases to bone have occurred.

11.16.11   *Multiple Myeloma*[165]

Renal disease is secondary to abnormal proliferation of plasma cells in the bone marrow and marked elevation of plasma protein levels due to the myeloma protein. Large amounts of urinary protein may deposit and cause concretions in the renal tubules. Associated hypercalcemia, hyperuricemia, and urinary tract infections are relatively common and may aggravate renal insufficiency. Occasionally amyloidosis may also be present associated with multiple myeloma.

Laboratory findings:

(a) Serum protein electrophoresis often demonstrates abnormal globulin fraction. Bence-Jones protein test may be positive in urine. This is best demonstrated by electrophoresis after the urine is concentrated by ultrafiltration.

(b) Renal function appears to frequently deteriorate, especially after tests involving dehydration.

11.17 RECAPITULATION

In practice certain tests have been selected as yielding the necessary information to guide the diagnosis and course of treatment with kidney disease without subjecting the patient to elaborate procedures.

For screening purposes, to decide whether kidney disease may exist, serum urea or nonprotein nitrogen levels as well as the serum creatinine are measured. A careful microscopic examination of the urinary sediment is also made. Structures in the sediment, such as oval fat bodies commonly seen in the nephrotic syndrome, heme granular casts found in acute ischemic nephropathy, and the presence of pyuria, hematuria,[162] or bacilluria, alert one to the presence of disease.

The values of the serum urea nitrogen and the creatinine levels are compared. Elevated creatinine levels suggest kidney parenchymal damage. An elevated serum urea simultaneous with a normal serum creatinine level suggests dehydration or decreased cardiac output, with reduced renal blood flow. A very high serum creatinine level with only moderate increase in the urea level suggests renal disease and decreased dietary protein intake.

If renal disease is suspected, the creatinine or urea clearance test is used to estimate *glomerular damage*. The PSP test can be used to evaluate the extent of *proximal tubular damage*. To estimate *distal tubule* activity, one can employ the concentration and dilution tests. As a measure of urine concentration, the urinary osmolarity determination is preferred over the specific gravity test, since it is less affected by proteinuria. Normal urine will yield values of 400–800 mosmol/kg of water.[136] After a 14 hr (overnight) fluid fast, the urine should contain at least 850 mosmol/kg, and the ratio of urine to serum osmolality should be 3 or higher. Urinary electrolyte levels, titratable acidity, and the pH of fresh urine are also examined.

The routine urine analysis on admission will indicate the presence of significant proteinuria. Presence of excessive amounts of protein in the urine suggests pathology in the glomerular basement membrane. Unfortunately, the amount of protein excreted in any casual specimen cannot be used as a measure of protein loss in the urine since negative results can be obtained on individual specimens even in severe proteinuria.[166] To estimate the extent of proteinuria, 24-hr urine specimens are analyzed for total protein on at least two occasions.

Marked proteinuria might indicate the nephrotic syndrome. Here the serum protein electrophoresis, total cholesterol, and total lipid determinations are useful. In those cases with gammopathies of the plasma, the abnormal proteins will tend to deposit in the tubules. In these cases urinary protein electrophoresis compared to serum protein electrophoresis is helpful.[167] For proteins other than albumin, such as Bence-Jones protein, the dip stick tests are negative.

The total volume of a 24-hr urine allows an estimate of the extent of urinary output and identifies the degree of oliguria or polyuria. Huge volumes are associated with certain disease states, such as diabetes insipidus, hyperparathyroidism, and alkaptonuria.

Urine ammonia is readily measured and is of value in estimating the rate of ammonia formation and thus the efficacy of the distal tubule in replacing Na with NH_3 and combination with hydrogen ion, especially in acidosis. Here one must use a fresh specimen. Presence of bacteria of the Proteus species, which are urea splitters, will result in false high values.

After the chemical tests have been completed, the geography of the urinary tract may be explored with dye contrast radiographs (intravenous pyelogram). The bladder dynamics may also be evaluated (cystometrogram, forced urinary flow rates).

In certain selected cases direct angiography of the renal vessels and percutaneous or open renal biopsy are of use to provide a *clinicopathological correlation with the kidney function tests.*

11.18 SELECTED READING—KIDNEY IN WATER AND ELECTROLYTE BALANCE

Rouiller, C. and Muller, A. F., Eds., *The Kidney*. Vol. III, *Morphology, Biochemistry, Physiology*, Academic Press, New York (1971).

Wesson, L. G., Jr., *Physiology of the Human Kidney*, Grune and Stratton, New York (1969).

Thurau, K., *Kidney and Urinary Tract Physiology*, University Park Press, Baltimore, Maryland (1973).

Fischer, J. W., Ed., *Kidney Hormones*, Academic Press, New York (1970).

Becker, E., *Structural Basis of Renal Disease*, Hoeber, New York (1968).

Pitts, R. F., *Physiology of the Kidney and Body Fluids*, 2nd ed., Yearbook, Chicago, Illinois (1968).

Deane, N., *Kidney and Electrolytes. Foundations of Clinical Diagnosis and Physiologic Therapy*, Prentice Hall, New York (1966).

DeWardener, H. E., *Kidney*, 3rd ed., Little, Brown, and Co., Boston, Massachusetts (1968).

Lippman, R. W., *Urine and the Urinary Sediment*, 2nd ed., C. C. Thomas, Springfield, Illinois (1967).

Smith, H. W., *Kidney, Structure and Functions in Health and Disease*, Oxford Univ. Press, New York (1951).

Reubi, F. C., *Clearance Tests in Clinical Medicine*, C. C. Thomas, Springfield, Illinois (1963).

Manuel, Y., Revillard, J. P., and Vetuel, H. *Proteins in Normal and Pathological Urine*, University Park Press, Baltimore, Maryland (1970).

Trueta, J. *Studies of the Renal Circulation*, C. C. Thomas, Springfield, Illinois (1947).

11.19 REFERENCES

1. Gilmore, J. P. and Michaelis, L. L., Influence of anesthesia on renal responses of the foxhound to intravascular volume expansion, *Am. J. Physiol.* **216**:1367–1369 (1969).

2. Griffith, L. D., Bulger, R. E., and Trump, B. F., The ultrastructure of the functioning kidney, *Lab. Invest.* **16**:220–246 (1967).

3. Vimtrup, B. J., On the number, shape, structure, and surface area of the glomeruli in the kidneys of man and mammals, *Am. J. Anat.* **41**:123–151 (1928).

4. Hollinshead, W. H., Renovascular anatomy, *Postgrad. Med.* **40**:241–246 (1966).

5. Jacobsen, N. O., Jørgensen, F., and Thomsen, A. C., An electron microscopic study of small arteries and arterioles in the normal human kidney, *Nephron* **3**:17–39 (1966).

6. Jørgensen, F., The ultrastructure of the normal human glomerulus, *Danish Med. Bull.* **14**:281–287 (1967).

7. Boyarsky, S. and Labay, P., Stimulation of ureteral peristalsis through the renal nerves, _Invest. Urol._ **5**:200–202 (1967).

8. Thurau, K., Renal hemodynamics, _Am. J. Med._ **36**:698–719 (1964).

9. Siegelman, S. S. and Goldman, A. G., Trueta phenomenon: angiographic documentation in man, _Radiology_ **90**:1084–1089 (1968).

10. Lilienfield, L. S., Rose, J. C., and Porfido, F. A., Evidence for a red cell shunting mechanism in the kidney, _Circ. Res._ **5**:64–68 (1957).

11. Brunner, F. P., Rector, F. C., and Selden, D. W., Mechanism of glomerulo-tubular balance. II. Regulation of proximal tubular reabsorption by tubular volume, as studied by stopped flow microperfusion, _J. Clin. Invest._ **45**:603–611 (1966).

12. Deen, W. M., Robertson, C. R., and Brenner, B. M., Glomerular ultra-filtration, _Fed. Proc._ **33**:14–20 (1974).

13. Robertson, C. R., Deen, W. M., Troy, J. L., and Brenner, B. M., Dynamics of glomerular ultrafiltration in the rat. III. Hemodynamics and autoregulation, _Am. J. Physiol._ **223**:1191–1200 (1972).

14. Renkin, E. M. and Gilmore, J. P., Glomerular filtration, in _Handbook of Physiology. Renal Physiology_, Am. Physiol. Soc., Washington, D.C. (1973), Chapter 9, pp. 185–248.

15. Moffat, D. B., The fine structure of the blood vessels of the renal medulla with particular reference to the control of the medullary circulation, _J. Ultrastruct. Res._ **19**:532–545 (1967).

16. Lewy, J. E. and Windhager, E., Peritubular control of proximal tubular fluid reabsorption in the rat kidney, _Am. J. Clin. Pathol._ **214**:943–954 (1968).

17. Bossert, W. H. and Schwartz, W. B., Relation of pressure and flow to control of sodium reabsorption in the proximal tubule, _Am. J. Physiol._ **213**:793–802 (1967).

18. Gottschalk, C. W., Renal tubular function lessons from micropuncture, in _The Harvey Lectures, 1962–3_, Academic Press, New York (1963), pp. 99–124.

19. Walker, A. M., Bott, P. A., Oliver, J., and MacDowell, M., The collection and analysis of fluid from single nephrons of the mammalian kidney, _Am. J. Physiol._ **134**:580–595 (1941).

20. Morgan, T. and Berliner, R. W., Permeability of the loop of Henle, vasa recta, and collecting duct to water, urea, and sodium, _Am. J. Physiol._ **15**:108–115 (1968).

21. Wirz, H., The location of the anti-diuretic action in the mammalian kidney, in _The Neurohypophysis_, Heller, H., ed., Butterworths, London (1957).

22. Keeler, R., Effect of hypothalamic lesions on renal excretion of sodium, _Am. J. Physiol._ **197**:847–849 (1959).

23. Gilman, A., and Brazeau, P., The role of the kidney in the regulation of acid–base metabolism, _Am. J. Med._ **15**:765–70 (1953).

24. Gottschalk, C. W., Lassiter, W. E., and Mylle, M., Localization of urine acidification in the mammalian kidney, _Am. J. Physiol._ **198**:581–585 (1960).

25. Wrong, O., and Davies, H. E. F., The excretion of acid in renal disease, _Q. J. Med._ **28**:259–313 (1959).

26. Barclay, J. A., Cooke, W. T., Kenney, R. A., and Nutt, M. E., The effect of

water diuresis and exercise on the volume and composition of urine, *Am. J. Physiol.* **148**:327–337 (1947).

27. Berliner, R. W., Kennedy, T. J., Jr., and Orloff, J., Relationship between acidification of the urine and potassium metabolism: effect of carbonic anhydrase inhibition on potassium excretion, *Am. J. Med.* **11**:274–282 (1951).

28. Smith, H. W., Finkelstein, N., Aliminosa, L., Crawford, B., and Graber, M., The renal clearance of substituted hippuric acid derivatives and other aromatic acids in dog and man, *J. Clin. Invest.* **24**:388–404 (1945).

29. Miles, B. E., Paxton, A., and deWardener, H. E., Maximum urine concentration, *B. Med. J.* **2**:901–905 (1954).

30. Ohta, M., Jarrett, R. J., and Field, F. B., Measurement of ATP in tissues with the use of $C^{14}O_2$ production from glucose-1-C^{14}, *J. Lab. Clin. Med.* **67**:1013–1024 (1966).

31. Forster, R. P. and Taggart, J. V., The use of isolated renal tubules for the examination of metabolic processes associated with active cellular transport, *J. Cell. Comp. Physiol.* **36**:251–270 (1950).

32. Hoffman, J. F., Ed., *Cellular Functions of Membrane Transport*, Prentice Hall, Englewood Cliffs, New Jersey (1964).

33. Edelman, I. S., Transport through biological membranes, *Ann. Rev. Physiol.* **23**:37–70 (1961).

34. Richardson, S. H., An ion translocase system from rabbit intestinal mucosa. Preparation and properties of the (Na^+–K^+) activated ATPase, *Biochim. Biophys. Acta* **150**:572–577 (1968).

35. Landon, E. J., Jazab, N., and Forte, L., Aldosterone and sodium–potassium dependent ATPase activity of rat kidney membranes, *Am. J. Physiol.* **211**:1050–1056 (1966).

36. Spitzer, A. and Windhager, E. E., Effect of peritubular oncotic pressure changes on proximal tubular fluid reabsorption, *Am. J. Physiol.* **218**:1188–1193 (1970).

37. Kokko, J. P., Membrane characteristics governing salt and water transport in the loop of Henle, *Fed. Proc.* **33**:25–30 (1974).

38. de Rouffignac, C. and Morel, F., The permeability to sodium of different segments of the nephron studied in the rat with the aid of intratubular microinjections of ^{22}Na as NaCl, *Nephron* **4**:92–118 (1967).

39. Bulger, R. E., Tisher, C. C., Myers, C. H., and Trump, F. F., Human renal ultrastructure. II. The thin limb of Henle's loop and the interstitium in healthy individuals, *Lab. Invest.* **16**:124–141 (1967).

40. Marsh, D. J., Solute and water flows in thin limbs of Henle's loop in the hamster kidney, *Am. J. Physiol.* **218**:824–831 (1970).

41. Lever, A. F., The vasa recta and countercurrent multiplication, *Acta Med. Scand.* **178**(Suppl. 434):1–43 (1966).

42. Lever, A. F. and Kriz, W., Countercurrent exchange between the vasa recta and the loop of Henle, *Lancet* **1**:1057–1060 (1966).

43. Jamison, R. L., Bennett, C. M., and Berliner, R. W., Countercurrent multiplication by the thin loops of Henle, *Am. J. Physiol.* **212**:357–366 (1967).

44. Rocha, A. S. and Kokko, J. P., Sodium chloride and water transport in the

medullary thick ascending limb of Henle: Evidence for active chloride transport, *J. Clin. Invest.* **52**:612–623 (1973).

45. Burg, M. B. and Green, N., Function of the thick ascending limb of Henle's loop, *Am. J. Physiol.* **224**:659–668 (1973).

46. Ullrich, K. J., Function of the collecting ducts, *Circulation* **21**:869–874 (1960).

47. Fourman, J. and Kennedy, G. C., An effect of antidiuretic hormone on the flow of blood through the vasa recta of the rat kidney, *J. Endocrinol.* **35**:173–176 (1966).

48. Carone, F. A., Everett, B. A., Blondeel, N. J., and Stolarczyk, J., Renal localization of albumin and its function in the concentrating mechanism, *Am. J. Physiol.* **212**:387–393 (1967).

49. Slotkoff, L. M. and Lilienfield, L. S., Extravascular renal albumin, *Am. J. Physiol.* **212**:400–406 (1967).

50. Wilde, W. S. and Vorburger, C., Albumin multiplier in kidney vasa recta analyzed by microspectrophotometry of T-1824, *Am. J. Physiol.* **213**:1233–1243 (1967).

51. Smith, H. W., The fate of sodium and water in the renal tubules, *Bull. N.Y. Acad. Med.* **35**:293–316 (1959).

52. Marsh, D. J. and Solomon, S., Relationship of electrical potential differences to net ion fluxes in rat proximal tubules, *Nature* **201**:714–715 (1964).

53. Wood, J. E. and Ahlquist, R. P., Eds., *Symposium on Angiotensin*, The American Heart Association, New York (1962).

54. Belleau, L. J. and Earley, L. E., Autoregulation of renal blood flow in the presence of angiotensin infusion, *Am. J. Physiol.* **213**:1590–1595 (1967).

55. Cook, W. F., The detection of renin in juxtaglomerular cells, *J. Physiol. (London)* **194**:73–74 (1968).

56. Skeggs, L. T., Lentz, K. E., Gould, A. B., Hochstrasser, H., and Kahn, J. R., Biochemistry and kinetics of the renin–angiotensin system, *Fed. Proc.* **26**:42–47 (1967).

57. Coon, J. W., Aldosteronism and hypertension, *Arch. Intern. Med.* **107**:813–828 (1961).

58. Cook, W. F. and Pickering, G. W., The location of renin in the kidney, *Biochem. Pharmacol.* **9**:165–171 (1962).

59. Goormaghtigh, N., Existence of an endocrine gland in the media of the renal arterioles, *Proc. Soc. Exp. Biol. Med.* **42**:688 (1939).

60. Latta, H. and Maunsbach, A. B., The juxtaglomerular apparatus as studied electron microscopically, *J. Ultrastruct. Res.* **6**:547 (1962).

61. Berkowitz, H. D., Miller, L. D., and Itskowitz, H. D., Renal function and the renin–angiotensin system in the isolated perfused kidney, *Am. J. Physiol.* **213**:928–934 (1967).

62. Britton, K. E., Renin and renal autoregulation, *Lancet* **2**:329–333 (1968).

63. Wolff, H. P., Aldosterone in clinical medicine, *Acta Endocrin. Suppl.* **124**:65–86 (1967).

64. Doyle, A. E., Jerums, G., Coghlan, J. P., and Scoggins, B., Renin, aldosterone, and sodium balance in hypertension, *Bull. Postgrad. Comm. Med. Univ. (Sydney)* **26**:102–110 (1971).

65. Laragh, J. H., Angers, M., Kelly, W. G., and Lieberman, S., Hypotensive agents and pressor substances, effect of epinephrine, norepinephrine, angiotensin II, and others on the secretory rate of aldosterone in man, *J. Am. Med. Assoc.* **174**:234–240 (1960).

66. Slater, J. D. H., Barbour, B. H., Henderson, H. H., Casper, A. G. T., and Bartter, F. C., Influence of the pituitary and the renin-angiotensin system on the secretion of aldosterone, cortisol, and corticosterone, *J. Clin. Invest.* **42**:1504–1520 (1963).

67. Gross, F., The regulation of aldosterone secretion by the reninangiotensin system under various conditions, *Acta Endocrinol. Suppl.* **124**:41–64 (1967).

68. Koch, K. M., Aynedjian, H. S., and Bank, N., Effect of acute hypertension on sodium reabsorption by the proximal tubule, *J. Clin. Invest.* **47**:1696–1709 (1968).

69. Tobian, L., Interrelationships of electrolytes, juxtaglomerular cells, and hypertension, *Physiol. Rev.* **40**:280–312 (1960).

70. Biron, P., Koiw, E., Nowaczynski, W., Brouillet, J., and Genest, J., The effects of intravenous infusions of valine-5-angiotensin II and other pressor agents on urinary electrolytes and corticosteroids, including aldosterone, *J. Clin. Invest.* **40**:338–47 (1960).

71. Frazier, H. S., Renal regulation of sodium balance, *New Eng. J. Med.* **279**:868–875 (1968).

72. Mills, J. H., The cardiovascular system and renal control of sodium excretion, *Can. J. Physiol. Pharmacol.* **46**:297–303 (1968).

73. Peart, W. S., Hypertension and the kidney. III. Experimental basis of renal hypertension, *Brit. Med. J.* **2**:1359 (1959); *Br. Med. J.* **2**(Suppl.):1429 (1959).

74. Boyd, G. W., Adamson, A. R., Fitz, A. E., and Peart, W. S., Radioimmunoassay determination of plasma–renin activity, *Lancet* **1**:213–218 (1969).

75. Goldblatt, H., *The Renal Origin of Hypertension*, C. C. Thomas, Springfield, Illinois (1948).

76. Ellis, A., Natural history of Bright's disease, II and III. *Lancet* **1942** (January 10):34–36; **1942** (January 17):72–76.

77. Eck, R. V. and Dayhoff, M. O., *Atlas of Protein Sequence and Structure*, National Biomed. Res. Found., Silverspring, Maryland (1966), p. 117.

78. Malnic, G., Klose, R. M., and Giebisch, G., Micropuncture study of renal potassium excretion in the rat, *Am. J. Physiol.* **206**:674–686 (1964).

79. Berliner, R. W., Renal mechanisms for potassium excretion, in *The Harvey Lectures*, Series 55 (1960), pp. 141–171.

80. Dickerman, H. W. and Walker, W. G., Effect of cationic amino acid infusion on potassium metabolism in vivo, *Am. J. Physiol.* **206**:403–408 (1964).

81. de Rouffignac, C. and Morel, F., Micropuncture study of water, electrolytes, and urea movements along the loops of Henle in Psammomys, *J. Clin. Invest.* **48**:474–486 (1969).

82. Duarte, C. G., Chomety, F., and Giebisch, G., Effects of amiloride, oubain and furosemide upon distal tubules function in the rat, *Clin. Res.* **17**:428 (1969).

83. Wilczewski, T. W., Olson, A. K., and Carrasquer, G., Effect of amiloride, furosemide, and ethacrynic acid on Na transport in the rat kidney, *Soc. Exp. Biol. Med.* **145**:1301–1305 (1974).

84. Ramsay, A. G., Stop-flow analysis of the influence of p_{CO_2} on renal tubular transport of K and H, *Am. J. Physiol.* **206**:1355–1360 (1964).

85. Wiederholt, M. and Wiederholt, B., The influence of dexamethason on the water and electrolyte excretion in adrenalectomized rats, *Arch. Ges. Physiol.* **302**:57–78 (1968).

86. Rector, F. C., Jr., Bloomer, H. A., and Seldin, D. W., Effect of potassium deficiency on the reabsorption of bicarbonate in the proximal tubule of the rat kidney, *J. Clin. Invest.* **43**:1976–1982 (1964).

86a. Rodriguez-Soriano, J. and Edelmann, C. M., Jr., Renal tubular acidosis, *Ann. Rev. Med.* **20**:363–382 (1969).

86b. Lightwood, R. and Butler, N., Decline in primary infantile renal acidosis. Aetological implications, *Br. Med. J.* **1**:855–857 (1963).

87. Giebisch, G. and Malic, G., in *Symposium of Renal Transport and Metabolism*, Springer-Verlag, New York (1969), pp. 123–138.

88. Rector, J. C., Jr., Micropuncture studies on the mechanism of urinary acidification, in *Renal Metabolism and Epidemiology of Some Renal Diseases*, Maple Press, York, Pennsylvania (1964), pp. 9–31.

89. Hayes, C. P., Jr., Owen, E. E., and Robinson, R. R., Renal ammonia excretion during acetazolamide or sodium bicarbonate administration, *Am. J. Physiol.* **210**:744–750 (1966).

90. Pitts, R. F. and Stone, W. J., Renal metabolism and excretion of ammonia, in *Proc. 3rd Intern. Cong. Nephrol.*, Washington, D.C. (1967), Vol. 1, pp. 123–135.

91. Pollak, V. E., Mattenheimer, H., DeBruin, H., and Weinman, K. J., Experimental metabolic acidosis: the enzymatic basis of ammonia production by the dog kidney, *J. Clin. Invest.* **44**:169–181 (1965).

92. Hills, A. G. and Reid, E. L., Renal ammonia balance. A kinetic treatment, *Nephron* **3**:221–256 (1966).

93. Pitts, R. F., Renal production and excretion of ammonia, *Am. J. Med.* **36**:720–742 (1964).

94. Rector, F. C., Carter, N. W., and Seldin, D. W., The mechanism of bicarbonate reabsorption in the proximal and distal tubules of the kidney, *J. Clin. Invest.* **44**:278–290 (1965).

95. Maren, T. H., Carbonic anhydrase: chemistry physiology and inhibition, *Physiol. Rev.* **47**:598–781 (1967).

96. Van Slyke, D. D., Phillips, R. A., Hamilton, P. B., Archibald, R. M., Futcher, P. H., and Hiller, A., Glutamine as a source material of urinary ammonia, *J. Biol. Chem.* **150**:481–482 (1945).

97. Reid, E. L. and Hills, A. G., Diffusion of carbon dioxide out of the distal nephron in man during antidiuresis, *Clin. Sci.* **28**:15–28 (1965).

98. Stone, W. J., Balagura, S., and Pitts, R. F., Diffusion equilibrium for ammonia in the kidney of the acidotic dog, *J. Clin. Invest.* **46**:1603–1608 (1967).

98a. Levitt, M. F., Goldstein, M. H., Lenz, P. R., and Wedeen, R., Mercurial diuretics, *Ann. N.Y. Acad. Sci.* **134**:375–387 (1966).

98b. Dirks, J. H., Cirksena, W. J., and Berliner, R. W., Micropuncture study of the effect of various diuretics on sodium reabsorption by the proximal tubules of the dog, *J. Clin. Invest.* **45**:1875–1885 (1966).

98c. Laragh, J. H., Mode of action and use of chlorothiazide and related compounds, *Circulation* **26**:121–132 (1962).

98d. Liddle, G. W., Aldosterone antagonists and triamterene, *Ann. N.Y. Acad. Sci.* **139**:466–470 (1966).

98e. Muth, R., Diuretic properties of furosemide in renal disease, *Ann. Intern. Med.* **69**:249–261 (1968).

99. Natelson, S. and Pochopien, D. J., Novel approach to the design of a pediatric microanalytical laboratory, *Microchem. J.* **18**:457–485 (1973).

100. Kaplan, A., Chaney, A. L., Lynch, R. L., and Meites, S., Urea nitrogen and urinary ammonia, in *Standard Methods of Clinical Chemistry*, Vol. 5, Academic Press (1965), pp. 245–256; also see Chaney, A. L. and Marbach, E. P., *Clin. Chem.* **8**:130–132 (1962).

101. Natelson, S., Crawford, W. L., and Munsey, F. A., in *Immature Infant*, Illinois Dept. Public Health, Division Prev. Med. (1952), p. 73.

102. Rapoport, A. and Husdan, H., Endogenous creatinine clearance and serum creatinine in clinical assessment of kidney function, *Can. Med. Assoc. J.* **99**:149–156 (1968).

103. Sasaki, M., Takahara, K., and Natelson, S., Urinary guanidinoacetate/guanidinosuccinate ratio as indicator of kidney dysfunction, *Clin. Chem.* **19**:315–321 (1973).

104. Bonas, J. E., Cohen, B. D., and Natelson, S., Separation and estimation of certain guanidino compounds. Application to human urine, *Microchem. J.* **7**:63–77 (1963).

105. Natelson, S., Stein, J., and Bonas, J. E., Improvements in the method of separation of guanidino organic acids by column chromatography. Isolation and identification of guanidinosuccinic acid from human urine, *Microchem. J.* **8**:371 (1964).

106. Stein, I. M., Cohen, B. D., and Kornhauser, R. S., Guanidino succinic acid in renal failure, experimental azotemia and inborn errors of the urea cycle, *New Eng. J. Med.* **280**:926–930 (1969).

107. Grof, J., Tanko, A., and Menyhart, J., New Method for measurement of guanidino succinic acid in serum and urine, *Clin. Chem.* **20**:574–575 (1974).

107a. Tamm, I. and Horsfall, F. L., A mucoprotein derived from human urine which reacts with influenza, mumps and Newcastle disease viruses, *J. Exp. Med.* **95**:71–79 (1952).

107b. Boyce, W. H., King, J. S., and Fielden, M. L., Total nondialyzable solids (TNDS) in human urine. IX. Immunochemical studies of the R-1 "uromucoid" fraction, *J. Clin. Invest.* **40**:1453–1465 (1961).

107c. Maxfield, M., Fractionation of the urinary mucoproteins of Tamm and Horsfall, *Arch. Biochem. Biophys.* **89**:281–288 (1960).

108. Chisholm, G. D., Evans, K., and Kulatilake, A. E., The quantitation of renal blood flow using I-125 hippuran. An experimental study in the perfusion of the isolated canine kidney, *Br. J. Urol.* **39**:50–57 (1967).

109. Bergstrom, J., Bucht, H., and Josephson, B., Determination of the renal blood flow in man by means of radioactive tracers. Diodrast and renal vein catheterization, *Scand. J. Clin. Lab. Invest.* **11**:71–81 (1959).

110. Alpert, L. K., A rapid method for the determination of diodrast iodine in blood and urine, *Bull. Johns Hopkins Hosp.* **68**:522–537 (1941).

111. Chasis, H., Redich, J., Goldring, W., Ranges, H. A., and Smith, H. W., Use of sodium *p*-amino hippurate for functional evaluation of human kidney. *J. Clin. Invest.* **24**:583–588 (1945).

112. Foa, P. P. and Foa, N. L., Simple method for determining effective renal blood flow and tubular excretory mass in man, *Proc. Soc. Exp. Biol.* **51**:375–378 (1942).

113. Calcagno, P. L. and Rubin, M. F., Renal extractions of *p*-aminohippurate in infants and children, *J. Clin. Invest.* **42**:1632–1639 (1963).

114. Tompsett, S. L., The determination of aminohippuric acid and aminobenzoic acids in urine, *Clin. Chim. Acta* **8**:308–311 (1963).

115. Natelson, S., in *Techniques of Clinical Chemistry*, 3rd ed., C. C. Thomas, Springfield, Illinois (1971), pp. 679–683.

116. Murphy, G. P., Palma, L. D., Moore, R. H., *et al.*, Rapid inulin clearance. Method for clinical use, *N.Y. State J. Med.* **69**:1735–1738 (1969).

117. Schreiner, G. E., Determination of inulin by means of resorcinol, *Proc. Soc. Exp. Biol. Med.* **74**:117–120 (1950).

118. Berger, E. Y., Farber, S. J., and Earle, D. P., Jr., Renal excretion of mannitol, *Proc. Soc. Exp. Biol.* **66**:62–66 (1947).

119. Kennedy, T. J., Jr. and Kleh, J., The relationship between the clearance and the plasma concentration of inulin in normal man, *J. Clin. Invest.* **32**:90–95 (1953).

120. Jeremy, D. and McIver, M., Inulin, ^{57}Co-labelled vitamin B_{12} and endogenous creatinine clearance in the measurement of glomerular filtration rate in man, *Aust. Ann. Med.* **15**:346–351 (1966).

121. Doolan, P. D., Alpen, E. L., and Theil, G. B., A clinical appraisal of the plasma concentration and endogenous clearance of creatinine, *Am. J. Med.* **32**:65–79 (1962).

122. Tobias, G. J., McLaughlin, R. R., Jr., and Hopper, J., Jr., Endogenous creatinine clearance, *New Eng. J. Med.* **266**:317–323 (1962).

123. Vere, D. W. and Walduck, A., Endogenous creatinine clearance and glomerular filtration rate, *Lancet* **2**:1299 (1964).

124. Tobias, G. J., McLaughlin, R. F., Jr., and Hopper, J., Jr., Endogenous creatinine clearance, a valuable clinical test of glomerular filtration and a prognostic guide in chronic renal disease, *New Eng. J. Med.* **266**:317 (1962).

125. Kim, K. E., Onesti, G., Ramirez, O., *et al.*, Creatinine clearance reappraisal, *Brit. Med. J.* **4**:11–14 (1969).

126. Moller, O. E., McIntosh, J. F., and Van Slyke, D. D., Studies of urea excretion. II. Relationship between urine volume and the rate of urea excretion by normal adults, *J. Clin. Invest.* **6**:427 (1928).

127. Dean, R. F. A. and McCance, R. A., Inulin, diodone, creatinine and urea clearances in newborn infants, *J. Physiol.* **106**:431–439 (1947).

128. Findley, T. and White, H. L., Measurement of diodrast and inulin clearances in man after subcutaneous administration, *Proc. Soc. Exp. Biol. Med.* **45**:623–625 (1950).

129. Landowne, M. and Alving, A. S., Method of determining specific renal functions of glomerular filtration maximal tubular excretion (or reabsorption) and "effective blood flow" using a single injection of a single substance, *J. Lab. Clin. Med.* **32**: 931–942 (1947).

130. Letteri, J. M. and Wesson, L. G., Jr., Glucose titration curves as an estimate of intrarenal distribution of glomerular filtrate in patients with congestive heart failure, *J. Lab. Clin. Med.* **65**:387–405 (1965).

131. Tomisek, A. J. and Natelson, S., Fluorometric assay of ultramicro quantities of glucose with Somogyi filtrate and hexokinase, *Microchem. J.* **19**:54–62 (1974).

132. Mitchell, A. D., Owens, R., and Valk, W. L., Clinical value of urea clearance and phenolsulfonephthalein test, *J. Urol.* **71**:230–236 (1954).

133. Dunea, G. and Freedman, P., Phenolsulfonphthalein excretion test, *J. Am. Med. Assoc.* **204**:621–622 (1968).

134. Gault, M. H. and Dossetor, J. B., The place of phenolsulfonphthalein (PSP) in the measurement of renal function, *Am. Heart. J.* **75**:723–727 (1968).

135. Bjerre-Christensen, K., The pitressin test of renal concentrating capacity. A comparative evaluation of the Addis-Shevsky and Pitressin test, *Acta Med. Scand.* **142**:215–230 (1952).

136. Jacobson, M. H., Levy, S. E., Kaufman, R. M., Gallinek, W. E., and Donnelly, O. W., Urine osmolality. A definite test of renal function, *Arch. Intern. Med.* **110**:83–89 (1962).

137. Earle, D. P. and Seegal, D., Eds., Symposium of glomerulonephritis, *J. Chronic Dis.* **5**:1–172 (1957).

138. Fischel, E. E. and Gajdusek, D.C., Serum complement in acute glomerulonephritis and other renal diseases, *Am. J. Med.* **12**:190–196 (1952).

139. Tidstrom, B., Urinary protein in glomerulonephritis and pyelonephritis (electrophoretic assay), *Acta Med. Scand.* **174**:385–391 (1963).

140. Dodge, W. F., Spargo, B. H., Bass, J. A., and Travis, L. B., The relationship between the clinical and pathologic features of poststreptococcal glomerulonephritis. A study of the early natural history, *Medicine* **47**:227–267 (1967).

141. Pollak, V. E., Rosen, S., Pirani, C. L., Muehrcke, R. C., and Kark, R. M., Natural history of lipoid nephrosis and membranous glomerulonephritis, *Ann. Intern. Med.* **69**:1171–1196 (1968).

142. Lewis, L. A., Plasma and urinary proteins in renal disease, *Med. Clin. N. Am.* **39**:1015–1026 (1955).

143. Allen, A. C., The clinicopathologic meaning of the nephrotic syndrome, *Am. J. Med.* **18**:277–314 (1955).

144. Baxter, J. H., Goodman, H. C., and Havel, R. J., Serum lipid and lipoprotein alterations in nephrosis, *J. Clin. Invest.* **39**:355–365 (1960).

145. Block, W. M., Jackson, R. L., Stearns, G., and Butsch, M. P., Lipoid nephrosis: clinical and biochemical studies of 40 children with 10 necropsies, *Pediatrics* **1**:733–752 (1948).

146. Burch, R. R., Pearl, M. A., and Sternberg, W. H., A clinicopathological study of the nephrotic syndrome, *Ann. Intern. Med.* **56**:54–67 (1962).

147. Jensen, H., Plasma protein and lipid pattern in the nephrotic syndrome, *Acta Med. Scand.* **182**:465–473 (1967).

148. Gitlin, D., Cornwell, D. G., Nakasato, D., Oncley, J. L., Huges, W. L., Jr., and Janeway, C. A., Studies on the metabolism of plasma proteins in the nephrotic syndrome, II. The lipoproteins, *J. Clin. Invest.* **37**:172–184 (1958).

149. Rifkin, H., and Peterman, M. L., Serum and urinary proteins in diabetic glomerulosclerosis, results of electrophoretic analysis, *Diabetes* **1**:28–32 (1952).

150. Brown, J. and Straatsma, B. R., Diabetes mellitus, current concepts and vascular lesions (renal and retinal), *Ann. Intern. Med.* **68**:634–661 (1968).

151. Faith, G. C., and Trump B. F., The glomerular capillary wall in human kidney disease: acute glomerulonephritis, systemic lupus erythematosus, and preeclampsia-eclampsia. Comparative electron microscopic observations and a review, *Lab. Invest.* **15**:1682–1719 (1966).

152. Pollak, V. E., Pirani, C. L., and Schwartz, F. D., The natural history of the renal manifestations of systemic lupus erythematosus, *J. Lab Clin. Med.* **63**:537–550 (1964).

153. Pesce, A. J., Mendoza, N., Boreisha, I., Gaizutis, M. A., and Pollack, V. E., Use of enzyme linked anti DNA antibody in systemic lupus erythematosis, *Clin. Chem.* **20**:353–359 (1974).

154. Holman, H. R. and Kunkel, H. G., Affinity between the lupus erythematosis factor and cell nucleic and nucleoprotein, *Science* **126**:162 (1957).

155. Schur, P. H. and Monroe, M., Antibodies to ribonucleic acid in systemic lupus erythematosis, *Proc. Nat. Acad. Sci.* **63**:1108–1112 (1969).

156. Kark, R. M. and Lannigan, R., Hypertension and some diseases of small vessels of the kidney, *Postgrad. Med.* **40**:270–281 (1966).

157. Womack, R. K. and Mathews, W. R., The renal manifestations of periaarteritis nodosa, *J. Urol.* **59**:733–747 (1948).

158. Quinn, E. L. and Kass, E. H., *Biology of Pyelonephritis*, Henry Ford Hospital, International Symposium, Little, Brown and Co., Boston, Massachusetts (1960).

159. Angell, M. E., Relman, A. S., and Robbins, S. L., "Active" chronic pyelonephritis without evidence of bacterial infection, *New Eng. Med. J.* **278**:1303–1308 (1968).

160. Flanigan, W. J., Renal function in chronic pyelonephritis, *South. Med. J.* **58**:1353–1358 (1965).

161. Lindeman, R. D., Scheer, R. L., and Raisz, L. G., Renal amyloidosis, *Ann. Intern. Med.* **54**:883–898 (1961).

162. Gueft, B. and Ghidoni, J. J., The site of formation and ultrastructure of amyloid, *Am. J. Pathol.* **43**:837–854 (1963).

163. Cohen, A. S., Amyloidosis, *New Eng. J. Med.* **277**:522–530, 574–583, 628–638 (1967).

164. Dorfman, L. E., Amador, E., and Wacker, W. E. C., Urinary lactic dehydrogenase activity. II. Elevated activities for the diagnosis of carcinomas of the kidney and bladder, *Biochem. Clin.* **2**:41–55 (1963).

165. Bryan, C. W. and Healy, J. K., Acute renal failure in multiple myeloma, *Am. J. Med.* **44**:128–133 (1968).

166. Lathem, W., The renal excretion of hemoglobin, regulatory mechanism and the differential excretion of free and protein-bound hemoglobin, *J. Clin. Invest.* **38**:652–658 (1959).

167. Altman, K. A. and Stellate, R., Variation of protein content of urine in a 24 hr period, *Clin. Chem.* **9**:63–69 (1963).

The Sweat Glands

12.1 FUNCTION OF THE EXOCRINE (ECCRINE) SWEAT GLANDS

The two major objectives of the function of the human sweat glands are:

1. To assist in the maintenance of body temperature through the evaporation of the sweat on the surface of the skin.
2. To aid in the maintenance of water, electrolyte, and nitrogen balance by excretion of excess water, electrolytes, and nitrogenous compounds.

In addition to the above two objectives, with which we are concerned in this volume, we should add the identification of sex and species by odor emanating from volatile organic compounds in the sweat. Among these volatile compounds are complex terpenes and volatile aliphatic acids, such as isovaleric, isobutyric, and caproic, with characteristic odors.[1]

In order to accomplish the first of the above objectives it is necessary to excrete a dilute solution since it is intended that the water evaporate rapidly in order to control body temperature. For this purpose the exocrine sweat gland (Figure 12.1) and the nephron of the kidney may be compared and contrasted.

(a) In both the nephron and sweat gland the starting material is blood plasma with an osmotic pressure approximating 300 mosmol/liter.
(b) A capillary network enmeshes the sweat gland. In the nephron this capillary network is the glomerulus. In both

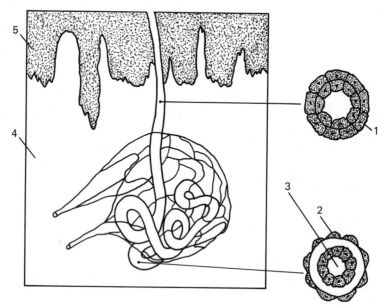

FIGURE 12.1   The eccrine sweat gland, composed of the secretory coil, which is surrounded by contractile myoepithelial cells, and a capillary network, is shown draining via the excretory duct. 1. Excretory duct. 2. Myoepithelial cell. 3. Secretory coil. 4. Subcutaneous tissue. 5. Epidermis.

cases blood pressure and flow through the capillaries control the filtration rate. This flow in both cases is controlled directly and indirectly by the central nervous system.

(c)   The objective of the nephron is to produce a *hyperosmotic* urine. The objective of the sweat gland is to produce *hypoosmotic* sweat.

(d)   While the kidney proceeds to concentrate the urine in the distal tubule, the straight distal tubule of the sweat gland does not have this property, as indicated by the hypo-osmotic nature of sweat.

For point (c) it is apparent that sodium and chloride excretion needs to be modulated in the sweat gland. For this purpose the secretory coil and proximal duct of the sweat gland (Figure 12.1) have an active Na–K-dependent ATPase system. Three sodium ions are transported per molecule of ATP hydrolyzed. The distal duct exhibits only one-tenth of the activity of the secretory coil and proximal duct and is therefore of minor value in sodium transport.[3]

The proximal coil may be compared to the thick loop of Henle of the nephron. After passing the thick loop of Henle the osmotic pressure of the uriniferous fluid has been reduced to 150 mosmol/liter. On passing the proximal duct of the sweat gland, sodium is reabsorbed and the osmotic pressure is reduced to less than 150 mosmol/liter.[2]

In studies comparing patients with essential and renal hypertension with control subjects, no difference was found in sweat sodium concentration in the three groups. In all cases a mean value of approximately 28 mEq/liter (an upper limit of 57 mEq/liter) of sodium concentration was found, which is within the normal range.[4] No inverse correlation between the sweat concentration and blood pressure was found, as occurs in urine. Apparently, aberration in the renin–angiotensin–aldosterone system in hypertensive patients does not affect sweat as it does urine formation.

From the above considerations one may draw the conclusion that the sweat gland behaves as though it were a nephron-type structure without the complete distal tubule or collecting duct function that exist in the kidney. In support of this concept is the observation that the antidiuretic hormone (ADH) has no effect on the composition of the sweat which is excreted.[5] The term *excretion* rather than *secretion* is used here since we are concerned with the function of the maintenance of constant pH and electrolyte concentrations in the blood of the human.

12.2 STRUCTURE OF THE EXOCRINE (ECCRINE) SWEAT GLANDS

The principal sweat-producing gland of the human is the *eccrine* (exocrine) sweat gland.[6] These glands are distributed over the entire body surface and on the palm of the hand. There are as many as 370 separate glands per cm².

The base of the gland consists of a coiled tube, referred to as the *secretory coil*, which lies in the subcutaneous tissue surrounded by contractile cells termed *myoepithelial* cells.[7–10] A duct drains to the surface of the skin (Figure 12.1). Other, more primitive glands, called *apocrine* sweat glands, are relatively few in number, and are found only in certain areas of the body, such as the axilla.

The coiled proximal tube of the gland is surrounded by a fine capillary network which encompasses the gland and brings the blood to and from the individual sweat gland.

12.3 *NEUROLOGICAL CONTROL OF SWEATING*

The great majority of the exocrine glands are stimulated by cholinergic nerve fibers of the autonomic nervous system. A few glands, notably those in the palm of the hand and sole of the foot, also receive adrenergic stimulation.[11] The apocrine glands are all activated by adrenergic nerve fibers. Excitation of all of these nerve fibers is caused by discharges from the anterior hypothalamus, and inhibited by stimuli arising in the posterior hypothalamus. Local reflexes to heat and cold also modify the nervous discharge.[12] Interruption of the autonomic nerve supply decreases sweat production. Thus sweating, directly and indirectly, is under continuous control of the central nervous system. Further evidence in this regard is that it is reponsive to psychological stimuli such as fear and anger.

12.4 *FACTORS INVOLVED IN DETERMINING THE VOLUME AND COMPOSITION OF THE SWEAT*

The following factors determine the rate and volume of perspiration.

1. *Number of functional sweat glands.* Although at birth a uniform number of glands is present, if an individual lives in a temperate climate for a long period of time, a certain number will atrophy. In addition, incident to blockage of the excretory duct in the superficial layers of skin, a certain percentage of sweat glands are periodically inactivated.[13]

2. *Environment and physical activity.* Both a warm environment and vigorous physical activity stimulate an increased sweat production.[14]

3. *Degree of acclimatization.* If an individual is maintained in a warm environment for longer than six weeks, his sweat production progressively increases in volume and decreases in electrolyte concentration up to a point. Acclimatization will raise the maximum quantity of sweat produced under the most severe heat conditions to a value as high as

1.5–3.5 liters/hr,[15–17] depending upon the intensity of the condition. In the past, stokers on ships and trains could excrete the higher value and collapse due to severe dehydration and salt loss.

4. *State of hydration and salt intake.* The volume of sweat is less in individuals who become dehydrated.[14] The electrolyte content will also decrease after salt depletion, although the total volume may be unchanged.[18] *The loss of water and salts through the sweat is thus an important mechanism for maintaining a normal degree of hydration and electrolyte concentration.*

12.5 EXCRETION OF ELECTROLYTES IN THE SWEAT GLANDS

In the adult, the sodium concentration in eccrine sweat can rise from a normal baseline of 10–30 mEq/liter,[19] to 100 mEq/liter following a sharp increase in sweat rate.[5] This occurs as the individual continues to perspire without water replacement, resulting in dehydration. In this manner the sweat glands serve to relieve the load on the kidney. In addition, nitrogenous compounds, such as urea, will be excreted in large amounts in the sweat.[20] In chronic renal failure, deposits of urea and salt crystals may be visible on the skin, giving a frost appearance. Increase in sweating in children also results in increased sodium loss, but increases are less steep than in the adult.

The sweat gland responds to increase in *aldosterone* concentration in the blood by decreasing the sodium output.[21,22] With *spironolactone* (an inhibitor of aldosterone secretion) sodium output is increased even on a low sodium diet.[23] The action of aldosterone is therefore to increase the quantity of sodium reabsorbed in the proximal duct. This duct is not permeable to water and is unresponsive to the antidiuretic hormone (ADH), as has been pointed out.[5] In this regard it resembles the ascending limb of the loop of Henle. Ouabain, which inhibits the Na–K-dependent ATPase, also inhibits transfer of sodium back to the blood stream in the proximal duct.[24] Since the Na/K ratio in the sweat is decreased with aldosterone administration, it has been proposed that this ratio be used as an index of deficiency or excess of mineralocorticoid concentration in the plasma.[25]

It should be pointed out that while sodium and chloride levels in sweat increase with the rate of sweating, the concentrations of potassium, calcium, lactic acid, urea, and creatinine continue to fall as sweating becomes more profuse. Glucose levels, on the other hand, are independent of the rate of sweating.[26]

While sodium and chloride levels are approximately 20–25% of those in the plasma, sweat potassium levels are usually somewhat higher than that found in the plasma. Apparently potassium is secreted in the duct partially in exchange for sodium ion.

The pH of the sweat approaches that in the urine, decreasing to levels as low as 4.0 but usually being at approximately 5.0. This suggests the presence of an acidification mechanism.

A substantial amount of the excreted hydrogen ion is in the form of ammonium ion. Thus, in acidosis, the perspiration mechanism participates in removing excess hydrogen ion.[27]

12.6 *CYSTIC FIBROSIS (MUCOVISCIDOSIS)*

Cystic fibrosis, also called mucoviscidosis, is a hereditary disease accompanied by dysfunction of the exocrine glands. Clinically the disease is characterized by persistent steatorrhea and repeated pulmonary infections.[28,29] Rarely will such individuals survive beyond puberty. There is poor digestion and absorption of starch, fats, and fat-soluble vitamins. Serum vitamin A and C levels are generally low. Vitamin A absorption is poor as measured by the vitamin A tolerance test.[40]

Pancreatic secretion is defective. The normal volume of pancreatic secretion is 2–3 ml/hr at two weeks of age and increases to 15 ml/hr at five years and to 45 ml/hr at seven years. In the individual with cystic fibrosis, levels as low as 1–2 ml/hr are observed even at seven years of age. Trypsin, amylase, and lipase activity are greatly reduced in duodenal juice.[41]

Cystic fibrosis is characterized by abnormally viscous secretions in a number of organs such as the pancreas, lungs, and intestinal tract. The newborn with cystic fibrosis may develop intestinal obstruction from failure of meconium dissolution in the fetal intestinal tract. This is due partly to decreased proteolytic activity but also to precipitation of a complex between albumin and an insoluble polysaccharide, containing substantial amounts of fucose. It has been suggested than an abnormality of glycoprotein metabolism is the basic cause of this disease.[34,35]

It was also demonstrated that glycosyl transferase was markedly hyperactive in lung tissue taken from nine patients with cystic fibrosis, as compared to the normal.[36] It is proposed that this results in the production of the abnormal glycoproteins of high viscosity.

It is also of interest to note that a protein is present in the serum of patients with cystic fibrosis with an isoelectric point of 8.41 which inhibits normal movement of cilia in rabbit trachea. Some have attempted to relate this to the accumulation of viscous mucous in the trachea and bronchioles of the child with cystic fibrosis. They claim that the improper action of the cilia caused by this protein in these patients results in stagnation and thickening of the mucus in the lung.[39]

10.7 *SWEAT AND CYSTIC FIBROSIS*

In the disease cystic fibrosis the sodium reabsorption mechanism in the proximal duct of the sweat gland is disturbed and large quantities of electrolytes are lost in the sweat.[28–31] These losses can be severe enough to induce large reductions in plasma volume, and hence bring about shock and circulatory failure. The defect appears to be in faulty function of the proximal duct.[31] Even at low sweat rates, sodium levels are above 70 mEq/liter. Sweat chloride level in excess of 60 mEq/liter is considered diagnostic of the disease.[32] It must be pointed out that significant elevation in sweat electrolyte levels also occurs in adrenal insufficiency, hereditary, pitressin-resistant, diabetes insipidus, and in hypothyroidism.[33]

If bromides are administered to normal individuals, the concentration in the sweat is substantially lower (approximately 25%) than that of the plasma. On the other hand, in the patient with cystic fibrosis they will rise to substantially higher values, exceeding 50% of the concentration in the plasma. This has been proposed as a test for cystic fibrosis. It is claimed that even heterozygotes can be detected by this test.[37,38]

12.8 *LABORATORY TEST FOR CYSTIC FIBROSIS*

In the *sweat test*, sweat sodium and chloride are assayed.[32,42] In one technique a Petri dish is taped to the back of the infant or child after the area has been thoroughly washed with distilled water, 70%

alcohol, and distilled water to remove evaporated salt. To stimulate sweating, the child is wrapped in a warm blanket. Beads of sweat can be seen collecting under the Petri dish. These are picked up with hematocrit capillaries, sealed, and sent to the laboratory for microassay for Na and Cl. As an alternative, 1-in. gauze squares are washed with distilled water and dried. These are placed in capped vials and weighed. The gauze is picked up with tweezers and placed on the back on the infant and covered with the Petri dish as before. The gauze, now wetted with sweat, is replaced in the bottle and weighed. Five milliliters of distilled water is added to each vial with mixing. An aliquot is taken and assayed for sodium and chloride and the concentration in the original sweat is calculated. For example, if 1 ml of sweat is obtained, the observed result is multiplied by five.

Most commonly sweating is induced by intradermal injection of mecholyl or by driving pilocarpine into the skin by an applied electrical voltage. This latter process is called *iontophoresis*, and equipment is available commercially for this purpose. (e.g., Sherwood Medical Industries, 1831 Olive Street, St. Louis, Missouri, 63103). The conductivity of the sweat is then measured and displayed on a meter as mEq/liter or as being normal or abnormal.[43] As an alternative, an ion-selective chloride electrode is applied to the sweat and the chloride concentration is measured directly.[44] An alternative is to place 5 μl into a micro-osmotic pressure measuring device (Wescor, Inc., Logan, Utah) and measure the osmotic pressure.

12.9 *NONPROTEIN NITROGEN AND TOTAL NITROGEN EXCRETION*

Amino acid excretion from the sweat reflects the composition of the blood. The essential amino acids, leucine, isoleucine, lysine, methionine, phenylalanine, threonine, and valine, are conserved and excreted only in very small amounts.[45,46] On the other hand, other amino acids are excreted in higher concentrations than they occur in the blood. Citrulline concentration in sweat is very high, as much as seven times the blood concentration. It has been suggested, for the above reasons, that secretion of amino acids is an active and selective process.[47]

In general, nonprotein nitrogen compounds are in higher concentration in the sweat than in the blood.[48] Ammonia levels are usually at least five times that in the blood, while urea levels are usually only slightly higher than the concentration in the blood.[20] As discussed earlier, the ammonia serves to react with excreted hydrogen ion.

Nitrogen output varies from approximately 112 mg/24 hr on a protein-free diet to 150 mg on a normal diet (75 g of protein daily). With a high protein diet (600 g of protein daily) 514 mg of nitrogen was excreted per day.[49]

The protein concentration in the sweat is low, usually less than 100 mg/100 ml. This is increased substantially in cystic fibrosis. Electrophoresis of sweat protein yields the proteins that normally are found in the urine, such as albumin, glycoproteins, and the gamma globulins.[50]

The meconium (first *postpartum* fecal discharge) from newborn infants with cystic fibrosis is high in protein. Determination of the protein in meconium after precipitation with trichloracetic acid or with an albumin dipstick is used to screen children for cystic fibrosis.[61]

Kallikrein is also found in sweat.[51] Kallikrein is a proteolytic enzyme found in the urine, resembling trypsin in its activity, which originates from various tissues including the pancreas. It has a

TABLE 12.1

Composition of Human Eccrine Sweat[54–60]

Component	Normal range	Component	Normal range
Volume/24 hr/ 1.73 m²	600–700 ml	Amino acid N	1.1–10.2 mg/100 ml
		Nonprotein N	17–64 mg/100 ml
pH	4.0–6.5	Ammonia N	5–9 mg/100 ml
Specific gravity	1.001–1.006	Urea N	8–30 mg/100 ml
Sodium	15–52 mEq/liter	Creatinine	0.4–1.3 mg/100 ml
Potassium	7–37 mEq/liter	Uric acid	0.5–2 mg/100 ml
Chloride	10–45 mEq/liter	Nitrogen (total)	20–90 mg/100 ml
Calcium	2–6 mg/100 ml		
Magnesium	0.15–4.0 mg/100 ml	Lactic acid	30–80 mg/100 ml
Phosphorus	1–3 mg/100 ml	Thiamine	0.2–1.0 μg/100 ml
Sulfur	7–7.4 μg/100 ml	Riboflavin	0.1–0.7 μg/100 ml
Iron	10–20 μg/100 ml	Folic acid	0.5–0.9 μg/100 ml
Iodine	0.15–0.6 μg/100 ml	Ascorbic acid	40–100 μg/100 ml
Copper	2–12 μg/100 ml	Inositol	15–36 μg/100 ml
Glucose	3–23 mg/100 ml	Phenols	1.5–4 mg/100 ml
Manganese	3–8 μg/100 ml	Corticoids	4–8 μg/100 ml

TABLE 12.2

Volatile Acids in Sweat[1]

Acid	μg/ml
Acetic	7.7 ± 6
Propionic	0.26 ± 0.1
Isobutyric	0.07 ± 0.06
Isovaleric	0.11 ± 0.11
Caproic	0.10 ± 0.09

vasodilating effect when injected intravenously since it hydrolyzes kininogen, an α globulin, to produce a hypotensive agent, kinin.

Other enzymes, such as amylase, are found in substantial amount in the sweat. Pancreatitis, resulting in high blood amylase levels, also results in high amylase levels in the sweat.

Sweat has antibacterial activity and in this regard serves to protect the skin from infection. The presence of immunoglobulins in the sweat apparently contributes to this activity.[52,53]

The concentrations of the most common components of sweat are summarized in Table 12.1 and Table 12.2.

12.10 SELECTED READING—HUMAN SWEAT

Rothman, S., *Physiology and Biochemistry of the Skin*, Univ. of Chicago Press, Chicago, Illinois (1954).

Johnson, S. A., Ed. *Skin and Internal Disease*, McGraw-Hill, New York (1968).

Kuno, Y., *Human Perspiration*, C. C. Thomas, Springfield, Illinois (1956).

Hurley, H. J. and Shelley, W. B., *The Human Apocrine Sweat Gland in Health and Disease*, C. C. Thomas, Springfield, Illinois (1960).

Carruthers, C., *Biochemistry of the Skin in Health and Disease*, C. C. Thomas, Springfield, Illinois (1962).

Sulzberger, M. B. and Herrmann, F., *The Clinical Significance of Disturbances in the Delivery of Sweat*, C. C. Thomas, Springfield, Illinois (1954).

Allen, A. C., Ed., *Skin*, 2nd ed., Grune and Stratton, New York (1966).

12.11 REFERENCES

1. Perry, T. L., Hansen, S., Diamond, S., Bullis, B., Mok, C., and Melancon, S. B., Volatile fatty acids in normal human physiological fluids, *Clin. Chim. Acta* **29**:369–374 (1970).

2. Cage, G. W. and Dobson, R. L., Sodium secretion and reabsorption in the human eccrine sweat gland, *J. Clin. Invest.* **44**:1270–1276 (1965).
3. Sato, K., Dobson, R. L., and Mali, J. W. H., Enzymatic basis for the active transport of sodium in the eccrine and sweat gland. Localization and characterization of Na K adenosine triphosphatase, *J. Invest. Derm.* **57**:10–16 (1971).
4. Levi, J., Rosenfeld, J. B., Hansel, Z., and Elian, E., Relationship of sweat sodium concentration to blood pressure in control subjects and in patients with essential and renal hypertension, *Israel J. Med. Sci.* **6**:665–668 (1970).
5. Ratner, A. C. and Dobson, R. L., The effect of antidiuretic hormone on sweating, *J. Invest. Derm.* **43**:379–381 (1964).
6. Dobson, R. L., The human eccrine sweat gland. Structural and functional interrelationships, *Arch. Environ. Health (Chicago)* **11**:423–429 (1965).
7. Dobson, R. L. and Abele, D. C., The correlation of structure and function in the human eccrine sweat gland, *Trans. Assoc. Am. Physicians* **75**:242–252 (1962).
8. Hurley, H. J. and Witkowski, J. A., The dynamics of eccrine sweating in man. I. Sweat delivery through myoepithelial contraction, *J. Invest. Derm.* **39**:329–338 (1962).
9. Hashimoto, K., Gross, B. G., and Lever, W. F., An electron microscopic study of the adult human apocrine duct, *J. Invest. Derm.* **46**:6–11 (1966); Electron microscopic study of the human adult eccrine gland. I. The duct, *J. Invest. Dermatol.* **46**:172–185 (1966).
10. Munger, B. L., The ultrastructure and histophysiology of human eccrine sweat glands, *J. Biophys. Biochem. Cytol.* **11**:385–402 (1961).
11. Kennard, D. W., The nervous regulation of the sweating apparatus of the human skin and emotive sweating in thermal sweating area, *J. Physiol. (London)* **165**:457–467 (1963).
12. Collins, J. K. and Weiner, J. S., Axon reflex sweating, *Clin. Sci.* **21**:333–344 (1961).
13. Gordon, B. I. and Maibach, H. I., On the mechanism of the inactive eccrine human sweat gland, *Arch. Dermatol.* **97**:66–68 (1968).
14. van Beaumont, W. and Bullard, R. W., Sweating: its rapid response to muscular work, *Science* **141**:643–646 (1963).
15. Gordon, R. S. Jr. and Cage, G. W., Mechanism of water and electrolyte secretions by the eccrine sweat glands, *Lancet* **1**:1246–1250 (1966).
16. Furman, K. I. and Beer, G., Dynamic changes in sweat electrolyte composition induced by heat stress as an indication of acclimatization and aldosterone activity, *Clin. Sci.* **24**:7–12 (1963).
17. Wyndham, C. H., Effect of acclimatization of the sweat rate/rectal temperature relationship, *J. Appl. Physiol.* **22**:27–30 (1967).
18. Sigal, C. B. and Dobson, R. L., The effect of salt intake on sweat gland function, *J. Invest. Dermatol.* **50**:451–455 (1968).
19. Coltman, C. A. Jr. and Atwell, R. J., The electrolyte composition of normal adult sweat, *Am. Rev. Resp. Dis.* **93**:62–69 (1966).
20. Komives, G. K., Robinson, S., and Roberts, J. T., Urea transfer across the sweat glands, *J. Appl. Physiol.* **21**:1681–1684 (1966).
21. Collins, K. J., The action of exogenous aldosterone on the secretion and composition of drug-induced sweat, *Clin. Sci.* **30**:207–221 (1966).

22. Shuster, S., Adrenal control of eccrine sweat secretion, *Proc. Roy. Soc. Med.* **55**:719–720 (1962).

23. Pazram, G., *et al.*, Modification of sweat electrolytes by spironolactones in secondary hyperaldosteronism, *Z. Ges. Inn. Med.* **20**:585–589 (1965).

24. Mangos, J., Transductal fluxes of Na, K, and water in the human eccrine sweat gland, *Am. J. Physiol.* **224**:1235–1240 (1973).

25. Grandchamp, A., Scherrer, J. R., Veyrat, R., and Muller, A. F., Assessment of the mineralocorticoid function by quantitative determination of Na and K in the sweat, *Helv. Med. Acta (Suppl.)* **48**:123 (1968).

26. Emrich, H. M., Stoll, E., Friolet, B., Colombo, J. P., Richterich, R., and Rossi, E., Sweat composition in relation to rate of sweating in patients with cystic fibrosis of the pancreas, *Pediat. Res.* **2**:464–478 (1968).

27. Brusilow, S. W. and Gordes, E. H., Ammonia secretion in sweat, *Am. J. Physiol.* **214**:513–517 (1968).

28. Johnston, M. C., Ed., Problems in cystic fibrosis, *Ann. N.Y. Acad. Sci.* **93**:485–624 (1962).

29. Anderson, D. H., Cystic fibrosis of pancreas and its relation to celiac disease. A clinical and pathological study, *Am. J. Dis. Child.* **56**:344–399 (1938).

30. Andrews, B. F., Bruton, O. C., and Knoblock, E. C., Sweat chloride concentration in children with allergy and with cystic fibrosis of the pancreas, *Pediatrics* **29**:204–208 (1962).

31. Cage, G. W., Dobson, R. L., and Waller, R., Sweat gland function is cystic fibrosis, *J. Clin. Invest.* **45**:1373–1378 (1966).

32. Shwachman, H. and Mahmoodian, A., The sweat test in cystic fibrosis. A comparison of overnight sweat collection versus the pilocarpine iontophoresis method, *J. Pediat.* **69**:285–287 (1966).

33. Madoff, L., Elevated sweat chlorides and hypothyroidism, *J. Pediat.* **73**:244–246 (1968).

34. Seutter, E. and Mali, J. W., Mucopolysaccharides in sweat, *Clin. Chim. Acta* **12**:17–21 (1965).

35. Biserte, G., Havez, R., and Cuvelier, R., The glycoproteins of bronchial secretions, *Expos. Ann. Biochim. Med.* **24**:85–120 (1963).

36. Louisot, P. and Levrat, C., A new pathogenic hypothesis for cystic fibrosis hyperactivity of glycosyl transferases at microsome level, *Clin. Chim. Acta* **48**:373–376 (1973).

37. Hager, Malecka, B. and Szezepanski, Z., Detection of mucoviscidosis heterozygotes by the bromide sweat test, *Pol. Tyg. Lek.* **24**:792–794 (1969).

38. Szczepanski, Z. and Hager-Maleckal, B., Determination of bromide in the sweat after oral administration of bromide preparations, *Pol. Med. J.* **6**:1507–1511 (1967).

39. Wilson, G. B., John, T. L., and Fonseca, J. R., Demonstration of serum protein differences in cystic fibrosis by isoelectric focusing in thin layer polyacrylamide gels, *Clin. Chim. Acta* **49**:79–91 (1973).

40. Lewis, J. M., Bodansky, O., Birmingham, J., and Cohlan, S. Q., Comparative absorption, excretion and storage of oily and aqueous preparations of Vitamin A, *J. Pediat.* **31**:496–508 (1947).

41. Andersen, D. H. and Early, M. V., Method of assaying trypsin suitable for routine use in diagnosis of congenital pancreatic deficiency, *Am. J. Dis. Child.* **63**:891 (1942).

42. Elam, J. F. and Palmer, R. E., An automated procedure for the determination of sweat electrolytes, *Am. J. Clin. Pathol.* **46**:65–68 (1966).

43. Jirka, M., Techniques of sweat collection in localized sweating by pilocarpine iontophoresis, *Clin. Chim. Acta* **11**:78–81 (1965).

44. Szabo, L., Kenny, M. A., and Less, W., Direct measurement of chloride in sweat with an ion selective electrode, *Clin. Chem.* **19**:727–730 (1973).

45. Coltman, C. A., Jr., Rowe, N. J., and Atwell, R. J., The amino acid content of sweat in normal adults, *Am. J. Clin. Nutr.* **18**:373–378 (1966).

46. Hungerland, H. and Liappis, N., Study of the free amino acids in human exocrine sweat, *Klin. Wschr.* **50**:973–976 (1972).

47. Miklaszewska, M., Comparative studies of free amino acids eccrine in sweat and plasma, *Pol. Med. J.* **7**:1313–1318 (1968).

48. Prasad, A. S., Schulert, A. R., Sandstead, H. H., Miale, A., Jr., and Farid, Z., Zinc, iron, and nitrogen content of sweat in normal and deficient subjects, *J. Lab. Clin. Med.* **62**:84–89 (1963).

49. Calloway, D. H., Odell, A. C. F., and Margen, S., Sweat and miscellaneous nitrogen losses in human balance studies, *J. Nutr.* **101**:775–786 (1971).

50. Cier, J. F., Manuel, Y., and Lacour, J. R., Electrophoretic studies of the proteins of human sweat with paper, gel and immunoelectrophoresis, *Compt. Rend. Soc. Biol.* **157**:1623–1626 (1963).

51. Fraki, J. E., Jansen, C. T., and Hopsu Havu, V. K., Human sweat kallikrein. Biochemical demonstration and chromatographic separation from several other esteropeptidases in the sweat, *Acta Dermatol. Venereol.* **50**:321–326 (1970).

52. Page, C. O., Jr. and Remington, J. S., Immunologic studies in normal human sweat, *J. Lab. Clin. Med.* **69**:634–650 (1967).

53. Jirka, M. and Masopust, J., Immunochemical behavior of proteins in human sweat, *Biochem. Biophys. Acta* **71**:217–218 (1963).

54. Astrand, I., Lactate content in sweat, *Acta Physiol. Scand.* **58**:359–367 (1963).

55. Brown, G. and Dobson, R. L., Sweat sodium excretion in normal women, *J. Appl. Physiol.* **23**:97–99 (1967).

56. Consolazio, C. F., Matoush, L. O., Nelson, R. A., Isaac, G. J., and Canham, J. E., Comparisons of nitrogen, calcium and iodine excretion in arm and total body sweat, *Am. J. Clin. Nutr.* **18**:443–448 (1966).

57. Ahlmann, K. L., Eränkö, O., Karvonen, M. J., and Leppänen, V., Mineral composition of thermal sweat in healthy persons, *J. Clin. Endocrinol.* **13**:773–782 (1953).

58. Levin, S., Shapira, E., Garin, T., Seligson, S., and Strassberg, D., Effect of age, ethnic background and disease on sweat chlorides, *Israel J. Med. Sci.* **2**:333–337 (1966).

59. Danton, R. A. and Nyhan, W. L., Concentrations of uric acid in the sweat of control and mongoloid children, *Proc. Soc. Exp. Biol. Med.* **121**:270–271 (1966).

60. Coltman, C. A., Jr. and Rowe, N. J., The iron content of sweat in normal adults, *Am. J. Clin. Nutr.* **18**:270–274 (1966).

III

*Appendix
and
Index*

Appendix

BIBLIOGRAPHY ON ANALYTICAL CLINICAL CHEMISTRY

Bergmeyer, H. U., Ed., *Methods of Enzymatic Analyses.* Academic Press, New York (1963).

Blackburn, S., *Amino Acid Determination*, Marcel Dekker, New York (1968).

Cawley, L. P., *Electrophoresis and Immunoelectrophoresis*, Little, Brown, and Co., Boston, Massachusetts (1969).

Davidsohn, I. and Henry, J. B., Todd Sanford, *Clinical Diagnosis by Laboratory Methods*, 14th ed., W. B. Saunders, Philadelphia, Pennsylvania (1969).

Henry, R. J., (ed.), *Clinical Chemistry, Principles and Techniques*, 2nd ed., Hoeber, New York (1974).

Keleti, G. and Lederer, W. H., *Handbook of Micromethods for the Biological Sciences*, Van Nostrand, New York (1974).

Natelson, S., *Techniques of Clinical Chemistry.* 3rd ed., C. C. Thomas, Springfield, Illinois (1972).

O'Brien, D., Ibbott, F. A., and Rodgerson, D. O., *Laboratory Manual of Pediatric Micro-Biochemical Techniques*, Hoeber, New York (1968).

Wolf, P. L., Williams, D., Tsudaka, T., and Acosta, L., *Methods and Techniques in Clinical Chemistry*, Wiley–Interscience, New York (1972).

Stahl, E., ed., *Thin Layer Chromatography*, 2nd ed., Academic Press, New York (1969).

Tietz, N. W., ed., *Fundamentals of Clinical Chemistry*, W. B. Saunders, Philadelphia, Pennsylvania (1970).

TABLE A.1

Surface Area (m²) for Adults From Height and Weight from Equation, Surface Area $= W^{0.425}\ H^{0.725} \times 0.007184$

Height (top line: ft, in.; bottom line: cm)

Weight (lb)	Weight (kg)	3 11	4 1	4 3	4 5	4 7	4 9	4 11	5 1	5 3	5 5	5 7	5 9	5 11	6 1	6 3	6 5	6 7	6 9
		120	125	130	135	140	145	150	155	160	165	170	175	180	185	190	195	200	205
33	15	0.73	0.75	0.78	0.80	x	x	x	x	x	x	x	x	x	x	x	x	x	x
44	20	0.83	0.85	0.88	0.90	0.92	x	x	x	x	x	x	x	x	x	x	x	x	x
55	25	0.91	0.94	0.96	0.99	1.02	1.04	x	x	x	x	x	x	x	x	x	x	x	x
66	30	0.98	1.01	1.04	1.07	1.10	1.13	1.15	1.18	x	x	x	x	x	x	x	x	x	x
77	35	1.05	1.08	1.11	1.14	1.17	1.20	1.23	1.26	1.29	1.32	x	x	x	x	x	x	x	x
88	40	1.11	1.14	1.18	1.21	1.24	1.27	1.30	1.34	1.37	1.40	1.43	x	x	x	x	x	x	x
99	45	1.17	1.20	1.24	1.27	1.30	1.34	1.37	1.40	1.44	1.47	1.50	1.53	x	x	x	x	x	x
110	50	1.22	1.26	1.29	1.33	1.36	1.40	1.43	1.47	1.50	1.54	1.57	1.60	1.64	x	x	x	x	x
121	55	x	1.31	1.35	1.38	1.42	1.46	1.49	1.53	1.56	1.60	1.63	1.67	1.70	1.74	x	x	x	x
132	60	x	1.36	1.40	1.44	1.47	1.51	1.55	1.59	1.62	1.66	1.70	1.73	1.77	1.80	1.84	x	x	x
143	65	x	x	1.44	1.48	1.52	1.56	1.60	1.64	1.68	1.72	1.75	1.79	1.83	1.87	1.90	1.94	x	x
154	70	x	x	1.49	1.53	1.57	1.61	1.65	1.69	1.73	1.77	1.81	1.85	1.89	1.93	1.96	2.00	2.04	x
165	75	x	x	x	1.58	1.62	1.66	1.70	1.74	1.78	1.82	1.86	1.90	1.94	1.98	2.02	2.06	2.10	2.14
176	80	x	x	x	x	1.66	1.71	1.75	1.79	1.83	1.88	1.92	1.96	2.00	2.04	2.08	2.12	2.16	2.19
187	85	x	x	x	x	x	1.75	1.80	1.84	1.88	1.92	1.97	2.01	2.05	2.09	2.13	2.17	2.21	2.25
198	90	x	x	x	x	x	x	1.84	1.88	1.93	1.97	2.01	2.06	2.10	2.14	2.18	2.23	2.27	2.31
209	95	x	x	x	x	x	x	1.88	1.93	1.97	2.02	2.06	2.11	2.15	2.19	2.23	2.28	2.32	2.36
220	100	x	x	x	x	x	x	x	1.97	2.02	2.06	2.11	2.15	2.20	2.24	2.28	2.33	2.37	2.41
231	105	x	x	x	x	x	x	x	x	2.06	2.11	2.15	2.20	2.24	2.29	2.33	2.38	2.42	2.46
242	110	x	x	x	x	x	x	x	x	2.10	2.15	2.19	2.24	2.29	2.33	2.38	2.42	2.47	2.51
253	115	x	x	x	x	x	x	x	x	x	2.19	2.24	2.28	2.33	2.38	2.42	2.47	2.52	2.56
264	120	x	x	x	x	x	x	x	x	x	2.23	2.28	2.32	2.37	2.42	2.47	2.51	2.56	2.61
275	125	x	x	x	x	x	x	x	x	x	2.27	2.32	2.37	2.41	2.46	2.51	2.56	2.61	2.65
286	130	x	x	x	x	x	x	x	x	x	x	2.36	2.41	2.45	2.50	2.55	2.59	2.65	2.70
297	135	x	x	x	x	x	x	x	x	x	x	2.39	2.44	2.49	2.54	2.59	2.64	2.69	2.74
308	140	x	x	x	x	x	x	x	x	x	x	2.43	2.48	2.53	2.58	2.63	2.68	2.73	2.78
319	145	x	x	x	x	x	x	x	x	x	x	2.47	2.52	2.57	2.62	2.67	2.73	2.78	2.83
330	150	x	x	x	x	x	x	x	x	x	x	2.50	2.56	2.61	2.66	2.71	2.76	2.82	2.87

Weight — **Height (top line: in.; bottom line: cm)**

lb	kg	7.9 / 20	9.8 / 25	11.8 / 30	13.8 / 35	15.7 / 40	17.7 / 45	19.7 / 50	21.7 / 55	23.6 / 60	25.5 / 65	27.6 / 70	29.5 / 75	31.5 / 80	33.5 / 85	35.4 / 90	37.4 / 95	39.4 / 100	41.3 / 105	43.3 / 110	45.3 / 115	47.2 / 120
2.2	1.0	0.06	0.08	0.09	0.10	0.11	0.11	0.12	0.13	0.14	×	×	×	×	×	×	×	×	×	×	×	×
3.3	1.5	0.08	0.09	0.10	0.11	0.12	0.14	0.15	0.16	0.18	0.19	×	×	×	×	×	×	×	×	×	×	×
4.4	2.0	0.09	0.10	0.11	0.13	0.14	0.15	0.17	0.18	0.19	0.20	0.21	×	×	×	×	×	×	×	×	×	×
5.5	2.5	0.09	0.11	0.13	0.14	0.15	0.17	0.18	0.19	0.21	0.22	0.23	0.24	×	×	×	×	×	×	×	×	×
6.6	3.0	0.10	0.12	0.14	0.15	0.17	0.18	0.20	0.21	0.22	0.24	0.25	0.26	0.28	×	×	×	×	×	×	×	×
7.7	3.5	×	0.13	0.15	0.16	0.18	0.19	0.21	0.22	0.24	0.25	0.27	0.28	0.29	0.31	×	×	×	×	×	×	×
8.8	4.0	×	0.15	0.15	0.17	0.19	0.21	0.22	0.24	0.25	0.27	0.28	0.30	0.31	0.33	0.34	×	×	×	×	×	×
9.9	4.5	×	×	0.16	0.18	0.20	0.22	0.23	0.25	0.27	0.28	0.30	0.31	0.33	0.34	0.36	0.37	×	×	×	×	×
11.0	5.0	×	×	0.17	0.19	0.21	0.23	0.24	0.26	0.28	0.29	0.31	0.33	0.34	0.36	0.37	0.39	0.40	×	×	×	×
12.1	5.5	×	×	×	0.20	0.22	0.24	0.25	0.27	0.29	0.31	0.32	0.34	0.36	0.37	0.39	0.40	0.42	0.43	×	×	×
13.2	6.0	×	×	×	0.20	0.22	0.24	0.26	0.28	0.30	0.32	0.34	0.34	0.37	0.39	0.40	0.42	0.43	0.45	0.47	×	×
15.4	7.0	×	×	×	×	0.24	0.26	0.28	0.30	0.32	0.34	0.36	0.38	0.39	0.41	0.43	0.45	0.46	0.48	0.50	0.53	×
17.6	8.0	×	×	×	×	0.25	0.28	0.30	0.32	0.34	0.36	0.38	0.40	0.42	0.44	0.45	0.47	0.49	0.51	0.53	0.54	0.56
19.8	9.0	×	×	×	×	×	0.29	0.31	0.33	0.36	0.38	0.40	0.42	0.44	0.46	0.48	0.50	0.52	0.54	0.56	0.57	0.59
22.0	10	×	×	×	×	×	0.30	0.33	0.35	0.37	0.40	0.42	0.44	0.46	0.48	0.50	0.52	0.54	0.56	0.58	0.60	0.62
24.2	11	×	×	×	×	×	×	0.34	0.36	0.39	0.41	0.43	0.46	0.48	0.50	0.52	0.54	0.56	0.58	0.60	0.62	0.64
26.4	12	×	×	×	×	×	×	0.35	0.38	0.40	0.43	0.45	0.47	0.50	0.52	0.54	0.56	0.58	0.60	0.62	0.65	0.67
28.6	13	×	×	×	×	×	×	×	0.39	0.42	0.44	0.47	0.49	0.51	0.54	0.56	0.58	0.60	0.62	0.65	0.67	0.69
30.8	14	×	×	×	×	×	×	×	×	0.43	0.46	0.48	0.51	0.53	0.55	0.58	0.60	0.62	0.64	0.67	0.69	0.71
33.0	15	×	×	×	×	×	×	×	×	×	0.47	0.50	0.52	0.55	0.57	0.59	0.62	0.64	0.66	0.69	0.71	0.73
35.2	16	×	×	×	×	×	×	×	×	×	0.48	0.51	0.53	0.56	0.59	0.61	0.65	0.65	0.67	0.70	0.73	0.75
37.4	17	×	×	×	×	×	×	×	×	×	×	0.52	0.55	0.58	0.60	0.63	0.67	0.67	0.69	0.72	0.75	0.77
39.6	18	×	×	×	×	×	×	×	×	×	×	0.54	0.56	0.59	0.62	0.65	0.68	0.68	0.71	0.74	0.76	0.79
41.8	19	×	×	×	×	×	×	×	×	×	×	×	0.58	0.60	0.63	0.66	0.70	0.70	0.73	0.76	0.78	0.81
44.0	20	×	×	×	×	×	×	×	×	×	×	×	0.59	0.62	0.64	0.67	0.71	0.72	0.75	0.78	0.80	0.83
46.2	21	×	×	×	×	×	×	×	×	×	×	×	×	0.63	0.66	0.69	0.73	0.74	0.77	0.80	0.82	0.85
48.4	22	×	×	×	×	×	×	×	×	×	×	×	×	×	0.67	0.70	0.74	0.76	0.78	0.82	0.85	0.87
50.6	23	×	×	×	×	×	×	×	×	×	×	×	×	×	×	×	0.75	0.78	0.80	0.83	0.86	0.89
52.8	24	×	×	×	×	×	×	×	×	×	×	×	×	×	×	×	0.77	0.79	0.81	0.84	0.87	0.90
55.0	25	×	×	×	×	×	×	×	×	×	×	×	×	×	×	×	×	0.80	0.82	0.85	0.88	0.91

TABLE A.3

Plasma Bicarbonate Ion Concentration (mEq/liter) from p_{CO_2} and pH, for p_{CO_2} = 3-60 mmHg

pH:

p_{CO_2}	6.90	6.95	7.00	7.05	7.10	7.15	7.20	7.25	7.30	7.35	7.40	7.45	7.50	7.55	7.60	7.65	7.70	7.75
3	0.57	0.64	0.72	0.80	0.90	1.01	1.13	1.27	1.43	1.60	1.80	2.02	2.26	2.54	2.85	3.19	3.58	4.02
4	0.76	0.85	0.95	1.07	1.20	1.35	1.51	1.70	1.90	2.13	2.40	2.69	3.02	3.38	3.80	4.26	4.78	5.36
5	0.95	1.06	1.19	1.34	1.50	1.68	1.89	2.12	2.38	2.67	2.99	3.36	3.77	4.23	4.74	5.32	5.97	6.70
6	1.14	1.28	1.43	1.61	1.80	2.02	2.27	2.54	2.85	3.20	3.59	4.03	4.52	5.07	5.69	6.39	7.17	8.04
7	1.33	1.49	1.67	1.87	2.10	2.36	2.64	2.97	3.33	3.74	4.19	4.70	5.28	5.92	6.64	7.45	8.36	9.38
8	1.52	1.70	1.91	2.14	2.40	2.69	3.02	3.39	3.80	4.27	4.79	5.37	6.03	6.77	7.59	8.52	9.56	10.7
9	1.70	1.91	2.15	2.41	2.70	3.03	3.40	3.81	4.28	4.80	5.39	6.05	6.78	7.61	8.54	9.58	10.8	12.1
10	1.89	2.12	2.38	2.67	3.00	3.37	3.78	4.24	4.76	5.34	5.99	6.72	7.54	8.46	9.49	10.7	12.0	13.4
11	2.08	2.34	2.62	2.94	3.30	3.70	4.16	4.66	5.23	5.87	6.59	7.39	8.29	9.30	10.4	11.7	13.1	14.8
12	2.27	2.55	2.86	3.21	3.60	4.04	4.53	5.09	5.71	6.40	7.18	8.06	9.04	10.2	11.4	12.8	14.3	16.1
13	2.46	2.76	3.10	3.48	3.90	4.38	4.91	5.51	6.18	6.94	7.78	8.73	9.80	11.0	12.3	13.8	15.5	17.4
14	2.65	2.97	3.34	3.74	4.20	4.71	5.29	5.93	6.66	7.47	8.38	9.40	10.6	11.8	13.3	14.9	16.7	18.8
15	2.84	3.19	3.58	4.01	4.50	5.05	5.67	6.36	7.13	8.00	8.98	10.1	11.3	12.7	14.2	16.0	17.9	20.1
16	3.03	3.40	3.81	4.28	4.80	5.39	6.04	6.78	7.61	8.54	9.58	10.8	12.1	13.5	15.2	17.0	19.1	21.5
17	3.22	3.61	4.05	4.55	5.10	5.72	6.42	7.20	8.08	9.07	10.2	11.4	12.8	14.4	16.1	18.1	20.3	22.8
18	3.41	3.82	4.29	4.81	5.40	6.06	6.80	7.63	8.56	9.60	10.8	12.1	13.6	15.2	17.1	19.2	21.5	24.1
19	3.60	4.04	4.53	5.08	5.70	6.40	7.18	8.05	9.03	10.1	11.4	12.8	14.3	16.1	18.0	20.2	22.7	25.5
20	3.79	4.25	4.77	5.35	6.00	6.73	7.55	8.48	9.51	10.7	12.0	13.4	15.1	16.9	19.0	21.3	23.9	26.8
21	3.98	4.46	5.01	5.62	6.30	7.07	7.93	8.90	9.99	11.2	12.6	14.1	15.8	17.8	19.9	22.4	25.1	28.2
22	4.17	4.67	5.24	5.88	6.60	7.41	8.31	9.32	10.5	11.7	13.2	14.8	16.6	18.6	20.9	23.4	26.3	29.5
23	4.35	4.89	5.48	6.15	6.90	7.74	8.69	9.75	10.9	12.3	13.8	15.5	17.3	19.5	21.8	24.5	27.5	30.8
24	4.54	5.10	5.72	6.42	7.20	8.08	9.07	10.2	11.4	12.8	14.4	16.1	18.1	20.3	22.8	25.6	28.7	32.2
25	4.73	5.31	5.96	6.69	7.50	8.42	9.44	10.6	11.9	13.3	15.0	16.8	18.8	21.1	23.7	26.6	29.9	33.5
26	4.92	5.52	6.20	6.95	7.80	8.75	9.82	11.0	12.4	13.9	15.6	17.5	19.6	22.0	24.7	27.7	31.1	34.9
27	5.11	5.74	6.44	7.22	8.10	9.09	10.2	11.5	12.8	14.4	16.2	18.1	20.4	22.8	25.6	28.7	32.3	36.2

30	5.68	6.37	7.15	8.02	9.00	10.1	11.3	12.7	14.3	16.0	18.0	20.2	22.6	25.4	28.5	31.9	35.8	40.2
31	5.87	6.58	7.39	8.29	9.30	10.4	11.7	13.1	14.7	16.5	18.6	20.8	23.4	26.2	29.4	33.0	37.0	41.6
32	6.06	6.80	7.63	8.56	9.60	10.8	12.1	13.6	15.2	17.1	19.2	21.5	24.1	27.1	30.4	34.1	38.2	42.9
33	6.25	7.01	7.86	8.82	9.90	11.1	12.5	14.0	15.7	17.6	19.8	22.2	24.9	27.9	31.3	35.1	39.4	44.2
34	6.44	7.22	8.10	9.09	10.2	11.5	12.9	14.4	16.2	18.1	20.4	22.8	25.6	28.8	32.3	36.2	40.6	45.6
35	6.63	7.43	8.34	9.36	10.5	11.8	13.2	14.8	16.7	18.7	21.0	23.5	26.4	29.6	33.2	37.3	41.8	46.9
36	6.82	7.65	8.58	9.63	10.8	12.1	13.6	15.3	17.1	19.2	21.6	24.2	27.1	30.4	34.2	38.3	43.0	48.3
37	7.00	7.86	8.82	9.89	11.1	12.5	14.0	15.7	17.6	19.7	22.2	24.9	27.9	31.3	35.1	39.4	44.2	49.6
38	7.19	8.07	9.06	10.2	11.4	12.8	14.4	16.1	18.1	20.3	22.8	25.5	28.6	32.1	36.1	40.5	45.4	50.9
39	7.38	8.28	9.29	10.4	11.7	13.1	14.7	16.5	18.6	20.8	23.4	26.2	29.4	33.0	37.0	41.5	46.6	52.3
40	7.57	8.50	9.53	10.7	12.0	13.5	15.1	17.0	19.0	21.3	24.0	26.9	30.2	33.8	38.0	42.6	47.8	53.6
41	7.76	8.71	9.77	11.0	12.3	13.8	15.5	17.4	19.5	21.9	24.6	27.5	30.9	34.7	38.9	43.7	49.0	55.0
42	7.95	8.92	10.0	11.2	12.6	14.1	15.9	17.8	20.0	22.4	25.2	28.2	31.7	35.5	39.9	44.7	50.2	56.3
43	8.14	9.13	10.3	11.5	12.9	14.5	16.3	18.2	20.5	22.9	25.7	28.9	32.4	36.4	40.8	45.8	51.4	57.6
44	8.33	9.35	10.5	11.8	13.2	14.8	16.6	18.7	20.9	23.5	26.3	29.6	33.2	37.2	41.8	46.8	52.6	59.0
45	8.52	9.56	10.7	12.0	13.5	15.2	17.0	19.1	21.4	24.0	26.9	30.2	33.9	38.1	42.7	47.9	53.8	60.3
46	8.71	9.77	11.0	12.3	13.8	15.5	17.4	19.5	21.9	24.6	27.5	30.9	34.7	38.9	43.6	49.0	54.9	61.7
47	8.90	9.98	11.2	12.6	14.1	15.8	17.8	19.9	22.4	25.1	28.1	31.6	35.4	39.7	44.6	50.0	56.1	63.0
48	9.09	10.2	11.4	12.8	14.4	16.2	18.1	20.4	22.8	25.6	28.7	32.2	36.2	40.6	45.5	51.1	57.3	64.3
49	9.28	10.4	11.7	13.1	14.7	16.5	18.5	20.8	23.3	26.2	29.3	32.9	36.9	41.4	46.5	52.2	58.5	65.7
50	9.47	10.6	11.9	13.4	15.0	16.8	18.9	21.2	23.8	26.7	29.9	33.6	37.7	42.3	47.4	53.2	59.7	67.0
51	9.65	10.8	12.2	13.6	15.3	17.2	19.3	21.6	24.3	27.2	30.5	34.3	38.4	43.1	48.4	54.3	60.9	68.4
52	9.84	11.1	12.4	13.9	15.6	17.5	19.6	22.0	24.7	27.8	31.1	34.9	39.2	44.0	49.3	55.4	62.1	69.7
53	10.0	11.3	12.6	14.2	15.9	17.9	20.0	22.5	25.2	28.3	31.7	35.6	39.9	44.8	50.3	56.4	63.3	71.0
54	10.2	11.5	12.9	14.4	16.2	18.2	20.4	22.9	25.7	28.8	32.3	36.3	40.7	45.7	51.2	57.5	64.5	72.4
55	10.4	11.7	13.1	14.7	16.5	18.5	20.8	23.3	26.2	29.4	32.9	36.9	41.5	46.5	52.2	58.6	65.7	73.7
56	10.6	11.9	13.4	15.0	16.8	18.9	21.2	23.7	26.6	29.9	33.5	37.6	42.2	47.4	53.1	59.6	66.9	75.1
57	10.8	12.1	13.6	15.3	17.1	19.2	21.5	24.2	27.1	30.4	34.1	38.3	43.0	48.2	54.1	60.7	68.1	76.4
58	11.0	12.3	13.8	15.5	17.4	19.5	21.9	24.6	27.6	31.0	34.7	39.0	43.7	49.0	55.0	61.7	69.3	77.7
59	11.2	12.5	14.1	15.8	17.7	19.9	22.3	25.0	28.1	31.5	35.3	39.6	44.5	49.9	56.0	62.8	70.5	79.1
60	11.4	12.8	14.3	16.1	18.0	20.2	22.7	25.4	28.5	32.0	35.9	40.3	45.2	50.7	56.9	63.9	71.7	80.4

TABLE A.4

Plasma Bicarbonate Ion Concentration (mEq/liter) from p_{CO_2} and pH, for $p_{CO_2} = 60$–$250\ mmHg$

p_{CO_2}	pH: 6.90	6.95	7.00	7.05	7.10	7.15	7.20	7.25	7.30	7.35	7.40	7.45	7.50	7.55	7.60	7.65	7.70	7.75
60	11.4	12.8	14.3	16.1	18.0	20.2	22.7	25.4	28.5	32.0	35.9	40.3	45.2	50.7	56.9	63.9	71.7	80.4
65	12.3	13.8	15.5	17.4	19.5	21.9	24.6	27.6	30.9	34.7	38.9	43.7	49.0	55.0	61.7	69.2	77.6	87.1
70	13.3	14.9	16.7	18.7	21.0	23.6	26.4	29.7	33.3	37.4	41.9	47.0	52.8	59.2	66.4	74.5	83.6	93.8
75	14.2	15.9	17.9	20.1	22.5	25.3	28.3	31.8	35.7	40.0	44.9	50.4	56.5	63.4	71.2	79.8	89.6	101
80	15.2	17.0	19.1	21.4	24.0	26.9	30.2	33.9	38.0	42.7	47.9	53.7	60.3	67.7	75.9	85.2	95.6	107
85	16.1	18.1	20.3	22.7	25.5	28.6	32.1	36.0	40.4	45.4	50.9	57.1	64.1	71.9	80.6	90.5	102	114
90	17.0	19.1	21.5	24.1	27.0	30.3	34.0	38.1	42.8	48.0	53.9	60.5	67.8	76.1	85.4	95.8	108	121
95	18.0	20.2	22.6	25.4	28.5	32.0	35.9	40.3	45.2	50.7	56.9	63.8	71.6	80.3	90.1	101	114	127
100	18.9	21.2	23.8	26.7	30.0	33.7	37.8	42.4	47.6	53.4	59.9	67.2	75.4	84.6	94.9	107	120	134
105	19.9	22.3	25.0	28.1	31.5	35.4	39.7	44.5	49.9	56.0	62.9	70.5	79.1	88.8	99.6	112	126	141
110	20.8	23.4	26.2	29.4	33.0	37.0	41.6	46.6	52.3	58.7	65.9	73.9	82.9	93.0	104	117	131	148
115	21.8	24.4	27.4	30.8	34.5	38.7	43.4	48.7	54.7	61.4	68.8	77.2	86.7	97.2	109	123	137	154
120	22.7	25.5	28.6	32.1	36.0	40.4	45.3	50.9	57.1	64.0	71.8	80.6	90.4	102	114	128	143	161
125	23.7	26.6	29.8	33.4	37.5	42.1	47.2	53.0	59.4	66.7	74.8	84.0	94.2	106	119	133	149	168
130	24.6	27.6	31.0	34.8	39.0	43.8	49.1	55.1	61.8	69.4	77.8	87.3	98.0	110	123	138	155	174
135	25.6	28.7	32.2	36.1	40.5	45.5	51.0	57.2	64.2	72.0	80.8	90.7	102	114	128	144	161	181
140	26.5	29.7	33.4	37.4	42.0	47.1	52.9	59.3	66.6	74.7	83.8	94.0	106	118	133	149	167	188

145	27.5	30.8	34.6	38.8	43.5	48.8	54.8	61.5	69.0	77.4	86.8	97.4	109	123	138	154	173	194
150	28.4	31.9	35.8	40.1	45.0	50.5	56.7	63.6	71.3	80.0	89.8	101	113	127	142	160	179	201
155	29.3	32.9	36.9	41.5	46.5	52.2	58.6	65.7	73.7	82.7	92.8	104	117	131	147	165	185	208
160	30.3	34.0	38.1	32.8	48.0	53.9	60.4	67.8	76.1	85.4	95.8	108	121	135	152	170	191	215
165	31.2	35.1	39.2	44.1	49.5	55.5	62.3	69.9	78.5	88.0	98.8	111	124	140	157	176	197	221
170	32.2	36.1	40.5	45.5	51.0	57.2	64.2	72.0	80.8	90.7	102	114	128	144	161	181	203	228
175	33.1	37.2	41.7	46.8	52.5	58.9	66.1	74.2	83.2	93.4	105	118	132	148	166	186	209	235
180	34.1	38.2	42.9	48.1	54.0	60.6	68.0	76.3	85.6	96.0	108	121	136	152	171	192	215	241
185	35.0	39.3	44.1	49.5	55.5	62.3	69.9	78.4	88.0	98.7	111	124	140	157	176	197	221	248
190	36.0	40.4	45.3	50.8	57.0	64.0	71.8	80.5	90.3	101	114	128	143	161	180	202	227	255
195	36.9	41.4	46.5	52.1	58.5	65.6	73.7	82.6	92.7	104	117	131	147	165	185	208	233	261
200	37.9	42.5	47.7	53.5	60.0	67.3	75.5	84.8	95.1	107	120	134	151	169	190	213	239	268
205	38.8	43.5	48.9	54.8	61.5	69.0	77.4	86.9	97.5	109	123	138	155	173	195	218	245	275
210	39.8	44.6	50.1	56.2	63.0	70.7	79.3	89.0	99.9	112	126	141	158	178	199	224	251	282
215	40.7	45.7	51.2	57.5	64.5	72.4	81.2	91.1	102	115	129	144	162	182	204	229	257	288
220	41.7	46.7	52.4	58.8	66.0	74.1	83.1	93.2	105	117	132	148	166	186	209	234	263	295
225	42.6	47.8	53.6	60.2	67.5	75.7	85.0	95.4	107	120	135	151	170	190	214	240	269	302
230	43.5	48.9	54.8	61.5	69.0	77.4	86.9	97.5	109	123	138	155	173	195	218	245	275	308
235	44.5	49.9	56.0	62.8	70.5	79.1	88.8	99.6	112	125	141	158	177	199	223	250	281	315
240	45.4	51.0	57.2	64.2	72.0	80.8	90.7	102	114	128	144	161	181	203	228	256	287	322
245	46.4	52.0	58.4	65.5	73.5	82.5	92.5	104	117	131	147	165	185	207	233	261	293	328
250	47.3	53.1	59.6	66.9	75.0	84.2	94.4	106	119	133	150	168	188	211	237	266	299	335

DERIVATION OF THE EQUATION FOR CALCULATING THE AMOUNT OF ALKALI REQUIRED TO BRING PATIENT TO NORMAL BLOOD pH

The Henderson–Hasselbalch equation can be written, for the bicarbonate system,

$$\text{pH} = 6.1 + \log \frac{x}{y}; \quad x = [HCO_3{}^-], \quad y = 0.03 p_{CO_2} \quad \text{(A.1)}$$

and $x + y = T$, where T = total CO_2, which has been measured. Then x/y = antilog(pH − 6.1). If we represent antilog(pH − 6.1) by B, we have

$$x + y = T; \quad x/y = B \quad \text{(A.2)}$$

Since the patient's pH is measured, B is known, and we have two equations and two unknowns. Solving, we have

$$x = TB/(1 + B) = [HCO_3{}^-] \text{ in mmol/liter} \quad \text{(A.3)}$$

and

$$y = T/(1 + B) = 0.03 p_{CO_2} \quad \text{(A.4)}$$

Now we are to add to, or subtract, from the $HCO_3{}^-$ of the patient's blood to bring to pH 7.40. *We assume that this will be added or subtracted while the patient is in anuria*, that is, no changes have occurred due to excretion of anions or cations. We also assume that in acidosis, the $NaHCO_2$ added is completely ionized, so that the number of millimoles of sodium bicarbonate added is equal to the bicarbonate ion added. This is a reasonable assumption in that, on the basis of it, the pK value of 6.1 was obtained experimentally.

Let us represent the amount of bicarbonate added by Q. In alkalosis Q will be negative. Then, when the pH of 7.4 is reached, we can represent the Henderson–Hasselbalch equation by

$$7.4 = 6.1 + \log \frac{x + Q}{y} \quad \text{(A.5)}$$

and

$$\frac{x + Q}{y} = \text{antilog } 1.3 = 19.95, \quad x + Q = 19.95y$$

Substituting values for x and y from equations (A.3) and (A.4), we have

$$\frac{TB}{B + 1} + Q = \frac{19.95\,T}{B + 1}$$

Solving for Q, we arrive at the equation

$$\text{mEq } HCO_3^- \text{ to be added/liter} = Q = \frac{T(19.95 - B)}{1 + B}$$

or Q, *the mmol of HCO_3^- to be added per liter of extracellular fluid* to bring to pH 7.4, is given by

$$Q = \frac{(\text{total } CO_2)[19.95 - \text{antilog } (pH - 6.1)]}{1 + \text{antilog } (pH - 6.1)} \qquad (A.6)$$

Since the total CO_2 and pH of the patient's plasma are measured, the HCO_3^- deficit or excess to bring to normal pH can be readily calculated.

Sample calculations are shown in Table A.5. In the calculation, values for the fraction $[19.95 - \text{antilog}(pH - 6.1)]/[1 + \text{antilog} (pH - 6.1)]$ are determined (A/B in the table). Multiplying this by the total CO_2, we can obtain the bicarbonate deficit. For pH values greater than 7.4, there is a bicarbonate excess (negative number), and this can also be calculated.

Multiplying the number given as A/B by the CO_2 content gives the bicarbonate deficit. For example, with a pH of 7.2 and a CO_2 content of 15 mmol/liter, the bicarbonate deficit is $0.542 \times 15 = 8.13$ mmol of bicarbonate per liter of extracellular fluid. With a pH of 7.6, the bicarbonate excess is $0.358 \times 15 = 5.37$ mmol/liter.

TABLE A.5

Sample Calculations for Preparing Table 9.3

pH	pH − 6.1	antilog (pH − 6.1)	19.95 − antilog (pH − 6.1) (A)	1 + antilog (pH − 6.1) (B)	A/B
6.9	0.8	6.31	13.64	7.31	1.87
7.0	0.9	7.94	12.01	8.94	1.34
7.1	1.0	10.0	9.95	11.0	0.905
7.2	1.1	12.59	7.36	13.59	0.542
7.3	1.2	15.85	4.10	16.85	0.243
7.4	1.3	19.95	0	20.95	0
7.5	1.4	25.12	− 5.17	26.12	−0.198
7.6	1.5	31.62	− 11.67	32.62	−0.358
7.7	1.6	39.81	− 19.86	40.81	−0.487

TABLE A.6

Normal Range for components of Major Importance in Fluid and Electrolyte Balance

Component	Plasma or serum	Urine	Sweat
Amino acid nitrogen	4–8 mg/100 ml	50–200 mg/24 hr	1.1–10.2 mg/100 ml
Ammonia (nitrogen)	70–180 μg/100 ml	200–900 mg/24 hr	5–9 mg/100 ml
Bicarbonate	22–26 mEq/liter	2–5 mEq/liter	2–4 mEq/liter
Buffer base cations — fixed anions, e.g., Cl	39–43	—	—
Calcium (total)	9.7–10.2 mg/100 ml (4.8–5.1 mEq/liter)	42–300 mg/24 hr (2.1–15 mEq/24 hr)	2–6 mg/100 ml
Calcium (ionized)	2.4–2.8 mEq/liter	—	—
Chloride	91–103 mEq/liter	—	12–35 mEq/liter
Citric acid	1.7–2.5 mg/100 ml	100–700 mg/24 hr	—
Carbon dioxide content (plasma or serum)	Ven: 25–30 mmol/liter Art: 23–27 mmol/liter	2–5 mmol/liter	2–4 mEq/liter
Copper	70–150 μg/100 ml	10–30 μg/24 hr	2–12 μg/100 ml
Creatinine	0.7–0.9 mg/100 ml	18–28 mg/kg/24 hr	0.4–1.3 mg/100 ml
Creatine	0.1–0.4 mg/100 ml	M: 30–112 mg/24 hr F: 80–200 mg/24 hr	0.1–0.5 mg/100 ml
Glucose	70–110 mg/100 ml	40–80 mg/100 ml	3–23 mg/100 ml
Hemoglobin	M: 13–18 g/100 ml F: 11–16 g/100 ml	—	—
Iron	65–150 μg/100 ml	100–500 μg/24 hr	10–20 μg/100 ml
Lactic acid	Art: 3–7 mg/100 ml Ven: 5–20 mg/100 ml	140–350 mg/24 hr	30–80 mg/100 ml

TABLE A.6.—*continued*

Component	Plasma or serum	Urine	Sweat
Magnesium	1.9–2.5 mg/100 ml (1.6–2.1 mEq/liter)	0.7–10.9 mg/100 ml (0.5–9.0 mEq/liter)	0.15–4.0 mg/100 ml (0.12–3.3mEq liter)
Nitrogen (total)	1.0–1.2 g/100 ml	10–20 g/24 hr	20–90 mg/100 ml
Nonprotein nitrogen	20–35 mg/100 ml	10–20 g/24 hr	17–64 mg/100 ml
O_2 capacity	18–24 vol %	—	—
O_2 content	Art: 93–96% cap. Ven.: 56–69% cap.	—	—
Osmotic pressure	285–302 mosmol/kg H_2O	400–1200 mosmol/kg H_2O	120–165 mosmol/kg H_2O
p_{CO_2}	Art: 33–48 mmHg Ven: 38–52 mmHg	—	—
pH	7.38–7.42	4.5–5.5	4.5–5.5
Phosphorus (inorganic)	2.5–4.2 mg/100 ml	300–1000 mg/24 hr	1–3 mg/100 ml
p_{O_2}	Art: 92–100 mmHg Ven: 35–45 mmHg	—	—
Potassium	4.1–5.0 mEq/liter	50–150 mEq/24 hr	7–37 mEq/liter
Protein (total)	6.6–7.1 g/100 ml	49–112 mg/100 ml	40–100 mg/100 ml
Albumin	3.6–4.8 g/100 ml	40–100 mg/100 ml	30–80 mg/100 ml
Pyruvic acid	0.3–1.0 mg/100 ml	4–9 mg/24 hr	—
Sodium	138–142 mEq/liter	50–225 mEq/24 hr	15–52 mEq/liter
Standard bicarbonate	21–25 mEq/liter	—	—
Sulfur (inorganic)	0.5–1.2 mg/100 ml	240–1120 mg/24 hr	7–7.4 μg/100 ml
Urea nitrogen	8–18 mg/100 ml	6–16 g/24 hr	8–30 mg/100 ml
Uric acid	3.7–6.4 mg/100 ml	250–750 mg/24 hr	0.5–2 mg/100 ml
Zinc	55–150 μg/100 ml	0.15–1.3 mg/24 hr	—

Index